Tree-Based Machine Learning Methods in SAS® Viya®

Sharad Saxena

S.sas.

sas.com/books

The correct bibliographic citation for this manual is as follows: Saxena, Sharad. 2022. *Tree-Based Machine Learning Methods in SAS® Viya®*. Cary, NC: SAS Institute Inc.

Tree-Based Machine Learning Methods in SAS® Viya®

Copyright © 2022, SAS Institute Inc., Cary, NC, USA

ISBN 978-1-954846-71-5 (Hardcover)
ISBN 978-1-954846-63-0 (Paperback)
ISBN 978-1-954846-64-7 (Web PDF)
ISBN 978-1-954846-65-4 (EPUB)
ISBN 978-1-954846-66-1 (Kindle)

SAS Institute Inc., SAS Campus Drive, Cary, NC 27513-2414

February 2022

Contents

About This Book

Preface

Decision trees are popular machine learning models. They are very intuitive and natural for us humans because in a way, they mimic the way we make decisions in our day-to-day lives. Several decision trees can be combined to produce ensemble models for better predictive performance than using a single decision tree. Decision trees and tree-based ensembles are supervised learning models used for problems involving classification and regression. They are largely used for prediction, but they also have several other uses in the modeling process and beyond.

Why you might want to use tree-based predictive modeling? Tree-based models are designed to handle categorical variables without you needing to create dummy variables. Tree-based models directly consider categorical levels when determining a split. Tree-based models can also handle observations with missing values instead of ignoring an observation with the missing value or being forced to impute. A tree can assign the missing value to one of those branches. This strategy can also be used to score an unknown level that occurs at scoring time but was not seen at training time. Tree-based models are also able to detect nonlinear relationships. Finally, a single decision tree can be interpreted as a set of rules or can easily be visualized in a tree plot.

SAS® Visual Data Mining and Machine Learning software on SAS® Viya® has three tree-based predictive modeling techniques – decision trees, forests, and gradient boosted trees. A decision tree is visually appealing and highly explainable. A forest model is a bagged ensemble of trees and has high predictive power but is harder to interpret than a single tree. A gradient boosting model is a boosted ensemble of trees and is very hard to interpret but has high predictive power. In SAS Visual Data Mining and Machine Learning, it is easy to start building one type of tree-based model and transition to building another. The output is similar so that you can understand the results more easily from one model to the next.

The analytics for all three tree-based predictive modeling techniques can be found in the Decision Tree Action Set in SAS Viya. These techniques can be accessed in the programming interface of SAS Visual Data Mining and Machine Learning when you use the TREESPLIT procedure, the FOREST procedure, or the GRADBOOST procedure. You can access these procedures through the SAS® Studio HTML client. SAS Studio also has tasks that can help you generate code. SAS® Visual Analytics has objects that can help you create these models interactively using a point-and-click interface. The actions in SAS Viya are also available to be called from other clients such as the SAS Visual Analytics client, Java, or Python, and more. For this book, we will primarily consider point-and-click visual interfaces that include mostly the pipelines in Model Studio, a bit of SAS Studio and SAS Visual Statistics. A little exposure of procedures in the SAS programming interface to the SAS Viya actions is also covered in the book.

The book includes discussions of tree-structured predictive models and the methodology for growing, pruning, and assessing decision trees, forests, and gradient boosted trees. You will acquire knowledge not only of how to use tree-based models for classification and regression, and some of their limitations, but also how the respective algorithms that shape them work. Each demonstration introduces a new data concern and then walks you through tweaking the modeling approach, modifying the properties, and changing the hyperparameters, thus building a right tree-based machine learning model. Along the way, you will gain experience making decision trees, forests, and gradient boosted trees that work for you.

The book also explains isolation forest (an unsupervised learning algorithm for anomaly detection), deep forest (an alternative for neural network deep learning), and Poisson and Tweedy gradient boosted regression trees. In addition, many of the auxiliary uses of trees, such as exploratory data analysis, dimension reduction, and missing value imputation are examined and running open source in SAS and SAS in open source is demonstrated.

What Does This Book Cover?

This book covers everything from using a single tree to more advanced bagging and boosting ensemble methods in SAS Viya.

Chapter 1 provides an introduction to tree-structured models and the ones that are available in SAS Viya. Decision trees are powerful predictive and explanatory modeling tools. They are flexible in that they can model interval, ordinal, nominal, and binary targets, and they can accommodate nonlinearities and interactions. They are simple to understand and present. Model Studio is also introduced in this chapter.

Chapter 2 discusses the types of decision trees. The tree can be either a classification tree, which models a categorical target, or a regression tree, which models a continuous target. The model is expressed as a series of if-then statements. For each tree, you specify a target (dependent, response) variable and one or more input (independent, predictor) variables. The input variables for tree models can be categorical or continuous. The initial node is called the root node, and the terminal nodes are called leaves. Partitioning is done repeatedly, starting with the root node, which contains all the data, and continuing to split the data until a stopping criterion is met. At each step, the parent node is split into two or more child nodes by selecting an input variable and a split value for that variable. As with other modeling tools, decision tree models generate scoring code and can be used to generate predictions on new data through the Score node in Model Studio.

Chapter 3 talks about how to grow a decision tree. Recursive partitioning is the method whereby a tree grows from the root node to its maximal size. Splitting criteria are used to assess and compare splits within and between inputs. Those criteria measure reduction in variability in child versus parent nodes. The goal is to reduce variability and thus increase purity. Various measures, such as the Gini index, entropy, and residual sum of squares, can be used to assess candidate splits for each node. The process of building a decision tree begins with growing a large, full tree. The full tree can overfit the training data, resulting in a model that does not adequately generalize to new data. Exhaustive split searches can be computationally expensive. Decision tree algorithms use shortcuts to reduce the split search primarily through restrictions on the number of child nodes (for example, binary splitting as used in CART trees) or using agglomerative clustering (as in CHAID).

Decision tree modeling with interval targets is like modeling with categorical targets. The split search considerations remain the same, and analogous (or equivalent) split criteria are used. Violation of the equal variance assumption from regression and ANOVA can lead to spurious split selection in decision trees. Target transformations, such as the logarithm, might improve split selection. The interactive training facility in SAS Viya enables the user to select variables and split points for model building. Entire models can be built by the user. Alternatively, initial splits can be user-defined, and subsequent splits can be chosen automatically by the tree algorithm.

Chapter 4 covers common advantages and disadvantages of decision trees along with some of their prominent secondary uses. Decision trees can directly incorporate missing values without the need for imputation. Methods include placing missing values in child nodes that lead to the greatest improvement in purity, or in nodes with the most cases. Alternatively, inputs that show highest agreement with primary split variables can be used to assign case membership in child nodes.

Variables' importance can be calculated for inputs involved in modeling, either through a measure of their contributions to impurity reduction or through a measure that compares a given fit statistic (for example, average squared error) for trees with the input included versus a tree in which the input is rendered uninformative.

Compared with other regression and classification methods, decision trees have the advantage of being easy to interpret and visualize, especially when the tree is small. Tree-based methods scale well to large data, and they offer various methods of handling missing values, including surrogate splits. However, decision trees do have limitations. Regression trees fit response surfaces that are constant over rectangular regions of the predictor space, so they often lack the flexibility needed to capture smooth relationships between the predictor variables and the response. Another limitation of decision trees is that small changes in the data can lead to very different splits, and this undermines the interpretability of the model. Random forest models address some of these limitations.

Decision trees have many uses beyond modeling. They are useful as multivariate data exploration tools because of their easily interpreted visual output and because they do not require much in the way of input preparation to use. In some cases, they can be used to aid in identifying interactions to add to regression models. Trees can be used to select important variables for subsequent models such as regression and neural networks because they are resistant to the curse of dimensionality and contain methods to calculate variable importance. Trees can also reduce the number of dimensions for subsequent models by collapsing levels of categorical inputs. Levels that have similar average target values are grouped into the same leaf. Trees can be used to discretize interval inputs to aid subsequent model interpretation or to help accommodate nonlinear patterns between inputs and the target. They might also help identify broad segments within which regression models can be fit. Trees can be used for imputation. Missing values are imputed based on predictive tree models in which variables with missing values are treated as the target and all other variables as model inputs. Tree imputation is available in the Imputation node.

Chapter 5 describes how to tune a decision tree. The "right-sized" tree can be found by using top-down or bottom-up criteria. Top-down criteria limit tree growth. CHAID algorithms tend to feature such criteria, and rely on statistical tests, alpha values, and p-value adjustments (Bonferroni), among others, to restrict the size of the maximal tree. In bottom-up pruning, a large tree is grown and then branches are removed in a backward fashion using some model assessment selection criterion. This strategy originated with cost-complexity pruning as used by CART decision trees. Model selection criteria include accuracy and average square error depending on the type of target variable and are applied to validation data. During bottom-up pruning, tree models are selected based on validation data, if present. Cross validation is a less expensive approach to model validation in which all data is used for both training and validation. Cross validation is available in the Decision Tree node in Model Studio.

When fitting multiple decision trees, you can refer to CART and CHAID methodologies to guide your choice of tree property settings. If you are unsure about what values of hyperparameters to be used for building a decision tree, it is best to use SAS® Viya® autotuning functionality.

Chapter 6 describes ensemble of trees and then focuses on the forest model. Tree-based ensemble models take advantage of the inherent instability of decision trees, whereby small changes in the training data can result in large changes in tree structure due to large numbers of competing splits. Perturb and combine methods generate multiple models by manipulating the distribution of the data or altering the construction method and then averaging the results. Bagging resamples training data with replacement and then fits models to each sample. Boosting adaptively reweights subsequent samples based on mistakes that are made in target classification from previous samples. Ensemble models might help smooth the prediction surface that is generated by individual tree models and thereby improve model performance.

A forest is just what the name implies. It's a bunch of decision trees – each with a randomly selected subset of the data – all combined into one result. Using a forest helps address the problem of overfitting inherent to an individual decision tree. The forest model in SAS Viya creates a random forest using literally hundreds of decision trees. A forest consists of several decision trees that differ from each other in two ways. First, the training data for a tree is a sample with replacement from all available observations. Second, the input variables that are considered for splitting a node are randomly selected from all available inputs. Among these randomly selected variables, the input variable that is most often associated with the target is used for the splitting rule. In other respects, trees in a forest are trained like standard trees. The training data for an individual tree excludes some of the available data that is used to assess the fit of the model.

The forest model in SAS Viya accepts interval and class target variables. For an interval target, the prediction in a leaf of an individual tree equals the average of the target values among the bagged training observations in that leaf. For a class target, the posterior probability of a target category equals the proportion of that category among the bagged training observations in that leaf. Predictions or posterior probabilities are then averaged across all the trees in the forest. Averaging over trees with different training samples reduces the dependence of the predictions on a training sample. Increasing the number of trees does not increase the risk of overfitting the data and can decrease it. However, if the predictions from different trees are correlated, then increasing the number of trees makes little or no improvement.

Chapter 7 covers some additional forest models. An isolation forest is a specially constructed forest that is used for anomaly detection instead of target prediction. When an isolation forest is created, it writes anomaly scores in the scored data table.

The recently proposed deep forest, which is also termed gcForest (multi-Grained Cascade Forest), is a novel decision tree ensemble approach. This method generates a deep forest ensemble with a cascade structure, which enables gcForest to do representation learning, which can find out the better features by end-to-end training. Its representational learning ability can be further enhanced by multi-grained scanning when the inputs are with high dimensionality, potentially enabling gcForest to be contextual or structural aware. The performance of deep forest claims to be more competitive than that of deep neural network (DNN).

Chapters 8 explains tree-based gradient boosting models. Gradient boosted trees create an ensemble model of weak decision trees in a stage-wise, iterative, sequential manner. Each tree uses a subsample of the data. Gradient boosting algorithms convert weak learners to strong learners. One advantage of gradient boosting is that it can reduce bias and variance in supervised learning. All points begin with the same weight. Points classified correctly are given a lower weight and those classified incorrectly are given a higher weight. Now the model focuses on high weight points and classifies them correctly. However, others that were classified correctly in the first iteration are now misclassified. This process continues for many iterations. In the end, all models are given a weight depending on their accuracy, and the model results are combined into one consolidated result.

Gradient boosting models can be fit in SAS Viya using several ways including the Gradient Boosting node in Model Studio, the GRADBOOST procedure, and so on. A decision tree in a gradient boosting model trains on new training data that are derived from the original training. Using different data to train different trees during the boosting process reduces the correlation of the predictions of the trees, which in turn improves the predictions of the boosting model.

The algorithm samples the original data without replacement to create the training data for an individual tree. It performs the action of sampling multiple times throughout a run, and each set of training data created is referred to as a subsample. In some cases, gradient boosting models can be overtrained and thus perform poorly on validation or test data. One method to combat overtraining in gradient boosting is early stopping. An additional advantage to early stopping is reduced training time in cases where the stopping criterion is met well before the specified maximum number of iterations occurs.

Chapter 9 explores some additional gradient boosting models. Gradient boosting can be used for transfer learning. You can use the auxiliary data to increase the number of observations for training the model. To prevent the model from being overly biased toward the auxiliary data, the gradient boosting algorithm down-weights auxiliary observations that are dissimilar from the training observations.

You can specify the distribution of the objective function in a gradient boosting model. The default distribution is Gaussian, binary, or multinomial for an interval, binary, or nominal target, respectively. The Poisson distribution is appropriate for count data. The Tweedie distribution is useful for modeling total losses in insurance.

Hyperparameter tuning is available in the three tree-based models, decision tree, forest, and gradient boosting, to find the best values for various options. These include the splitting criterion, maximum depth, and number of bins in decision trees; the fraction of training data to sample, maximum tree depth, number of trees, and number of variables to consider for each split in forest; and the L1 and L2 regularization parameters, learning rate, fraction of training data to sample, and number of variables to consider for each split in gradient boosting. There are several objective functions to choose from for the optimization algorithm as well as search methods, including one based on a genetic algorithm.

At the end, a practice case study is provided.

Is This Book for You?

Building representative tree-based machine learning models that generalize well on new data requires careful consideration of both the data used for the model to train, and the assumptions about the various tree-based algorithms. It is important to choose the right options and best hyperparameter values for both the data that you will be modeling and the business problem at hand.

If you want to gain insights about tree-based models and all the nitty-gritty details involved in decision trees, random forests, and gradient-boosted trees, then this book is for you! The discussion can help you quickly get value out of tree-based models from intermediate to advanced level. This book is most suitable for predictive modelers and data analysts who want to build decision trees and ensembles of decision trees using SAS Visual Data Mining and Machine Learning in SAS Viya.

What Are the Prerequisites for This Book?

Before reading this book, you should have the following:

- An understanding of basic statistical concepts. You can gain this knowledge from the course "SAS® Visual Statistics on SAS® Viya®: Interactive Model Building".
- Familiarity with SAS Visual Data Mining and Machine Learning software. You can gain this knowledge from the course "Machine Learning Using SAS® Viya®".

What Should You Know about the Examples?

This book includes worked demonstrations and practices for you to follow to gain hands-on experience with SAS Visual Data Mining and Machine Learning software including, but not limited to, Model Studio.

Software Used to Develop the Book's Content

SAS Visual Data Mining and Machine Learning 8.5 is a comprehensive visual – and programming – interface supports the end-to-end data mining and machine learning process. Analytics team members of all skill levels are empowered to handle all analytics life cycle tasks in a simple, powerful and automated way. You can solve complex analytical problems with a comprehensive visual interface that handles all tasks in the analytics life cycle. SAS Visual Data Mining and Machine Learning, which runs in SAS Viya 3.5, combines data wrangling, exploration, feature engineering, and modern statistical, data mining, and machine learning techniques in a single, scalable in-memory processing environment. The software also includes SAS Visual Statistics 8.5 and SAS Visual Analytics 8.5.

Model Studio is included in SAS Viya. It is an integrated visual environment that provides a suite of analytic data mining tools that enable you to explore and build models. It is part of the Discovery phase of the analytic life cycle. The data mining tools provided in Model Studio enable you to deliver and distribute analytic model data mining champion models, score code, and results.

Example Code and Data

The data sets used in the book's demonstrations and practices are provided to download.

You can access the example code and data for this book by linking to its author page at https://support.sas.com/saxena. The Solutions to the Case Study can also be found on the author page.

We Want to Hear from You

SAS Press books are written *by* SAS Users *for* SAS Users. We welcome your participation in their development and your feedback on SAS Press books that you are using. Please visit sas.com/books to do the following:

- Sign up to review a book
- Recommend a topic
- Request information about how to become a SAS Press author
- Provide feedback on a book

Do you have questions about a SAS Press book that you are reading? Contact the author through saspress@sas.com or https://support.sas.com/author_feedback.

SAS has many resources to help you find answers and expand your knowledge. If you need additional help, see our list of resources: sas.com/books.

About the Author

Dr. Sharad Saxena is Principal Analytical Training Consultant based at the SAS R&D center in Pune, India. He has been working in the field of statistics and analytics since 2000. He has been doing education consulting in the area of advanced analytics and machine learning across the globe including the UK, USA, Singapore, Italy, Australia, Netherlands, Middle East, China, Philippines, Nigeria, Hong Kong, Malaysia, Indonesia, Mexico, and India for a variety of SAS customers in banking, insurance, retail, government, health, agriculture, and telecommunications. Dr. Saxena earned a bachelor's degree in mathematics with statistics and economics minors, a master's degree in statistics, and a Ph.D. in statistics from the School of Studies in Statistics at Vikram University, India. He has also completed some short-term courses in allied areas from the Indian Statistical Institute, Kolkata, and the Indian Institute of Management, Ahmedabad. He is recipient of the prestigious U.S. Nair Young Statistician Award in 2001–02, and his biography has been included in *Who's Who in Science and Engineering* 2006–2007 for his outstanding contributions in the field of statistics. Prior to joining SAS, he worked as a Product Head–Business Analytics at TimesPro (The Times of India Group) where he started and managed the vocational training program in business analytics. Overall, he has more than two decades of rich experience in research, teaching, training, consulting, writing, and education product design, more than 14 years of which have been with SAS and the remaining in academia as a faculty member with some top-notch institutes in India like the Institute of Management Technology, Ghaziabad; Institute of Management, Nirma University, and more. Dr. Saxena has more than 35 publications including research papers in journals such as the *Journal of Statistical Planning and Inference*, *Communications in Statistics–Theory and Methods*, *Statistica*, *Statistical Papers*, and Vikalpa. He is a co-author of the book titled *Randomness and Optimal Estimation in Data Sampling* published by American Research Press. Apart from various talks delivered at national and international conferences, he has written some case studies, technical notes, white papers, book reviews, and popular columns in newspapers. He is an active member of many professional and academic bodies.

Learn more about this author by visiting his author page at http://support.sas.com/saxena. There you can download free book excerpts, access example code and data, read the latest reviews, get updates, and more.

Acknowledgements

This book is based on the SAS training course, "Tree-Based Machine Learning Methods in SAS® Viya®," developed by the author of this book. Some of the content of this training course was based on material originally developed by William E. Potts, and additional contributions were made by Manoj Singh, Padraic Neville, and Jordan Bakerman. The technical review of the book was done by Mike Patetta.

Additional content and editing were supplied by Sian Roberts and Catherine Connolly. Design, editing, and production support was provided by Robert Harris and Suzanne Morgen.

Foreword

"Take chances, make mistakes, get messy!" – Valerie Frizzle, The Magic School Bus

Anyone who was a nerdy science-loving kid understands the thrill of that statement. When Ms. Frizzle took her students on magic field trips, Arnold was always worried. Maybe that's because he didn't get to choose his adventure. That is where you, dear reader, are special. The best thing about adult life is getting to choose what to study and enjoying the new things that you can learn every day. Today, you will journey to the fascinating world of tree models.

I've been fortunate to call Sharad Saxena my colleague for over 10 years, and it is always a treat to be in his statistics lectures. Dr. Saxena's passion for learning and for sharing his knowledge with others comes through in his writing and his examples throughout the book. After devoting months of his career to making this book – and its accompanying course, "Tree-Based Machine Learning Methods in SAS Viya"—the fruits of his labor are yours to enjoy, with clear explanations, interesting history, and compelling use cases.

The appeal of this book, and of this topic, is that it all starts with a simple decision split. For example, what is the best way to predict which shoppers are likely to spend over $100 on their next visit? Let's call those who make purchases over $100 "Big Spenders." Among historical shopper data, you discover that if the previous shopping trip was more than 30 days ago, then on the next visit, 15% of shoppers were Big Spenders. Otherwise, about 12% of shoppers were Big Spenders. So, your decision split is: if days since last purchase >30, then predict Big Spender. Else, predict Not Big Spender. It is so simple. I bet you've already perceived that one split is not very useful; if customer patterns were so simple and predictable, then we wouldn't need statistical models, would we? You need lots of splits, and lots of variables, to construct a prediction surface that accounts for the nuanced differences among shoppers and their behaviors.

Applying this straightforward decision splitting method across many variables, recursively, can result in a model with hundreds or thousands of if-then rules. Theoretically, you could keep splitting until every case (except ties) has its own leaf. But this isn't useful; there is a limit to how complex a model can be and still generalize well to new cases. Pruning the tree can be just as important as splitting branches.

The first half of the book provides a strong foundation in tree models, how they are grown and pruned, what the predictions look like, and how to train them in SAS Viya. The second half of the book is dedicated to ensembles of trees and making sense of the output of these complex models.

If you follow along with the examples in the book, then by the end of this book, you will find that you can do so much with tree models. More than that, you will have expanded how you think about predictive models, whether it's the ability of trees to produce simple, elegant if-then explanations, the ability of forest models to smooth a choppy prediction surface, or the ability of isolation forests to detect anomalies in very large data. Along the way, you will come to appreciate just how powerful and flexible these models can be, starting with just a humble two-way decision split.

Catherine Truxillo, PhD

Director, Advanced Analytics Education

SAS

December 2021

Chapter 1: Introduction to Tree-Structured Models

Introduction

> "Sometimes you make the right decision, sometimes you make the decision right."
>
> *–Phil McGraw*

A decision tree has many analogies in real life. In decision analysis, a tree can be used to represent decisions and decision making visually and explicitly. As the name suggests, it uses a tree-like model of decisions.

The adjective *decision* in decision trees is a curious one, and misleading. In the 1960s, originators of the tree approach described the splitting rules as decision rules. The terminology remains popular. This is ill-fated because it inhibits the use of ideas and terminology from decision theory. The term decision tree is used in decision theory to depict a series of decisions for choosing alternative activities. You create the tree and specify probabilities and benefits of outcomes of the activities. Software, including SAS, finds the most beneficial path. The project follows a single path and never performs the unchosen activities. The decider follows a path based on a set of criteria.

Decision theory is not about data analysis. The choice of a decision might be made without reference to data. The trees in this book are only about data analysis. A tree is fit to a data set to enable interpretation and prediction of data. An apt name would be *data-splitting trees* that would be used for *supervised learning* also called *predictive modeling*.

In supervised learning, a set of *input* variables (predictors) is used to predict the value of one or more *target* variables (outcome). The mapping of the inputs to the target is a *predictive model*. The goal is to create a model that predicts the value of a target variable by learning simple decision rules inferred from the input variables. The data used to estimate a predictive model is a set of *cases* (observations, examples) consisting of values of the inputs and target. The fitted model is typically applied to new cases where the target is unknown.

Decision Tree – What Is It?

There are several tree-structured models that include one or more decision trees. Decision trees are a fundamental machine learning technique that every data scientist should know. Luckily, the construction and implementation of decision trees in SAS Viya is straightforward and easy to produce.

A *decision tree* represents a grouping of the data that is created by applying a series of simple rules. Each rule assigns an observation to a group based on the value of one input. One rule is applied after another, resulting in a hierarchy of groups within groups. The hierarchy is called a *tree*, and each group is called a *node*. The original group contains the entire data set and is called the *root node* of the tree. A node with all its successors forms a branch of the node that created it. The final nodes are called *leaves*. For each leaf, a decision is made and applied to all observations in the leaf. The type of decision depends on the context. In supervised learning, the decision is the predicted value.

You use the decision tree to do one of the following tasks:

- classify observations based on the values of nominal, binary, or ordinal targets
- predict outcomes for interval targets
- predict the appropriate decision when you specify decision alternatives

The tree depicts the first split into groups as branches emanating from a root and subsequent splits as branches emanating from nodes on older branches. Figure 1.1 is an example decision tree predicting a nominal target *Cause of Death* using two binary inputs *Weight Status* and *Smoking Status*. The decision nodes include a bar chart related to the node's sample target values and other details. The leaves of the tree are the final groups, the unsplit nodes. For some perverse reason, trees are always drawn upside down, like an organizational chart. For a tree to be useful, the data in a leaf must be similar with respect to some target measure so that the tree represents the segregation of a mixture of data into purified groups.

Types of Decision Trees

Decision trees are a nonparametric supervised learning method used for both classification and regression tasks. A classification tree models a categorical response, and a regression tree models a continuous response. See Figure 1.2. Both types of trees are called decision trees because the model is expressed as a series of if-then statements. For each type of tree, you specify a response variable

Figure 1.1: A Simple Decision Tree

Decision Tree **Cause of Death**   Observations Used **1,968**
Unused **3,241**

Tree

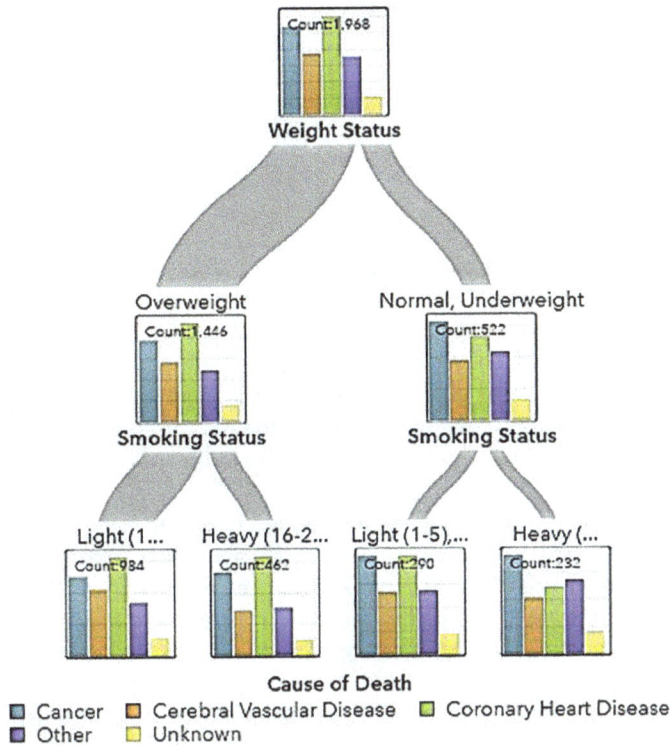

Cause of Death
- ■ Cancer ■ Cerebral Vascular Disease ■ Coronary Heart Disease
- ■ Other ☐ Unknown

Figure 1.2: Classification and Regression Trees

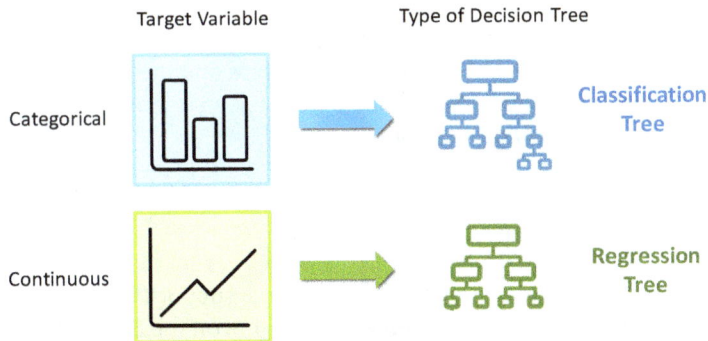

(also called a target variable), whose values you want to predict, and one or more input variables (called predictor variables), whose values are used to predict the values of the target variable.

The predictor variables for tree models can be categorical or continuous. The set of all combinations of the predictor variables are called the *predictor space*. The model is based on partitioning the predictor space into nonoverlapping groups, which correspond to the leaves of the tree. Partitioning is done repeatedly, starting with the root node, which contains all the data, and continuing until a stopping criterion is met. At each step, the parent node is split into child nodes by selecting a predictor variable and a split value for that variable that minimize the variability according to a specified measure (or the default measure) in the response variable across the child nodes. Various measures, such as the Gini index, entropy, and residual sum of squares, can be used to assess candidate splits for each node. The selected predictor variable and its split value are called the primary *splitting rule*.

Tree-structured models are built from training data for which the response values are known, and these models are subsequently used to score (classify or predict) response values for new data. For classification trees, the most frequent response level of the training observations in a leaf is used to classify observations in that leaf. For regression trees, the average response of the training observations in a leaf is used to predict the response for observations in that leaf. The splitting rules that define the leaves provide the information that is needed to score new data; these rules consist of the primary splitting rules, surrogate rules, and default rules for each node.

The process of building a decision tree begins with growing a large, full tree. The full tree can overfit the training data, resulting in a model that does not adequately generalize to new data. To prevent overfitting, the full tree is often pruned back to a smaller subtree that balances the goals of fitting training data and predicting new data. Two commonly applied approaches for finding the best subtree are cost-complexity pruning and C4.5 pruning.

Compared with other regression and classification methods, tree-structured models have the advantage that they are easy to interpret and visualize, especially when the tree is small. Tree-based methods scale well to large data, and they offer various methods of handling missing values, including surrogate splits.

However, tree-structured models have limitations. Regression tree models fit response surfaces that are constant over rectangular regions of the predictor space, so they often lack the flexibility needed to capture smooth relationships between the predictor variables and the response. Another limitation of tree models is that slight changes in the data can lead to quite different splits, and this undermines the interpretability of the model.

Tree-Based Models in SAS Viya

SAS Viya is a cloud-enabled, analytic run-time environment with several supporting services, including SAS Cloud Analytic Services (CAS). CAS is the in-memory engine on the SAS Viya Platform.

SAS Viya builds tree-based statistical models for classification and regression. You can build three tree-based models in SAS Viya starting from a single tree to more complex ensembles of trees like forest and gradient boosting.

A random forest is just what the name implies. It is a bunch of decision trees – each with a randomly selected subset of the data – all combined into one result. Using a random forest helps address the problem of overfitting inherent to an individual decision tree.

Gradient boosting creates an ensemble model of weak decision trees in a stage-wise, iterative, sequential manner. Gradient boosting algorithms convert weak learners to strong learners. One advantage of gradient boosting is that it can reduce bias and variance in supervised learning.

Analytics Platform from SAS

The SAS Analytics Platform is a software foundation that is engineered to address today's business challenges and to generate insights from your data in any computing environment. SAS Viya is the latest extension of the SAS Analytics Platform, which is designed to orchestrate your entire analytic ecosystem, connecting and accelerating all analytics life cycle – from data, to discovery, to deployment. SAS Viya seamlessly scales to data of any size, type, speed, and complexity, and is interoperable with SAS 9. As an integrated part of the SAS Analytics Platform, SAS Viya is a cloud-enabled, in-memory analytics engine.

The SAS Viya Platform architecture is illustrated in Figure 1.3. At the heart of SAS Viya is SAS Cloud Analytic Services (CAS), an in-memory, distributed analytics engine. It uses scalable, high-performance, multi-threaded algorithms to rapidly perform analytical processing on in-memory data of any size.

SAS Viya contains microservices. A microservice is a small service that runs in its own process and communicates with a lightweight mechanism (hypertext transfer protocol (HTTP)). Microservices

Figure 1.3: SAS Viya Platform Architecture

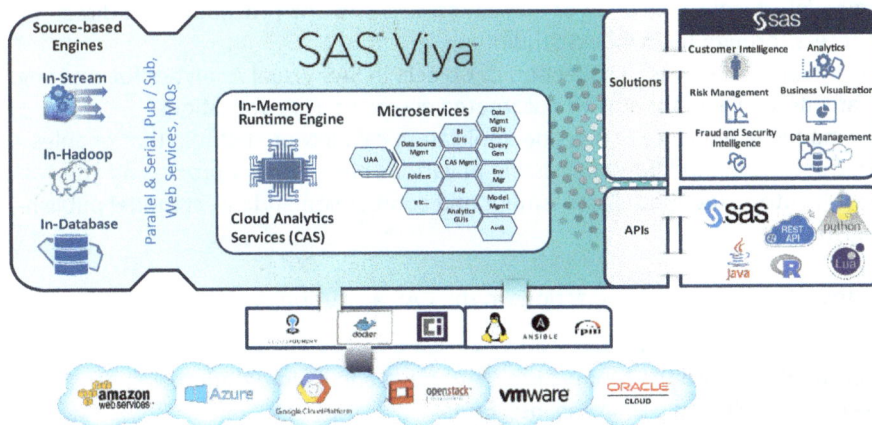

are a series of containers that define all the different analytic life cycle functions, sometimes described as "actions" that fit together in a modular way. The in-memory engine is independent from the microservices and allows for independent scalability.

On the left of Figure 1.3 you see a series of source-based data engines.

SAS Viya has a middle tier implemented on a micro-services architecture, deployed and orchestrated through the industry standard cloud Platform as a Service also known as Cloud Foundry. Through Cloud Foundry, SAS Viya can be deployed, managed, monitored, scaled, and updated. Cloud Foundry enables SAS Viya to support multiple cloud infrastructure allowing customers to deploy SAS in a hybrid cloud environment spanning multiple clouds including the combination of on-premises cloud infrastructure and public cloud infrastructure.

You can choose to use other platforms like Docker and the open container initiative. You can operate on private infrastructure such as OpenStack or VMware, or open infrastructure such as Amazon Web Services, Azure, and so on.

Existing SAS solutions and new ones are being built on SAS Viya. In addition, you can use REST API to include SAS Viya actions in your existing applications. A REST API is an application programming interface that conforms to the constraints of representational state transfer (REST) architectural style and allows for interaction with RESTful web services.

SAS Visual Data Mining and Machine Learning

SAS Visual Data Mining and Machine Learning is a product offering in SAS Viya that contains the underlying CAS actions and SAS procedures for data mining and machine learning applications, and graphical user interface (GUI)-based applications for various levels and types of users.

These applications are as follows:

- *Programming interface:* a collection of CAS action sets and SAS procedures for direct coding or access through tasks in SAS Studio.
- *Interactive modeling interface:* a collection of objects in SAS Visual Analytics for creating models in an interactive manner with automated assessment visualizations.
- *Automated modeling interface:* a pipeline application called Model Studio that enables you to construct automated flows consisting of various nodes for preprocessing and modeling with automated model assessment and comparison and direct model publishing and registration.

Each of these executes the same underlying actions in the CAS execution environment.

In this book, you primarily explore the Model Studio interface and its integration with other SAS Visual Data Mining and Machine Learning interfaces.

You can use the SAS Visual Data Mining and Machine Learning web client to assemble, configure, build, and compare tree-based models visually and programmatically.

SAS Viya provides two programming run-time servers for processing data that is not performed by the CAS server. Which server is used is determined by your SAS environment. When your SAS environment includes the SAS Viya visual and programming environments, your SAS administrator determines the server. The SAS Workspace Server and the SAS Compute Server support the same SAS code and produce the same results.

There are several interfaces and ways of executing analyses in SAS Viya. This includes the CAS actions, SAS procedures, and visual applications shown in Figure 1.4.

The Decision Tree Action Set

Decision Tree Action Set (Table 1.1) provides actions for modeling and scoring with tree-based models that include decision trees, forests, and gradient boosting.

Figure 1.4: Interfaces and Ways of Executing Analyses in SAS Viya

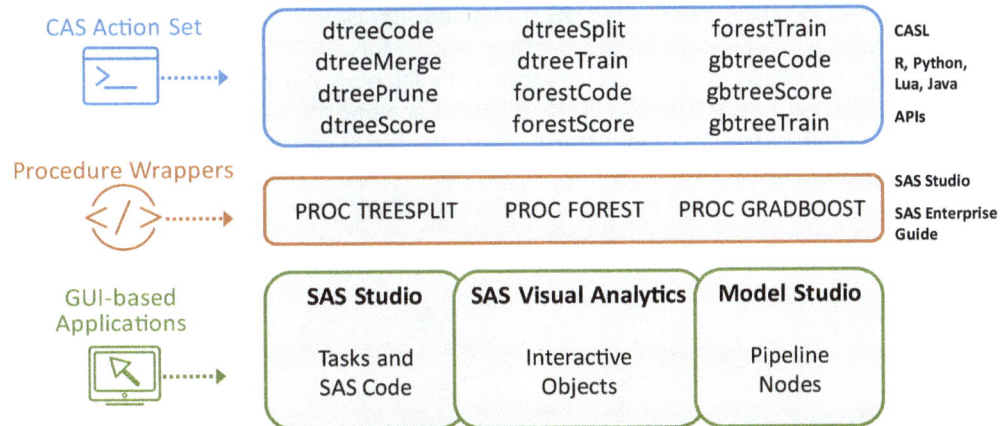

Table 1.1 Decision Tree Action Set

Action Name	Description
dtreeCode	Generates DATA step scoring code from a decision tree model
dtreeMerge	Merges decision tree nodes
dtreePrune	Prunes a decision tree
dtreeScore	Scores a table using a decision tree model
dtreeSplit	Splits decision tree nodes

(Continued)

Table 1.1 Decision Tree Action Set

Action Name	Description
dtreeTrain	Trains a decision tree
forestCode	Generates DATA step scoring code from a forest model
forestScore	Scores a table using a forest model
forestTrain	Trains a forest
gbtreeCode	Generates DATA step scoring code from a gradient boosting tree model
gbtreeScore	Scores a table using a gradient boosting tree model
gbtreeTrain	Trains a gradient boosting tree

SAS Viya also supports new analytic methods that can be accessed from SAS and other programming languages that include R, Python, Lua, and Java, as well as public REST APIs.

TREESPLIT, FOREST, and GRADBOOST Procedures

The TREESPLIT procedure builds tree-based statistical models for classification and regression in SAS Viya. The procedure produces a classification tree, which models a categorical response, or a regression tree, which models a continuous response. For each type of tree, you specify a target variable whose values you want PROC TREESPLIT to predict and one or more input variables whose values the procedure uses to predict the values of the target variable.

The following statements and options are available in the TREESPLIT procedure:

```
PROC TREESPLIT <options>;
        AUTOTUNE <options>;
        CLASS variables;
        CODE <options>;
        FREQ variable;
        GROW criterion <options>;
        MODEL response = variable. . .;
        OUTPUT OUT=CAS-libref.data-table output-options;
        PARTITION <partition-options>;
        PRUNE prune-method <(prune-options)>;
        VIICODE <options>;
        WEIGHT variable;
```

The PROC TREESPLIT statement and the MODEL statement are required.

The FOREST procedure creates a predictive model called a forest (which consists of several decision trees) in SAS Viya. The FOREST procedure creates an ensemble of decision trees to predict a single target of either interval or nominal measurement level. An input variable can have an interval or nominal measurement level.

The following statements are available in the FOREST procedure:

```
PROC FOREST <options>;
      AUTOTUNE <options>;
      CODE <options>;
      CROSSVALIDATION <options>;
      GROW criterion;
      ID variables;
      INPUT variables </ LEVEL=NOMINAL | INTERVAL>;
      OUTPUT OUT=CAS-libref.data-table <option>;
      PARTITION partition-option;
      SAVESTATE RSTORE=CAS-libref.data-table;
      TARGET variable </ LEVEL=NOMINAL | INTERVAL>;
      VIICODE <options>;
      WEIGHT variable;
```

The PROC FOREST, INPUT, and TARGET statements are required. The INPUT statement can appear multiple times.

The GRADBOOST procedure creates a predictive model called a *gradient boosting model* in SAS Viya. Based on the boosting method in Hastie, Tibshirani, and Friedman (2001) and Friedman (2001), the GRADBOOST procedure creates a predictive model by fitting a set of additive trees.

The following statements are available in the GRADBOOST procedure:

```
PROC GRADBOOST <options>;
      AUTOTUNE <options>;
      CODE <options>;
      CROSSVALIDATION <options>;
      ID variables;
      INPUT variables </ options>;
      OUTPUT OUT=CAS-libref.data-table <option>;
      PARTITION partition-option;
      SAVESTATE RSTORE=CAS-libref.data-table;
      TARGET variable </ LEVEL=NOMINAL | INTERVAL>;
      TRANSFERLEARN variable </ options>;
      VIICODE <options>;
      WEIGHT variable;
```

The PROC GRADBOOST, INPUT, and TARGET statements are required. The INPUT statement can appear multiple times.

Decision Tree, Forest, and Gradient Boosting Tasks and Objects

Shown in Figure 1.5 are SAS Studio tasks (left) and SAS Visual Analytics objects (right) relevant to tree-based models.

Figure 1.5: SAS Studio Tasks and SAS Visual Analytics Objects

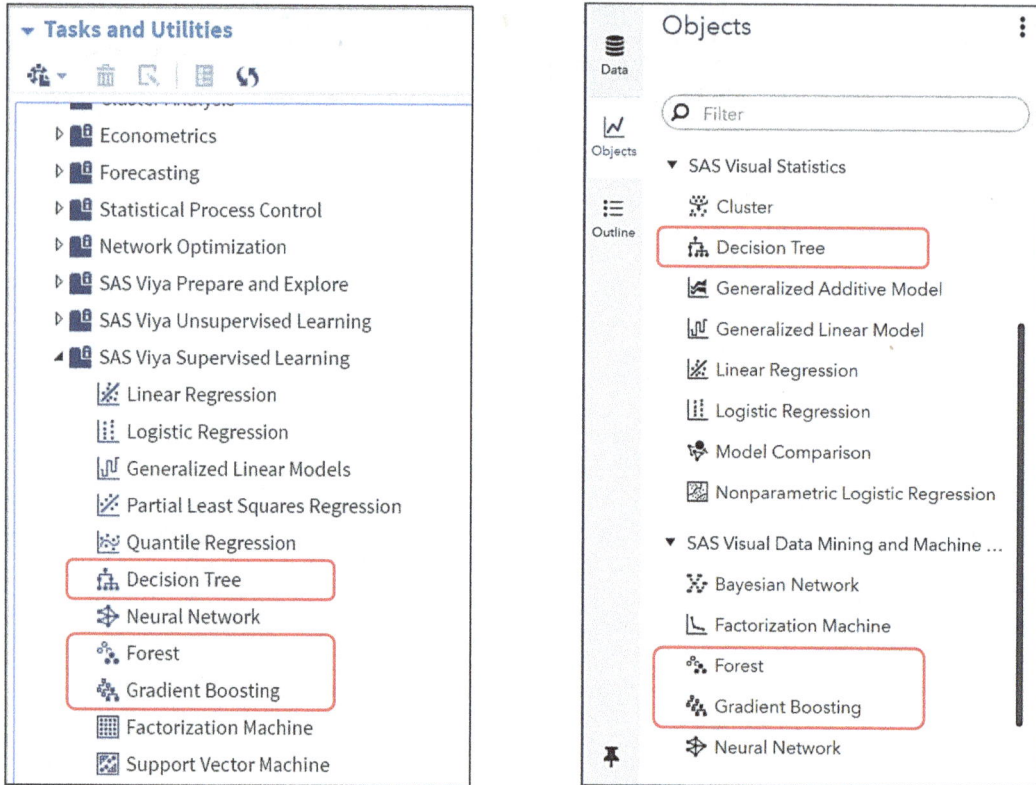

SAS Studio is more than just an editor. It is familiar to SAS programmers who just want to write code – no point and click required to start writing in SAS. If you are not familiar with SAS code, SAS Studio includes visual point-and-click tasks that generate code so that you do not have to code. SAS Studio comes with code snippet libraries for frequently used operations, as well as interactive assistance for defining code that works.

> Decision trees are available in SAS Visual Analytics without adding SAS Visual Statistics. However, SAS Visual Statistics does augment the decision tree functionality. The decision tree in SAS Visual Statistics uses a modified version of the C4.5 algorithm. If SAS Visual Data Mining and Machine Learning is licensed at your site (and you have permission), the Forest and Gradient Boosting objects can be accessed under SAS Visual Data Mining and Machine Learning.

SAS Viya enables you to develop, deploy, and manage enterprise-class analytical assets throughout the analytics life cycle (data, discovery, and deployment) with a single platform with the underlying engine called CAS.

SAS Viya delivers a single, consolidated, and centralized analytics environment. Customers no longer need to stitch together different analytic code bases.

It natively supports programming in SAS and access to SAS from other languages such as R, Python, Java, and Lua. This means that data scientists and coders not familiar with SAS can use SAS Viya, but they do not need to learn SAS code.

It supports access to SAS from third-party applications with public REST APIs, so developers can easily include SAS Analytics in their applications.

Regardless of which interface is used, the same CAS actions are applied behind the scenes for the same procedure. This provides important consistency.

A *CAS action*, the smallest unit of functionality in CAS, sends a request to the CAS server. The action parses the arguments of the request, invokes the action function, returns the results, and cleans the resources.

Introducing Model Studio

Model Studio enables you to explore ideas and discover insights by preparing data and building models. Model Studio is a central, web-based application that includes a suite of integrated data mining tools. The data mining tools supported in Model Studio are designed to take advantage of the SAS Viya programming and cloud processing environments to deliver and distribute data mining champion models, score code, and results.

Demo 1.1: Model Studio Introductory Flow

In this demonstration, you create a project and define metadata in Model Studio based on the **insurance_part** data set. This data is about a target marketing campaign for a bank that was undertaken to identify segments of customers who are likely to respond to a variable annuity (an insurance product) marketing campaign. The data set contains banking customers and 48 inputs that describe each customer. The 48 input variables represent other product usage in a three-month period and demographics. Two of the inputs are nominally scaled; the others are interval or binary. A binary target variable, **Ins**, indicates whether the customer bought the variable annuity product or not.

1. From the Windows taskbar, launch Google Chrome. When the browser opens, select **SAS Viya** ⇨ **SAS Drive** from the bookmarks bar or from the link on the page.
2. Log on using your user ID and password.
   Note: Use caution when you enter the user ID and password because values can be case sensitive.
3. Click **Sign In**.

4. Select **Yes** in the Assumable Groups window. The SAS Drive home page appears.

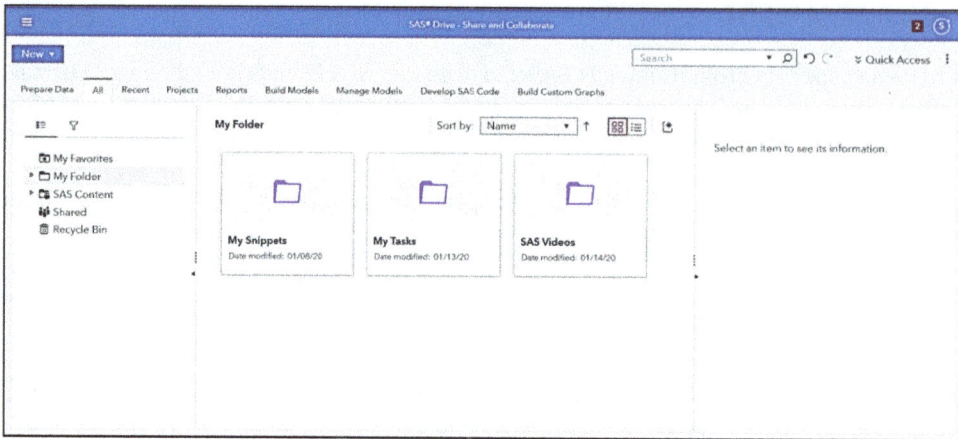

Note: The SAS Drive page on your computer might not have the same tiles as the image above.

5. Click the **Applications** menu (☰) in the upper left corner of the SAS Drive page. Select **Build Models**.

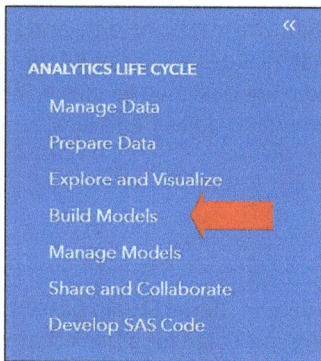

The Model Studio Projects page is now displayed.

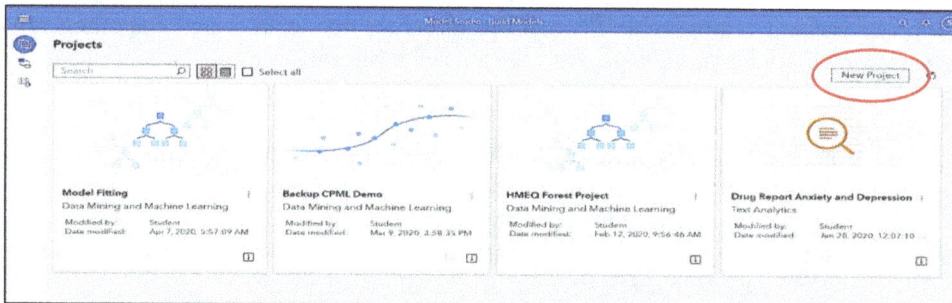

6. Select **New Project** in the upper right corner of the Projects page.

7. Enter **Insurance_ClassTree** as the name in the New Project window. Leave the default Type of **Data Mining and Machine Learning**.
Select **Browse** in the **Data** field.

8. Import a SAS data set into CAS.
   a. In the Choose Data window, click **Import**.

   b. Under Import, expand **Local Files** and then select **Local File**.

  c. Navigate to the data folder.

  d. Select the **insurance_part.sas7bdat** table. Click **Open**.

  e. Select **Import Item**. Model Studio parses the data set and populates the window with data set configurations.

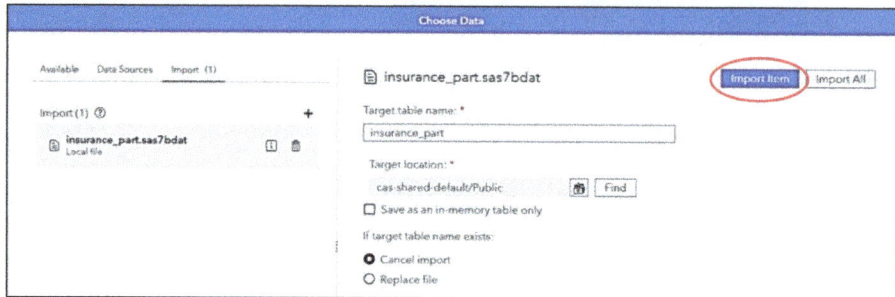

  Note: When the data is in memory, it is available for other projects through the Available tab.

  f. Click **OK** after the table is imported.

9. Click **Advanced** in the New Project window.

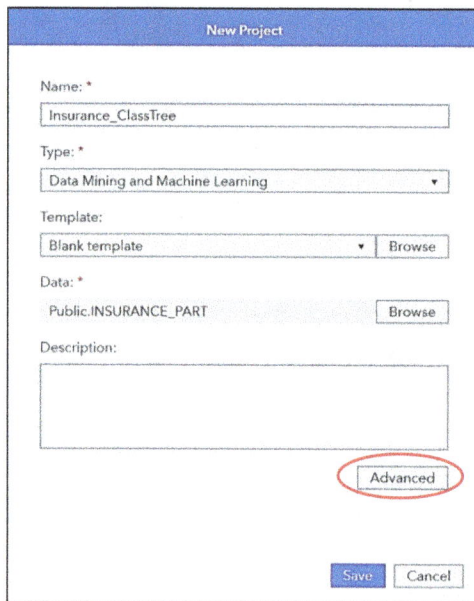

10. The Advanced project settings appear, and **Advisor Options** is one of the selections. Change **Interval cutoff** from 20 to **4**.

This is the threshold for a numeric input to be assigned as an interval variable, which means that if a numeric input has more than four distinct values or a nominal variable has more levels than four, it will be interval; otherwise, it will be nominal.

11. Click **Partition Data**. Uncheck the **Create partition variable** box as the data is already partitioned into training (70%) and validation (30%).

Click **Save** to return to the New Project window.

12. Click **Save** again.

When the project is created, you need to assign a target variable to run a pipeline.

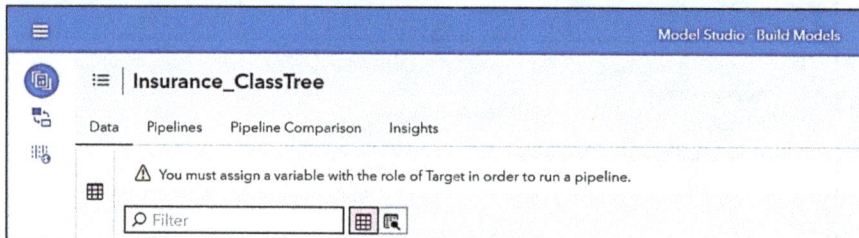

13. In the variables window, scroll down and select **INS** (Step 1). Then in the right pane, select **Target** under the **Role** property (Step 2).

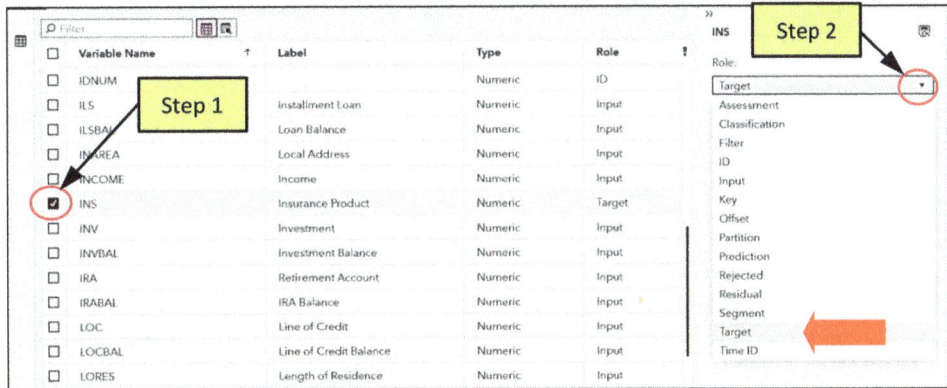

The right pane enables you to specify several properties of the variables, which includes **Role**, **Level**, **Order**, **Transform**, **Impute**, **Lower Limit**, and **Upper Limit**.

14. In the variables window, deselect **INS** and select **IDNUM**. In the right pane, select **Nominal** under the Level property.

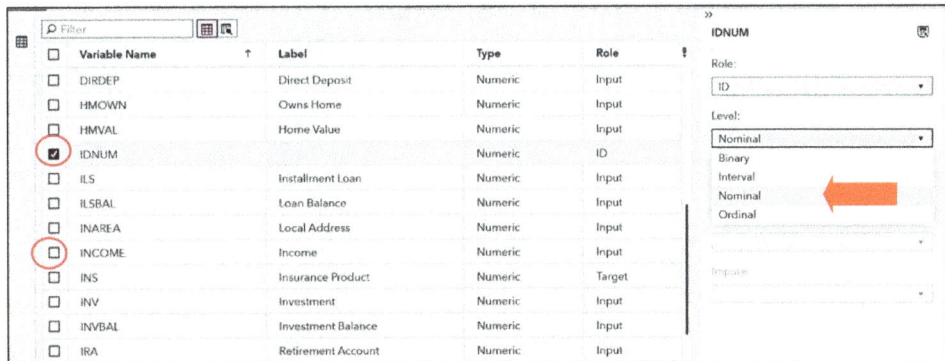

IDNUM is a customer identification variable in which each customer has a unique value.

15. Deselect **IDNUM**. Click the **Level** column to sort as per variable roles. Hide the right pane by clicking the **>>** icon to be able to see the additional columns.

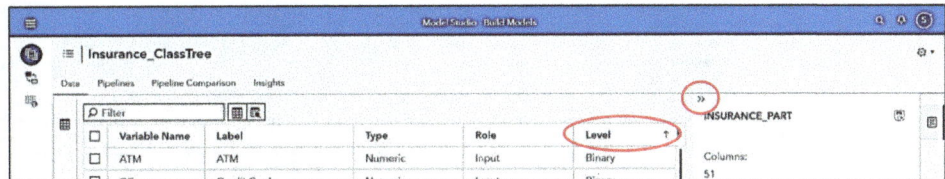

Columns' numerical details are shown.

Variable Name	Label	Type	Role	Level	↑	Order	C...	Nu...	Missing	Minimum	Maximum	Mean
PartInd	Partition Indicator	Numeric	Partition	Binary		Default		2	0.0000	0.0000	1.0000	0.7000
ATM	ATM	Numeric	Input	Binary		Default		2	0.0000	0.0000	1.0000	0.4098
CC	Credit Card	Numeric	Input	Binary		Default		2	12.7344	0.0000	1.0000	0.4804
CD	Certificate of Deposit	Numeric	Input	Binary		Default		2	0.0000	0.0000	1.0000	0.1299
DDA	Checking Account	Numeric	Input	Binary		Default		2	0.0000	0.0000	1.0000	0.8144
DIRDEP	Direct Deposit	Numeric	Input	Binary		Default		2	0.0000	0.0000	1.0000	0.2957
HMOWN	Owns Home	Numeric	Input	Binary		Default		2	17.2444	0.0000	1.0000	0.5445
ILS	Installment Loan	Numeric	Input	Binary		Default		2	0.0000	0.0000	1.0000	0.0506
INAREA	Local Address	Numeric	Input	Binary		Default		2	0.0000	0.0000	1.0000	0.9599
INS	Insurance Product	Numeric	Target	Binary		Default		2	0.0000	0.0000	1.0000	0.3464
INV	Investment	Numeric	Input	Binary		Default		2	12.7344	0.0000	1.0000	0.0301
IRA	Retirement Account	Numeric	Input	Binary		Default		2	0.0000	0.0000	1.0000	0.0634
LOC	Line of Credit	Numeric	Input	Binary		Default		2	0.0000	0.0000	1.0000	0.0628
MM	Money Market	Numeric	Input	Binary		Default		2	0.0000	0.0000	1.0000	0.1156
MOVED	Recent Address Change	Numeric	Input	Binary		Default		2	0.0000	0.0000	1.0000	0.0290
MTG	Mortgage	Numeric	Input	Binary		Default		2	0.0000	0.0000	1.0000	0.0500
NSF	Number Insufficient Fund	Numeric	Input	Binary		Default		2	0.0000	0.0000	1.0000	0.0876
SAV	Saving Account	Numeric	Input	Binary		Default		2	0.0000	0.0000	1.0000	0.4636
SDB	Safety Deposit Box	Numeric	Input	Binary		Default		2	0.0000	0.0000	1.0000	0.1076
BRANCH	Branch of Bank	Character	Input	Nominal		Default		19	0.0000			
IDNUM		Numeric	ID	Nominal		Default		>254	0.0000	1.0000	19,357.0000	9,679.0000
RES	Area Classification	Character	Input	Nominal		Default		3	0.0000			

Focus on all the categorical variables first. All the binary variables are lumped together at the top and all the nominal inputs are lumped together at the bottom (scroll down to see). There are 17 binary inputs. Three of them contain missing values (**CC, HMOWN, INV**). Approximately 35% of the customers bought the insurance product. **BRANCH** has 19 levels, and **RES** has three (**Rural, Urban, Suburban**). Notice that you have a partition indicator variable, **_PartInd_**.

16. Readjust your scroll bar to see all the interval inputs now.

Variable Name	Label	Type	Role	Level	↑	Order	Comment	Number of Levels	Missing	Minimum	Maximum	Mean
ACCTAGE	Age of Oldest Account	Numeric	Input	Interval		Default		>254	6.4111	0.3000	61.5000	5.8820
AGE	Age	Numeric	Input	Interval		Default		79	19.7241	16.0000	94.0000	47.9001
ATMAMT	ATM Withdrawal Amount	Numeric	Input	Interval		Default		>254	0.0000	0.0000	427,731.2600	1,262.8780
CASHBK	Number Cash Back	Numeric	Input	Interval		Default		4	0.0000	0.0000	4.0000	0.0155
CCBAL	Credit Card Balance	Numeric	Input	Interval		Default		>254	12.7344	1,903.9900	1,593,580.2400	9,064.0477
CCPURC	Credit Card Purchases	Numeric	Input	Interval		Default		6	12.7344	0.0000	5.0000	0.1513
CDBAL	CD Balance	Numeric	Input	Interval		Default		>254	0.0000	0.0000	386,700.0000	2,431.6041
CHECKS	Number of Checks	Numeric	Input	Interval		Default		39	0.0000	0.0000	49.0000	4.2573
CRSCORE	Credit Score	Numeric	Input	Interval		Default		>254	2.0509	509.0000	820.0000	666.5161
DDABAL	Checking Balance	Numeric	Input	Interval		Default		>254	0.0000	-774.8300	278,093.8300	2,135.5186
DE		Numeric	Input	Interval		Default		>254	0.0000	0.0000	16,000.0000	5,197.1115
DEP	Checking Deposits	Numeric	Input	Interval		Default		19	0.0000	0.0000	28.0000	2.1261
DEPAMT	Amount Deposited	Numeric	Input	Interval		Default		>254	0.0000	0.0000	359,618.4700	2,236.7032
HMVAL	Home Value	Numeric	Input	Interval		Default		173	18.0142	67.0000	754.0000	110.6684
ILSBAL	Loan Balance	Numeric	Input	Interval		Default		>254	0.0000	0.0000	24,634.6300	528.2789
INCOME	Income	Numeric	Input	Interval		Default		192	18.0142	0.0000	233.0000	40.3670
INVBAL	Investment Balance	Numeric	Input	Interval		Default		>254	12.7344	-2,214.9200	8,323,746.0200	1,996.8655
IRABAL	IRA Balance	Numeric	Input	Interval		Default		>254	0.0000	0.0000	285,339.7700	577.9684
LOCBAL	Line of Credit Balance	Numeric	Input	Interval		Default		>254	0.0000	-71.6100	523,147.2400	1,204.2149
LORES	Length of Residence	Numeric	Input	Interval		Default		39	18.0142	0.5000	19.5000	7.0244
MMBAL	Money Market Balance	Numeric	Input	Interval		Default		>254	0.0000	0.0000	68,965.9200	1,884.2266
MMCRED	Money Market Credits	Numeric	Input	Interval		Default		6	0.0000	0.0000	5.0000	0.0374
MTGBAL	Mortgage Balance	Numeric	Input	Interval		Default		>254	0.0000	0.0000	1,628,532.3800	7,626.0502
NSFAMT	Amount NSF	Numeric	Input	Interval		Default		>254	0.0000	0.0000	666.8500	2.3472
PHONE	Number Telephone Banking	Numeric	Input	Interval		Default		19	12.7344	0.0000	30.0000	0.4040
POS	Number Point of Sale	Numeric	Input	Interval		Default		35	12.7344	0.0000	54.0000	1.0747
POSAMT	Amount Point of Sale	Numeric	Input	Interval		Default		>254	12.7344	0.0000	2,408.4300	48.6339
SAVBAL	Saving Balance	Numeric	Input	Interval		Default		>254	0.0000	0.0000	609,587.7200	3,138.5250
TELLER	Teller Visits	Numeric	Input	Interval		Default		26	0.0000	0.0000	25.0000	1.3512

Of the 29 interval variables, 12 contain missing values. Unlike parametric models like regression and neural network, decision trees handle missing values quite well. This is discussed later in the book.

17. Click the **Pipelines** tab in the **Insurance_ClassTree** project.

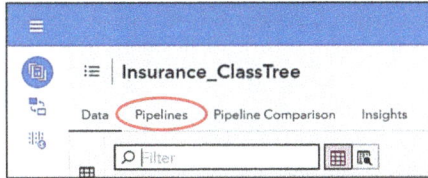

A blank template pipeline was created with only a Data node.

End of Demonstration

Quiz

1. Model Studio is a central, web-based application that includes a suite of integrated data mining and machine learning tools.
   a.   True
   b.   False

2. Decision trees can be created for nominal targets as well as for the interval targets.
   a.   True
   b.   False

3. Which of the following approaches can you use to build decision trees and tree-based models in SAS Viya? (Select all that apply.)
   a.   Python, Java, Lua, and R languages
   b.   CASL language
   c.   SAS procedure wrappers
   d.   GUI-based applications like Model Studio and SAS Visual Statistics

Answers

1. True	2. True	3. a, b, c, and d.

Chapter 2: Classification and Regression Trees

Classification Trees

A classification tree is a predictive model where the target variable is categorical. The decision tree algorithm is then used to identify the "class" or "label" within which a target variable would most likely fall. An example of a classification-type problem would be determining who will or will not default to a bank loan or who will or will not churn from a telecom service provider.

Consider the case of a nominal target variable. Handwriting recognition is a classic application of supervised prediction.

The example data set is a subset of the pen-based recognition of handwritten digits data, available from the UCI repository (Blake et al. 1998). The cases are digits written on a pressure-sensitive tablet. The input variables measure the position of the pen. They are scaled to be between 0 and 100. Two of the original 16 inputs are shown (**X1** and **X10**) in Figure 2.1. The target is the true written digit (0-9). This subset contains the 1064 cases corresponding to the three digits 1, 7, and 9. Each case represents a point in the input space. (The data were jittered for display because many of the points overlap.)

Figure 2.1: Handwriting Recognition Example

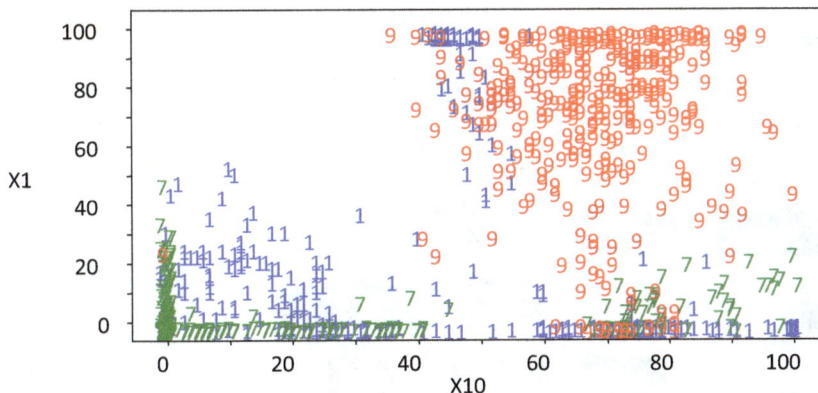

In the pen-digits data, the inputs have an interval measurement scale and the target has a nominal measurement scale. The generic, supervised, prediction problem places no restrictions on the scales of the inputs or the target.

A classification tree built on the handwriting recognition data is shown in Figure 2.2.

A decision tree is read from the top down starting at the *root node*. Each internal node represents a split based on the values of one of the inputs. The inputs can appear in any number of splits throughout the tree. Cases move down the branch that contains its input value.

In a binary tree with interval inputs, each internal node is a simple inequality. A case moves left if the inequality is true and right otherwise. The terminal nodes of the tree are called *leaves*. The leaves represent the predicted target. All cases reaching a particular leaf are given the same predicted value. When the target is categorical, the model is a called a *classification tree*. The leaves give the predicted class as well as the probability of class membership.

Decision Regions

The leaves of a classification tree partition the input space into rectilinear regions, shown in Figure 2.3. The leaves give the predicted target class. The class assignments are determined by the posterior probability of each class. That is, the predicted classification is determined by the class that is most prevalent in the terminal (leaf) node.

A classification tree can be thought of as defining several multivariate step functions. Each function corresponds to the posterior probability of a target class shown in Figure 2.4.

Figure 2.2: Classification Tree

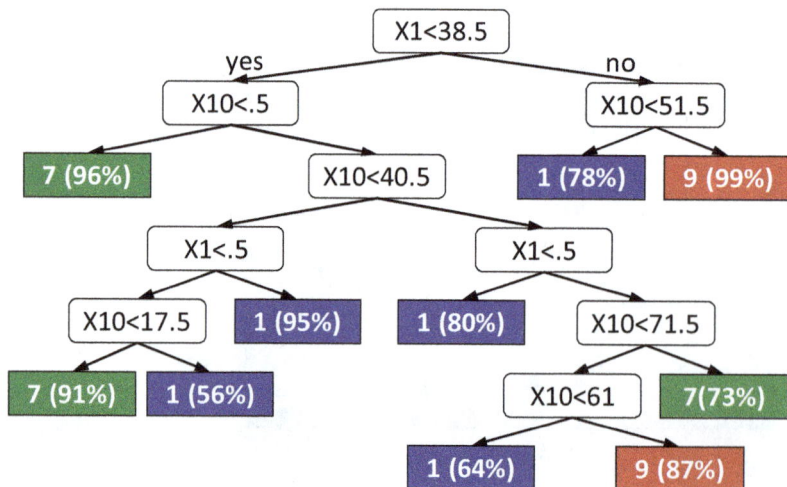

Figure 2.3: Rectilinear Regions

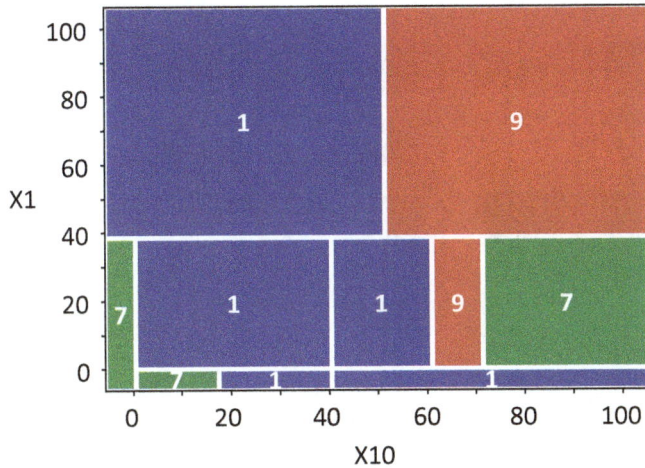

Figure 2.4: Leaves of Classification Tree

Leaf	Pr(1|x)	Pr(7|x)	Pr(9|x)	Decision
1	.03	.96	.01	7
2	.09	.91	.00	7
3	.56	.44	.00	1
4	.95	.05	.00	1
5	.80	.10	.10	1
6	.64	.09	.27	1
7	.00	.13	.87	9
8	.10	.73	.17	7
9	.78	.01	.21	1
10	.01	.00	.99	9

Posterior Probabilities

A classification tree can also provide a measure of confidence that the classification is correct.

For a trichotomous target, there are three posterior probabilities that are constrained to sum to 1. An allocation rule assigns class labels to regions based on the relative magnitudes of the posterior probabilities. The calculation shown in Figure 2.5, for example, is corresponding to the posterior probabilities of leaf 10 (last row shown in Figure 2.4). Likewise, you can observe the posterior probabilities corresponding to other leaves.

Figure 2.5: Posterior Probabilities

100·Pr(DIGIT=**1**|**x**)   100·Pr(DIGIT=**7**|**x**)   100·Pr(DIGIT=**9**|**x**)

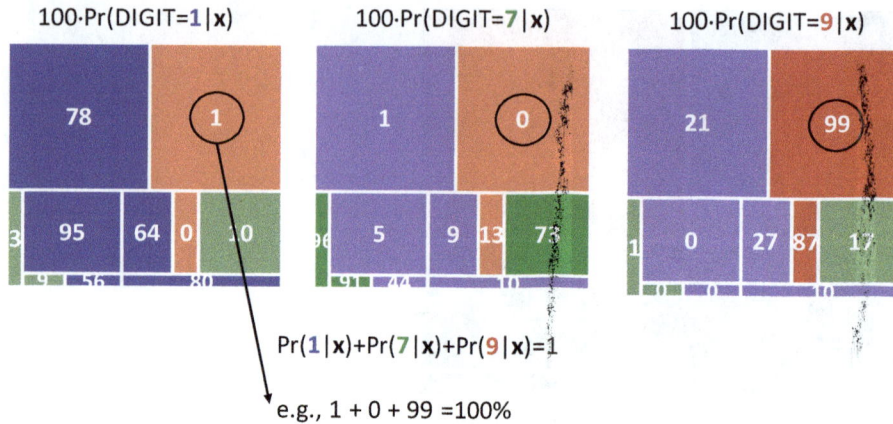

$$Pr(\mathbf{1}|\mathbf{x})+Pr(\mathbf{7}|\mathbf{x})+Pr(\mathbf{9}|\mathbf{x})=1$$

e.g., 1 + 0 + 99 =100%

Figure 2.6: Multi-Way Splits

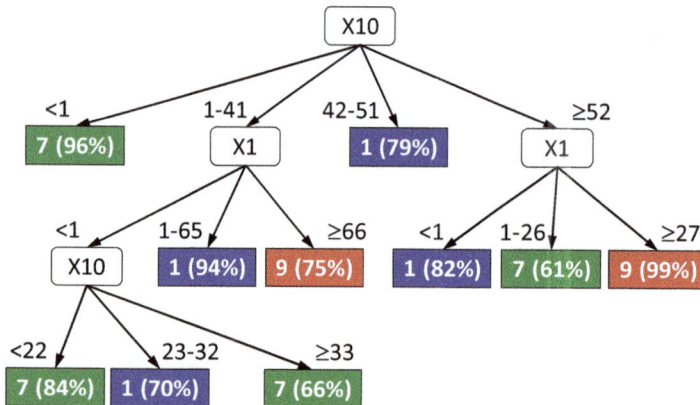

Decision trees can have multi-way splits where the values of the inputs are partitioned into disjoint ranges. See Figure 2.6.

In decision trees in which numeric attributes are split, several branches are generally more coherent than the usual binary splits because attributes rarely appear more than once in any path from root to leaf.

Demo 2.1: Building a Default Classification Tree

In this demonstration, you create an introductory pipeline in Model Studio based on the **insurance_part** data set. This pipeline creates a classification tree to be used to predict insurance purchases in a fictitious retail insurance company.

1. Ensure that the **Insurance_ClassTree** project is open in Model Studio and that you are on the **Pipelines** tab with **Pipeline1** open.
2. Right-click the Data node and select **Add child node** ⇨ **Supervised Learning** ⇨ **Decision Tree**.

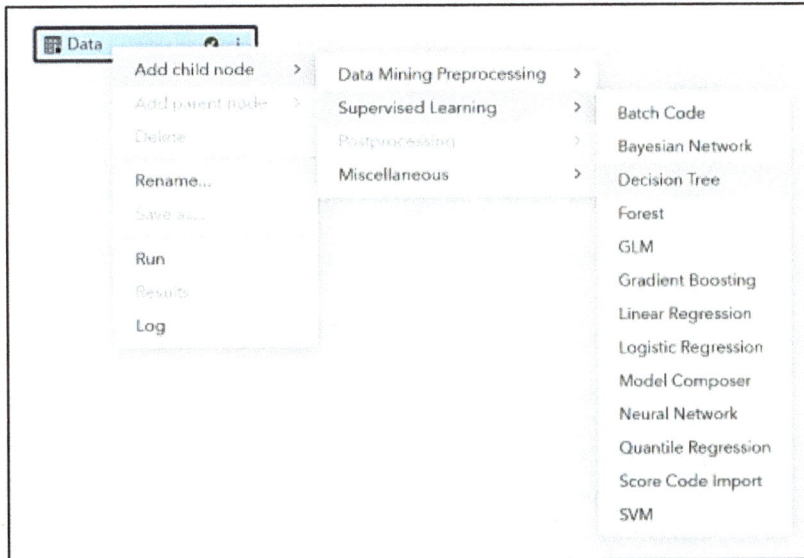

3. Click the **Run Pipeline** button to run the pipeline.

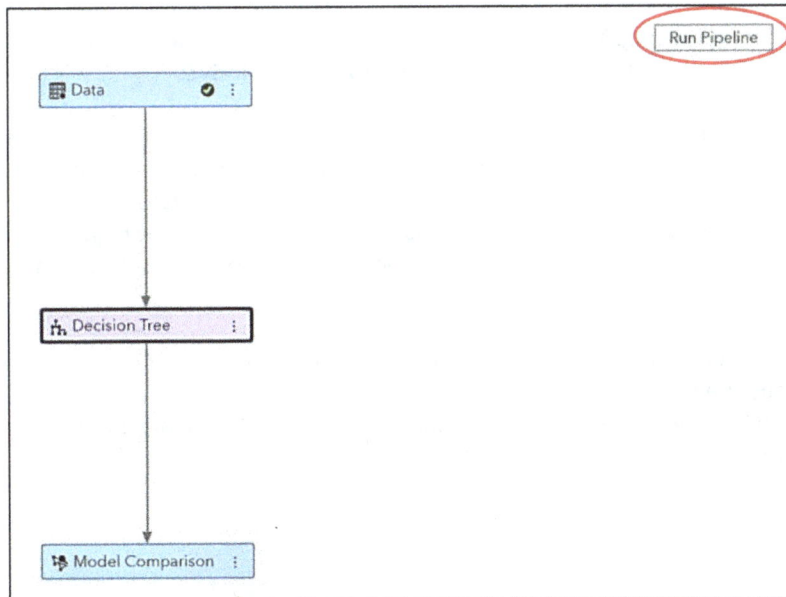

4. When the pipeline finishes running, right-click the **Decision Tree** node and select **Results**.

   Click ↗ to expand the **Tree Diagram** and examine the decision tree created.

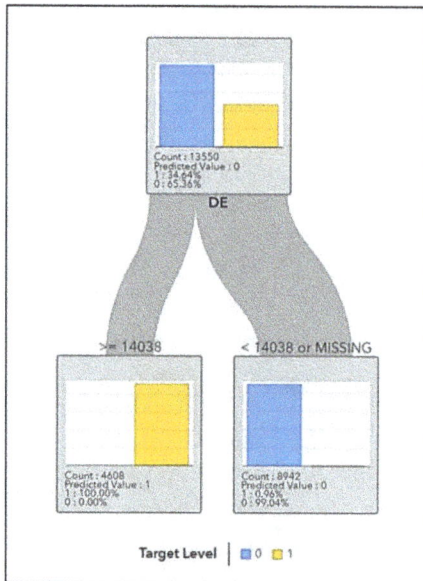

The results are at first surprising. The tree split only once. The split is almost perfect. The left child leaf has 100% **INS=1**, and the right child leaf has around 99% **INS=0**.

The root node has split using the variable **DE**. Overly simple trees with near perfect splits sometimes indicate the presence of inappropriate inputs. In this case, **DE** is an unformatted SAS date variable. It indicates the date of insurance purchase and is an outcome, rather than a predictor, of the target. Such *temporal infidelity* is more common than you might expect, particularly when working with large data sets and poorly understood inputs. Decision trees can be very useful tools for initial data exploration.

5. Click ✕ to exit the maximized view of the tree diagram. Click the **Close** button to discontinue the results.
6. Go back to the **Data** pane and click the check box of the **DE** variable. In the right pane, select **Rejected** under the **Role** property.

	Variable N... ↑	Label	Type	Role	Level	Order	
☐	CRSCORE	Credit Score	Numeric	Input	Interval	Default	
☐	DDA	Checking Account	Numeric	Input	Binary	Default	
☐	DDABAL	Checking Balance	Numeric	Input	Interval	Default	
☑	DE		Numeric	Rejected	Interval	Default	
☐	DEP	Checking Deposits	Numeric	Input	Interval	Default	
☐	DEPAMT	Amount Deposited	Numeric	Input	Interval	Default	
☐	DIRDEP	Direct Deposit	Numeric	Input	Binary	Default	
☐	HMOWN	Owns Home	Numeric	Input	Binary	Default	
☐	HMVAL	Home Value	Numeric	Input	Interval	Default	
☐	IDNUM		Numeric	ID	Nominal	Default	

Note: Remember to deselect the check box of this variable after you have changed the metadata.

7. Click the **Pipelines** tab. Notice that because of the change in metadata, the green check marks in the nodes in the pipeline have been changed to gray circles.
8. Click the **Run Pipeline** button to rerun the pipeline.
9. When the pipeline finishes running, right-click the **Decision Tree** node and select **Results**. There are several charts and plots to help you evaluate the model's performance.
   The first two plots are the tree diagram and treemap, which present the tree structure of the final decision tree, such as the depth of the tree and all terminal nodes (leaves).
   A tree diagram displays the decision tree. The width of the line that connects the node in the tree is proportionally given by the ratio of the number of observations in the branch to the number of observations in the root node.
   A treemap displays a compact graphical display of the tree. The rectangular regions that represent the tree nodes are colored by probability of event or proportion correctly classified for a class target or the average of an interval target.

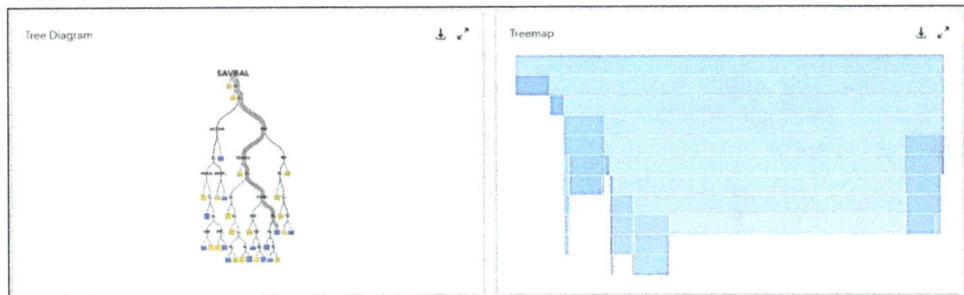

The treemap (also known as icicle plot in SAS Visual Analytics) is intended to be used in conjunction with a tree diagram to gauge the relative size of each leaf. Soon we will return to these two outputs.

The Pruning Error Plot displays a graph of the misclassification rate for a class target or ASE for an interval target. These values are given as a function of the number of leaves in a subtree for each data partition. There could be several subtrees. Each subtree will have a different number of leaves in it. A reference line is drawn to indicate how many leaf subtrees were selected as the final model (31 leaves in this case).

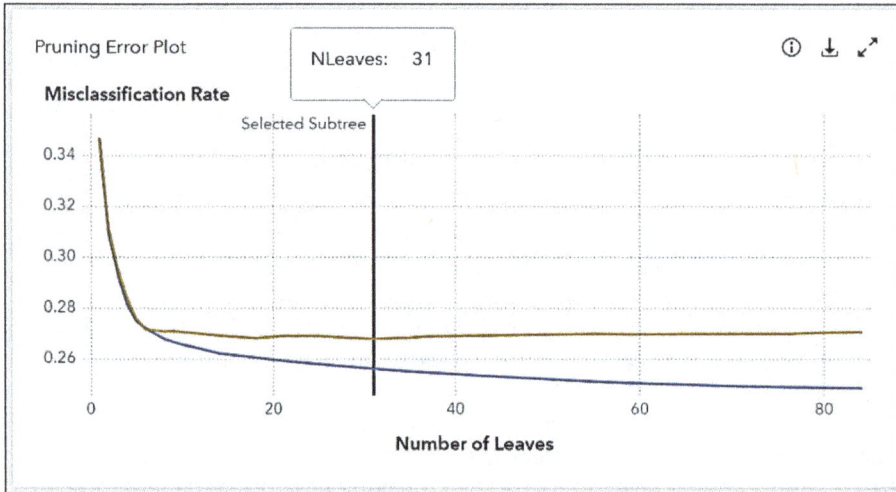

The Variable Importance table shows importance data for each variable (maximize the window to see).

Intuitively, the variables used in a tree have different levels of importance. What makes a variable important is the strength of the influence and the number of cases influenced.

You would typically use the selected variables as the inputs in a model such as logistic regression or neural network. In practice, trees often provide far fewer variables than seem appropriate for a regression. The sensible solution is to include some variables from another technique, such as correlation. No single selection technique is capable of fully prophesizing which variables will be effective in another modeling tool.

Variable Importance						
Variable Label	Role	Variable Name	Validation Imp...	I...	Relative Importance	Count
Saving Balance	INPUT	SAVBAL	222.0828	0	1	2
Money Market	INPUT	MM	91.0402	0	0.4099	1
CD Balance	INPUT	CDBAL	72.2555	0	0.3254	2
Checking Balance	INPUT	DDABAL	65.7168	0	0.2959	2
Checking Account	INPUT	DDA	36.4405	0	0.1641	1
IRA Balance	INPUT	IRABAL	5.5151	0	0.0248	1
Credit Card Purchases	INPUT	CCPURC	2.9355	0	0.0132	1
Age	INPUT	AGE	1.9427	0	0.0087	2
Branch of Bank	INPUT	BRANCH	0.9591	0	0.0043	2

Note: The definition and computation of each column in the variable importance table will be discussed in Lesson 4.

Model Studio gives three types of scoring code. The Node Score Code window shows the individual score code for a specific node that can be deployed in production. You will get this score code against every node in the Data Mining Preprocessing and the Supervised Learning groups that create DATA step score code. The nodes that create ASTORE score files will not generate this output.

```
Node Score Code

1      length _strfmt_ $12; drop _strfmt_;
2      _strfmt_ = ' ';
3
4      array _tlevname_46688142_{2} $12 _temporary_ ( '          1'
5      '          0');
6
7      array _dt_fi_46688142_{2} _temporary_;
8
9      _node_id_ = 0;
10     _new_id_ = -1;
11     nextnode_46688142:
12     if _node_id_ eq 0 then do;
13         _numval_ = SAVBAL;
14         if missing(_numval_) then do;
15             numval = 1 7076021240672200.
```

Note: An analytic store (ASTORE) is a binary file that contains the state from a predictive analytic procedure. This state from a predictive analytic procedure (such as a forest) is created using the results from the training phase of model development. The ASTORE file type works very well for scoring complex machine learning models. A key feature of an analytic store is that it can be easily transported from one host to another. This is in part because it is a compact and universal file form. The store names the one and only SAS component that can restore the state of the computation/memory that can restore the post training memory and go on from there. It is also called a *warm restart* for scoring.

After you add a model in the pipeline (a Supervised Learning node), that node will generate two additional score codes and a train code.

The Path Score Code is the flow score code, which includes score code for all the nodes until and including that modeling node.

```
Path Score Code                                                                    ↙↗
   14     ---------------------------------------------------------------  ,
   15     *Nodeid: _6IFIN4DH3SG8VUKIUKFADHUYQ;                                      |
   16     *-------------------------------------------------------------*;
   17       length _strfmt_ $12; drop _strfmt_;
   18       _strfmt_ = ' ';
   19
   20       array _tlevname_46688142_{2} $12 _temporary_ ( '           1'
   21                  0');
   22
   23       array _dt_fi_46688142_{2} _temporary_;
   24
   25       _node_id_ = 0;
   26       _new_id_ = -1;
   27       nextnode_46688142:
   28       if _node_id_ eq 0 then do;
```

Typically, the Path Score Code or Path EP Score Code is the score code that would be used in other SAS environments (for example, SAS Studio) for scoring. EP stands for *embedded process*, and perhaps it is the underlying engine for in-database scoring (via scoring accelerators). The Path EP Score code is what you will get when any of the nodes in the process flow create an ASTORE score file. Otherwise, you will get Path Score code. In addition, the DS2 Package Code (shown below) is score code packaged slightly different. As a DS2 package, this score code can be used for SAS Micro Analytic Services. Note that both the Path EP Score code and the DS2 Package Code are score codes written in DS2.

```
DS2 Package Code                                                                   ↙↗
   1     package MS_6dfebc1de7034587958e93cd4ed93c02_20JUN2019032425071 / overwrite=yes;
   2        dcl nchar(12) "I_INS";
   3        dcl double "P_INS0";
   4        dcl double "P_INS1";
   5        dcl double "_leaf_id_";
   6        dcl double "EM_EVENTPROBABILITY" having label n'Probability for INS =1';
   7        dcl nchar(32) "EM_CLASSIFICATION" having label n'Predicted for INS';
   8        dcl double "EM_PROBABILITY" having label n'Probability of Classification';
   9        varlist allvars [_all_];
  10
  11        method _6IFIN4DH3SG8VUKIUKFADHUYQ();
  12          dcl double _P_;
  13          dcl double _NEW_ID_;
  14          dcl double _NODE_ID_;
  15          dcl double _NUMVAL_;
```

On the other hand, the Training Code window shows the SAS training code that can be used to train the model based on different data sets or in different platforms (for example, when you scroll down in the Training code window, you can see the TREESPLIT procedure is used to train the decision tree model).

Training Code

```
1      *-------------------------------------------------------------*;
2      * Macro Variables for input, output data and files;
3        %let dm_datalib =;
4        %let dm_lib     = WORK;
5        %let dm_folder  = %sysfunc(pathname(work));
6      *-------------------------------------------------------------*;
7      *-------------------------------------------------------------*;
8       * Training for tree;
9      *-------------------------------------------------------------*;
10     *-------------------------------------------------------------*;
11      * Initializing Variable Macros;
12     *-------------------------------------------------------------*;
13  ⊖  %macro dm_unary_input;
14       %mend dm_unary_input;
15       %global dm_num_unary_input;
```

Note: There is a Score Data node in Model Studio that can be used to collect or accumulate score code at any point in the pipeline.

Finally, the Output window shows the final decision tree model parameters, the variable importance table, and the pruning iterations.

Output

The SAS System

The TREESPLIT Procedure

Model Information	
Split Criterion	IGR
Pruning Method	Cost Complexity
Max Branches per Node	2
Max Tree Depth	10
Tree Depth Before Pruning	10
Tree Depth After Pruning	10
Number of Leaves Before Pruning	103
Number of Leaves After Pruning	31

The TREESPLIT procedure is used to create a decision tree with 31 leaves. Information gain ratio is the default split criterion, and cost complexity is the default pruning method. You will learn about these topics in Lessons 3 and 5 respectively.

10. Click the **Assessment** tab.

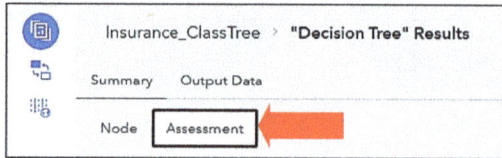

The first chart is **Lift Reports**, which displays the cumulative lift as a function of the depth for the model. The cumulative lift is given for each of the data roles. To examine other statistics as a function of depth, use the drop-down menu in the upper right corner. Other statistics include lift, gain, captured response percentage, cumulative captured response percentage, response percentage, and cumulative response percentage. This result is displayed only if the target is a class variable.

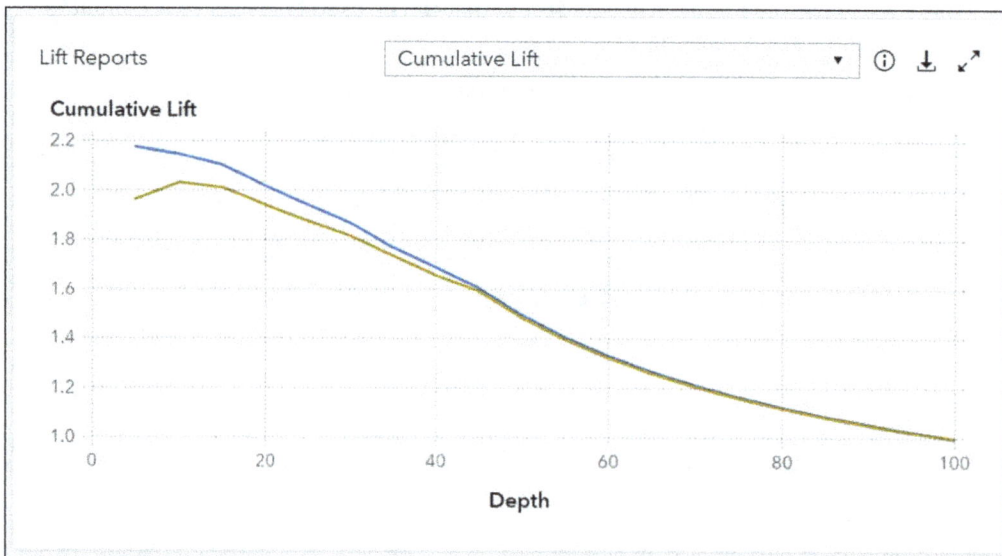

ROC Reports displays the ROC (receiver operating characteristic) chart for a model, giving the sensitivity as a function of 1-specificity. The sensitivity is given for each of the data roles. To examine other statistics, use the drop-down menu in the upper right corner. Other statistics include accuracy and F1 score. This result is displayed only if the target is a class variable.

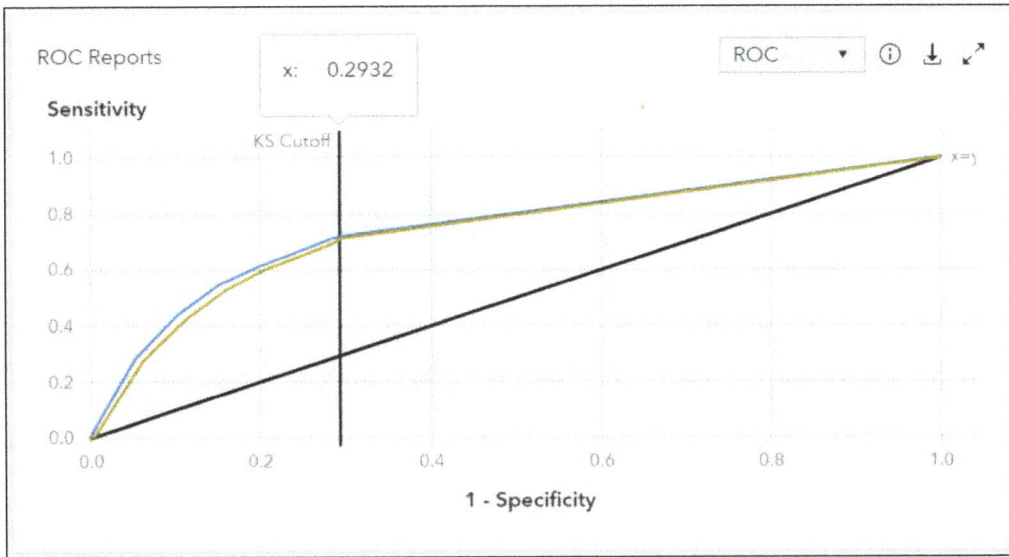

To help identify the best cutoff to use when scoring your data, the KS Cutoff reference line is drawn at the value of 1-specificity where the greatest difference between sensitivity and 1-specificity is observed for the VALIDATE partition. The KS Cutoff line is drawn at the cutoff value 0.2, where the 1-specificity value is 0.293 and the sensitivity value is 0.714 (hover your mouse to see).

The Fit Statistics output shows a table of the fit statistics for the model, broken down by data role (maximize the window to see).

				Fit Statistics						
Target ...	Data Role	Partitio...	Formatt...	Sum of ...	Averag...	Divisor ...	Root Av...	Misclas...	Multi-Cl...	KS (You...
INS	TRAIN	1	1	13,550	0.1792	13,550	0.4233	0.2561	0.5399	0.4289
INS	VALIDATE	0	0	5,807	0.1852	5,807	0.4303	0.2678	0.5680	0.4204

The Fit Statistics table shows an average squared error of 0.1852 and a misclassification rate of 0.2678 on the VALIDATE partition.

Finally, you have the Event Classification chart, which displays the confusion matrix at various cutoff values for each partition.

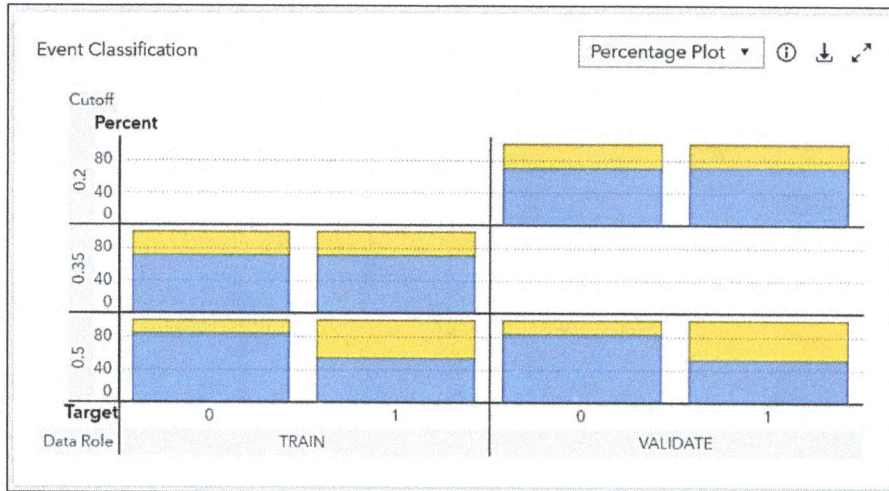

The Event Classification chart is a visual representation of the confusion matrix at various cutoff values for each partition. The classification cutoffs used in the plot are the default (0.5) and the KS value for the TRAIN partition (0.35), and VALIDATE partition (0.2).

The confusion matrix contains true positives (events that are correctly classified as events), false positives (non-events that are classified as events), false negatives (events that are classified as non-events), and true negatives (non-events that are correctly classified as non-events). Use the drop-down menu to view the information in the chart as percentages or counts. You can also view the information summarized in a table. This result is displayed only if the target is binary variable.

From the Event Classification chart, it is easy to see that for our example data and model, you get high numbers of correct (blue color) predictions in almost all the partitions that support the assumption of a low bias model.

11. Select **Node** and then scroll up to return to the Tree Diagram and Treemap result windows.

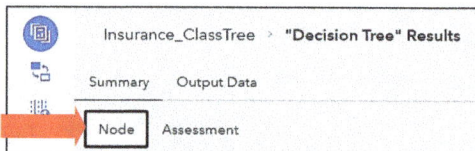

12. Click ↗ to expand the Tree Diagram and examine the decision tree created. When you move your mouse pointer over a node of a tree, a text box displays information about the node. You can also scroll to zoom in.
13. Click the left node after the first split of the top (root) node of the decision tree. Scroll to zoom in and view the characteristics of observations in this partition of the data.

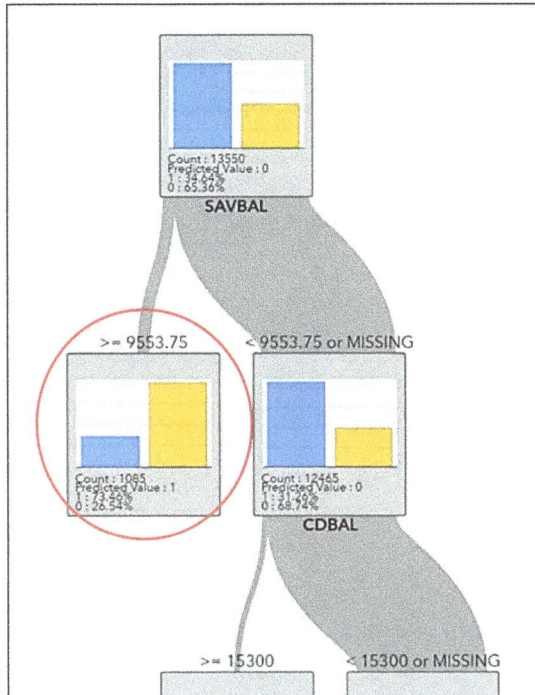

Trees notice relationships from the interaction of inputs. For example, buying an insurance annuity (INS) might not correlate with certificate of deposit (CDBAL) unless saving balance (SAVBAL) is less than 9553.75. The tree noticed both inputs. CDBAL has a split at 15300.

Moreover, trees discard superfluous inputs. Saving balance (SAVBAL) and money market balance (MMBAL) both highly correlate with buying insurance annuity (INS), but the tree needs only one of the inputs, saving balance in this case. Here superfluousness need not to be confused with redundancy. When saving balance (SAVBAL) and money market balance (MMBAL) are correlated with each other, they are said to be redundant. Decision trees have no built-in method for ignoring redundant inputs. Trees might arbitrarily select from a set of correlated inputs. It is therefore recommended that you reduce redundancy first before you use decision trees.

Finally, trees treat nominal and ordinal inputs on a par with interval ones. Instead of the categorical area classification (RES) input, the split could be at home value (HMVAL), an interval input.

A tree gives an impression that certain inputs uniquely explain the variations in the target. A completely different set of inputs might give a different explanation that is just as good.

By default, the growth process continues until the tree reaches a maximum depth of 10 (you can specify a different limit by using the **Maximum depth** property). The result is often a large tree that overfits the data and is likely to perform poorly in predicting future data. A recommended strategy for avoiding this problem is to prune the tree to a smaller subtree that minimizes prediction error, the concept discussed in lesson 5.

Note: The zoom functionality centers the diagram on your cursor. A good tip is to position the mouse pointer on the part of the tree that you are interested in investigating before zooming.

You can use a SAS Code node with the following code to compute the correlation coefficients.

Training code

```
1     %dmcas_register(dataset=&sysuserid);
2   ⊖ data &sysuserid;
3         set &dm_data;
4     run;
5
6     title 'Correlation with Target';
7   ⊖ proc corr nosimple data=&dm_data rank;
8         var %dm_interval_input ;
9         with %dm_dec_target;
10    run;
11
```

The ten most correlated inputs with the target are shown below.

INS Insurance Product	SAVBAL	MMBAL	CDBAL	DEP	DDABAL	PHONE	MMCRED	HMVAL	CCPURC	CHECKS
	0.18912	0.16882	0.15804	-0.14097	0.11322	-0.10628	0.08784	0.07933	0.07083	-0.07064
	<.0001	<.0001	<.0001	<.0001	<.0001	<.0001	<.0001	<.0001	<.0001	<.0001
	19357	19357	19357	19357	19357	16892	19357	15870	16892	19357

14. Click ⊠ to exit the maximized view of the tree diagram. Click ↗ to expand Treemap. Locate the same left node after the first split of the top (root) node of the decision tree.

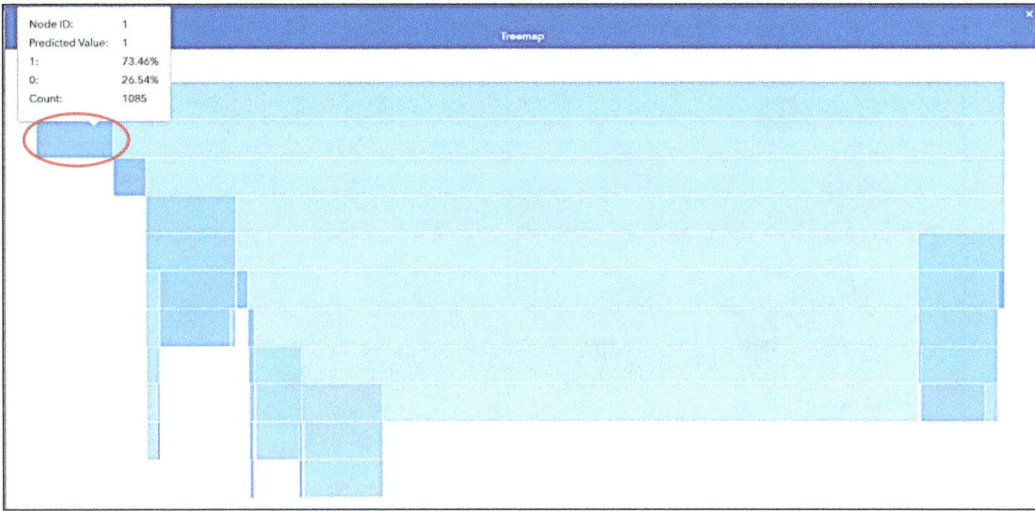

The darker the color, the higher the probability of event. The node width is proportional to the number of observations in the node. Selecting a node or hovering your pointer on a node displays information about the node.

The node shown above is one of the highest event probabilities, which contains almost three-fourths of the events.

15. Click ❎ to exit the maximized view of the treemap. Click **Close** to discontinue the results.

16. Under Tree Diagram Options, deselect Display embedded bar charts in tree diagram and change Class target node color to Proportion correctly classified.

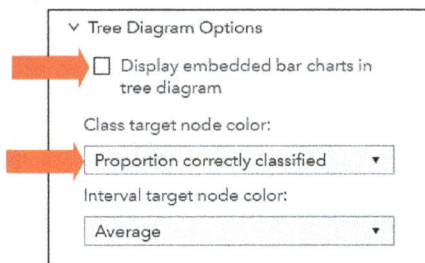

17. Click the **Run Pipeline** button.
18. Open the **Results** of the Decision Tree node again.
19. Observe that the embedded bar charts in the tree diagram have disappeared and that the nodes are colored per class target node color setting.

20. Observe the change in color composition in the treemap.

The darker the color, the purer the node. The node shown above is the purest one that has almost 17% versus 83% events and non-events into it.

21. Close the **Results** window.

End of Demonstration

Model Scoring

After building a decision tree model, or for that matter any other model, we put the model to use by scoring new data. The scoring process is sometimes referred to as *model deployment, model production,* or *model implementation*. All tasks performed earlier in the analytics life cycle lead to this task: generating predictions through scoring. An organization gets value from the model when it is in production. To maximize that value, it is necessary to monitor model performance over time and update the model as needed.

For scoring, the model is first translated into another format (typically, score code). In the SAS Viya environment, score code is a SAS program that you can easily run on a scoring data set. Then the model is applied to the scoring data set to obtain predicted outcomes. Based on the predictions from the model, the enterprise makes business decisions and takes action.

After you download the score code, there are several model deployment options in SAS Viya:

- The Score Data node in Model Studio enables you to score a data table with the score code that was generated by the predecessor nodes in the pipeline. The scored table is saved to a caslib. By default, the scored table is temporary, exists only for the duration of the run of a node, and has local session scope. The Score Data node enables you to save the scored table to disk in the location that is associated with the specified output library. After it is saved to disk, this table can be used by other applications for further analysis or reporting.

  Note: This approach is shown in this book.

- Models that create ASTORE code can be scored in SAS Studio using the ASTORE procedure.
- You can also run a scoring test in SAS Model Manager.
- Model Studio creates API for score code to be called from SAS or Python or through a REST API. Within a project in Model Studio, you can go to the Pipeline Comparison tab and click the **Project pipeline** menu (three vertical dots) in the upper right and select **Download score API**. The code that it provides can be used directly in other applications. (You just need to modify the CAS server and port that you are calling, the data source that you want to score, and the name of the output caslib and scored output table.) Even when you call the score code through the API from a Python program, the score code runs in CAS, not in Python.
- Score code can also be published to Micro Analytics Services (web service) and to SAS Event Stream Processing.

Demo 2.2: Scoring a Decision Tree Model in Model Studio

This demonstration builds on the **Insurance_ClassTree** project on the **insurance_part** data. You will score a decision tree model created in previous demonstration within Model Studio.

1. Ensure that the **Insurance_ClassTree** project is open in the Model Studio and that you are on the **Pipelines** tab with **Pipeline1** open.
2. Right-click the **Decision Tree** node and select **Add child node** ⇨ **Miscellaneous** ⇨ **Score Data**.

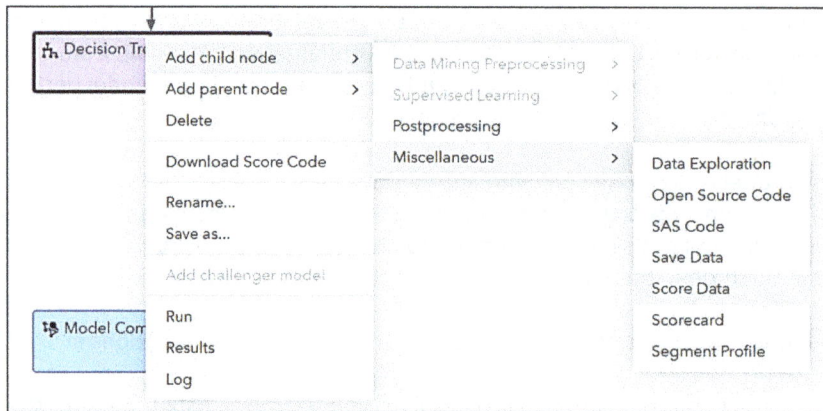

3. Select the **Score Data** node. The Score Data node enables you to score a data table with the score code that was generated by the predecessor nodes in the pipeline. The scored table is saved to a CAS library.

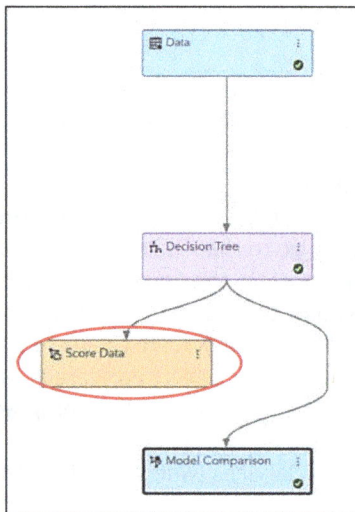

> **Note:** Nodes cannot be added after the Score Data node because it does not prepare data for successor nodes.

4. Under Score Data, select the input table that you would like to score. Click **Browse** in the **Table name** field.

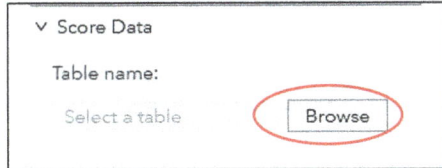

5. Import a SAS data set into CAS.

   a. In the Browse Data window, click **Import**.

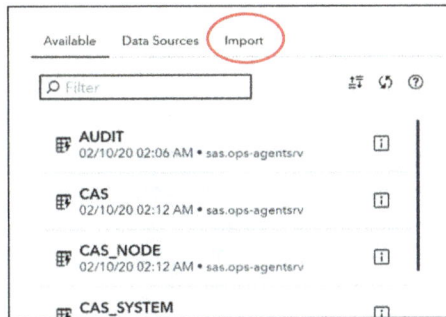

   b. Under Import, expand **Local files** and then select **Local file**.

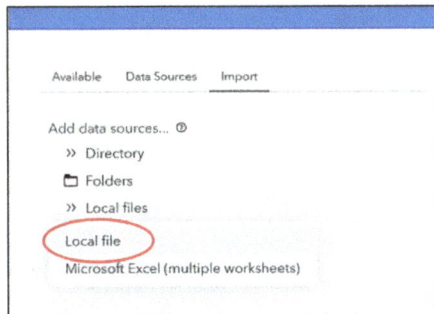

   Navigate to the data folder.

   c. Select the **score_insurance.sas7bdat** table. Click **Open**.

d. Select **Import Item**.

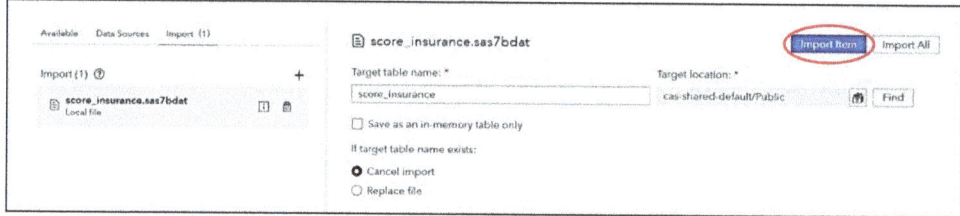

e. Click **OK** after the table is imported.

6. Under the Output Data property, click **Browse** in the **Output library** field.

7. Choose the **Public** library and then click **OK**.

    By default, the scored table is temporary, exists only for the duration of the run of a node, and has local session scope. The Score Data node enables you to save the scored table to disk in the location that is associated with the specified output library. After it is saved to disk, this table can be used by other applications for further analysis or reporting.

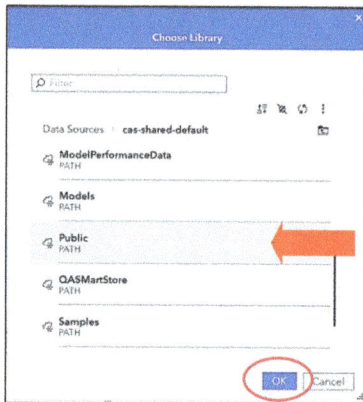

8. Type **Ins_ScoredData** in the **Table name** field. Check the **Replace existing table** box.

9. Right-click the **Score Data** node and click **Run**.
10. Right-click the **Score Data** node and click **Results**.
11. The scored input data that you specified in the Table name property is available on the Output Data tab. Click the **Output Data** tab.

12. Click **View Output Data**.

13. Click the **Enable sampling** box. Click the **Percentage of rows** radio button and enter **10**. Click **View Output Data**.

14. Scroll right until the end to see the predicted columns.

_leaf_id_	Into: INS	Predicted: INS=1	Predicted: INS=0	Probability for INS =1	Predicted for INS	Probability of Classification
3	1	0.7426636569	0.2573363431	0.7426636569		0.7426636569
59	0	0.3904939422	0.6095060578	0.3904939422		0.6095060578
42	0	0.3575757576	0.6424242424	0.3575757576		0.6424242424
13	1	0.625	0.375	0.625		0.625
40	0	0.173866524	0.826133476	0.173866524		0.826133476
40	0	0.173866524	0.826133476	0.173866524		0.826133476
1	1	0.734562212	0.265437788	0.734562212		0.734562212
41	1	0.5390374332	0.4609625668	0.5390374332		0.5390374332
40	0	0.173866524	0.826133476	0.173866524		0.826133476

15. Close the **Results** window.

End of Demonstration

Regression Trees

Now consider the case of an interval target variable. A *regression tree* is another variant of decision trees, designed to approximate real-valued functions, instead of being used for classification methods.

Figure 2.7: Boston Housing Example (2-dimensions)

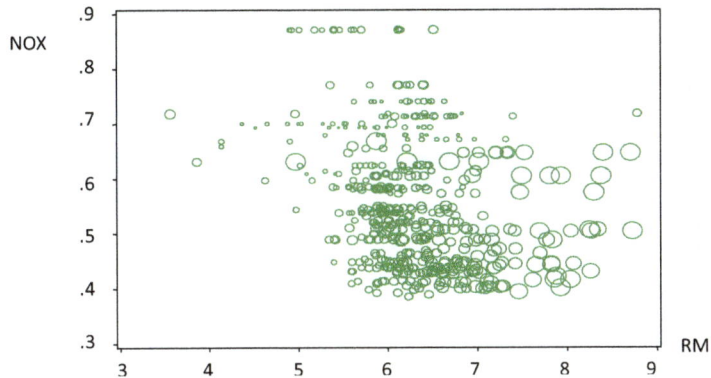

Figure 2.8: Boston Housing Example (3-dimensions)

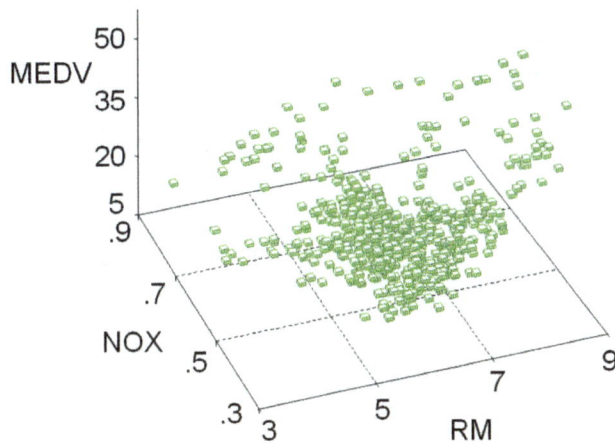

A regression tree refers to a predictive model where the target variable is not fixed or continuous. The decision tree algorithm is then used to predict the value of the target variable. An example of a regression type problem is to predict the selling prices of a residential house.

Figure 2.9: Regression Tree

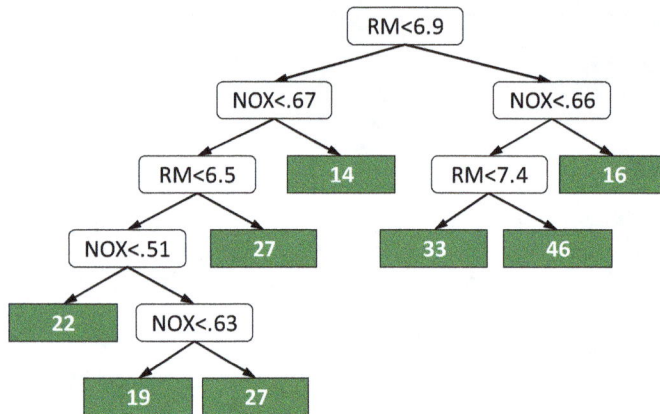

Figure 2.10: Leaves of Regression Tree as Boolean Rules

If RM ∈ {*values*} and NOX ∈ {*values*}, then MEDV=*value*.

Leaf	RM	NOX	Predicted MEDV
1	<6.5	<.51	22
2	<6.5	[.51, .63)	19
3	<6.5	[.63, .67)	27
4	[6.5, 6.9)	<.67	27
5	<6.9	≥.67	14
6	[6.9, 7.4)	<.66	33
7	≥7.4	<.66	46
8	≥6.9	≥.66	16

I will show how to build a regression tree using Boston housing data, which is available from the UCI repository (Blake et al. 1998). The cases are 506 census tracts in Boston. The target is the median home value (**MEDV**). Two of the 13 inputs are shown:

- the average number of rooms (**RM**)

- the nitrogen oxide concentration in the air (**NOX**)

The data are shown in a two-dimensional bubble plot in Figure 2.7. The radius of each bubble is proportional to the median home value (MEDV).

The same data are shown in a three-dimensional scatter plot in Figure 2.8.

Figure 2.11: Partitioned Input Space

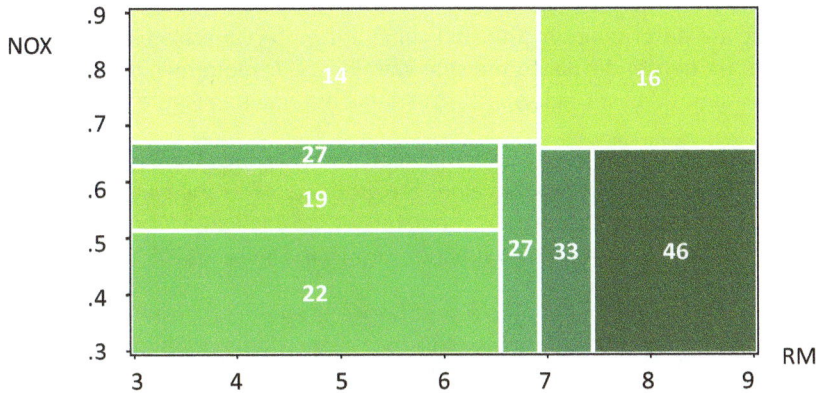

Figure 2.12: Multivariate Step Function

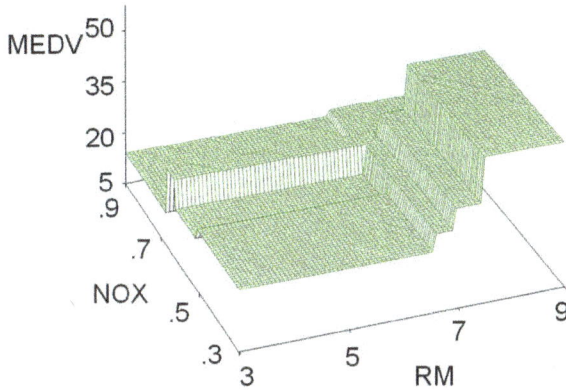

When the target is continuous, the model is a called a *regression tree*. Regression trees are piecewise constant models that provide concise summaries of how the predictors determine the predictions. These models are usually easier to interpret than linear regression models. A regression tree built on the Boston housing data is shown in Figure 2.9.

Predicted Values and Partitioned Input Space

The leaves in a regression tree give the predicted value of the continuous target. All cases that reach a particular leaf are assigned the same predicted value, see Figure 2.10.

The path to each leaf in a regression tree can be expressed as a Boolean rule. The rules take this form:

If the inputs ∈ {*region of the input space*}, then the predicted value = *value*.

The regions of the input space are determined by the split values. For interval-scaled inputs, the boundaries of the regions are perpendicular to the split variables. Consequently, the regions are intersections of subspaces defined by a single splitting variable as shown in Figure 2.11.

The leaves of the decision tree partition the input space into rectilinear regions. The partition is created such that observations with similar response values are grouped. After the partition is completed, a constant value of the response variable is predicted within each area. Consequently, the fitted regression tree model is a multivariate step function shown in Figure 2.12.

Since regression trees partition the predictor space in a set of regions and fit a constant value within each region they are sometimes called *piecewise constant regression models*. An important aspect of tree-based regression models is that they provide a propositional logic representation of these regions in the form of a tree. Each path from the root of the tree to a leaf corresponds to a region.

The surface is a piecewise constant and not joined continuously at the boundaries. A step function is highly flexible and is capable of modeling nonlinear trends. For this reason, decision trees are known as *universal approximators*.

Demo 2.3: Building a Regression Tree Pipeline from SAS Visual Statistics

This demonstration uses the Boston housing data to create a regression tree model using SAS Visual Analytics interface, explores the residuals, and then creates a pipeline in Model Studio.

1. Go to the **Applications** menu (≡) in the upper left side.
2. Click **Explore and Visualize** in the application shortcut area to launch SAS Visual Analytics.
3. Import the Housing data set into CAS.

   a.   Click **Start with Data**.

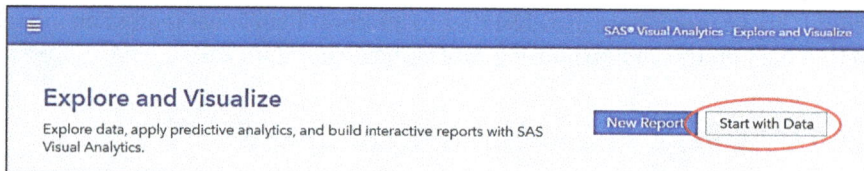

   b.   In the Choose Data window, click **Import**.
   c.   Under Import, expand **Local Files** and then select **Local File**.
   d.   Navigate to the data folder.
   e.   Select the **housing.sas7bdat** table. Click **Open**.

  f. Select the default for **Import Item**.
  g. Click **OK**.
  h. Close the **Start your tour of SAS Visual Analytics!** window if that pops up.

4. Click the **Objects** icon in the left pane. Under SAS Visual Statistics, either double-click or drag and drop **Decision tree** onto the canvas.

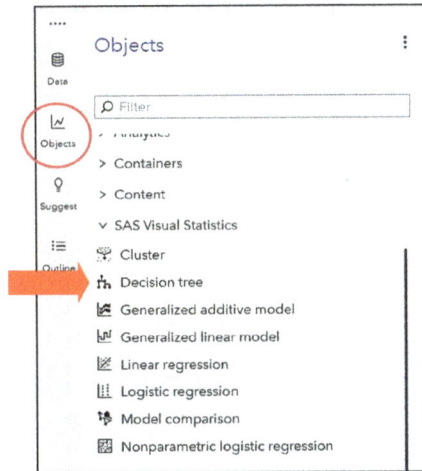

5. Click **Roles** icon in the right pane. Assign **MEDV** as the Response and all other variables as Predictors, except RM and PTRATIO.

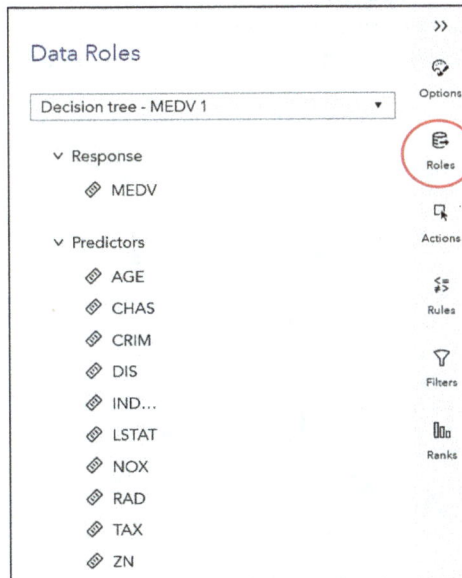

Note: When your category response variable contains more than two levels, SAS Visual Statistics treats all observations in the level of interest as an event and all other observations as nonevents. The **Event level** property enables you to choose the event level of interest.

6. In the Options pane, under Model Display, expand **General** properties. Change the plot layout to **Stack** to expand the fit summary on the canvas. After careful examination, only seven of the 10 input variables are used to build this regression tree model. To verify this, you can also examine the Variable importance chart.
7. Click the left node after the first split of the top (root) node of the decision tree.

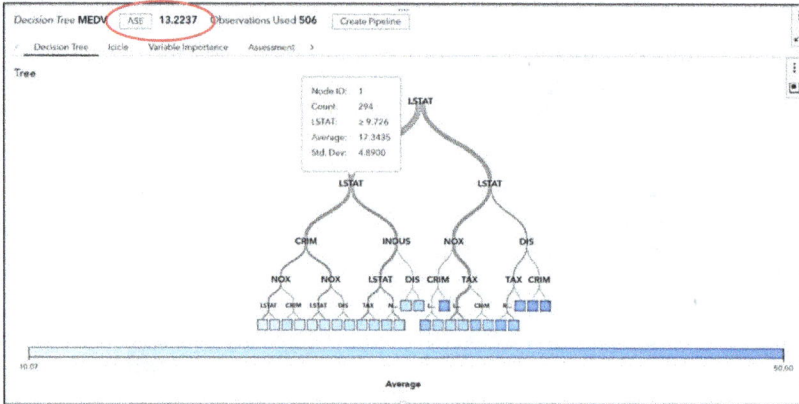

You can scroll to zoom in and view the characteristics of observations in this partition of the data. Observe that the average square error of this model is 13.2237.
8. Click the **Icicle** tab. This plot is similar to what we had referred to as treemap in Model Studio.

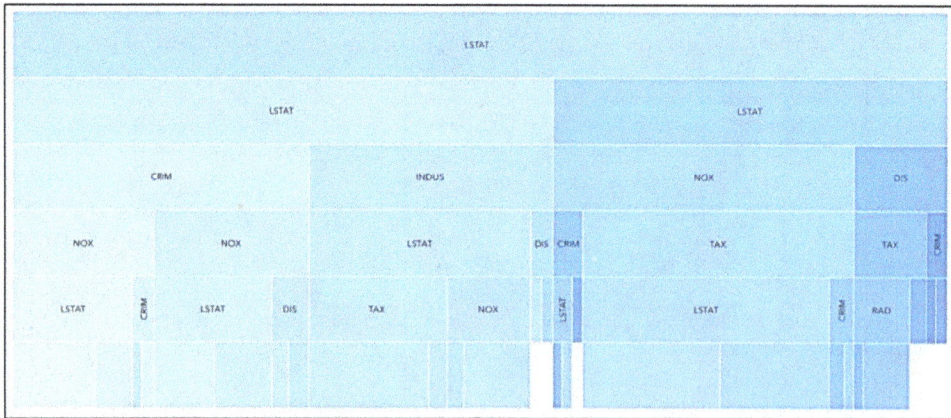

9. Click the **Variable Importance** tab.

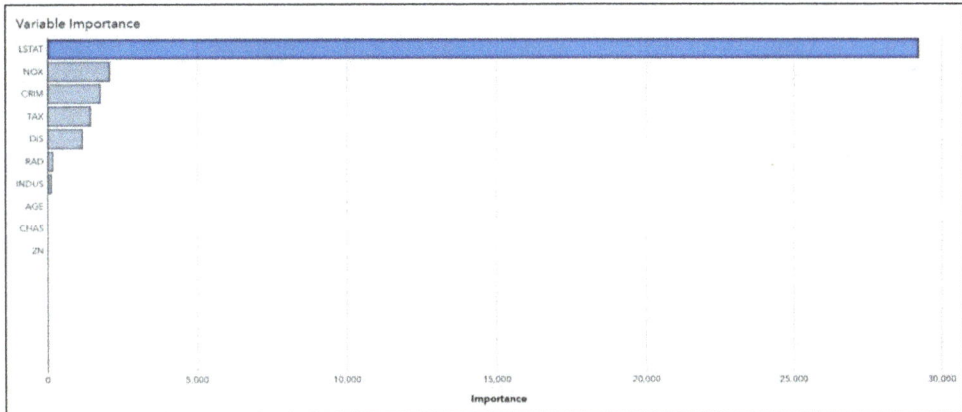

The two most important variables in this model are **LSTAT** (percent low socioeconomic status population) and **NOX** (nitric oxide concentration).

10. Right-click in the variable importance plot and select **Leaf Statistics**.

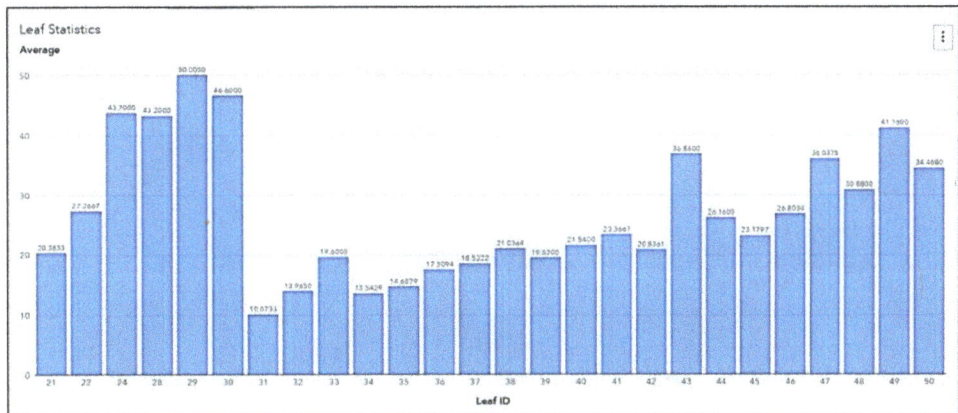

The Leaf Statistics panel displays a bar chart of simple statistics for each leaf. This enables you to do a quick visual comparison of leaf purity. In the plot above, node 29 has the highest **MEDV** average at nearly 50.00.

11. In the report, click ⬉ (**Maximize**) on the object toolbar to enter maximize mode and see the Details table.

12. Click **Node Rules** tab in the bottom pane and adjust the view to see the English rules.

Node Statistics | Node Rules | Variable Importance | Assessment | Assessment Statistics

Node ID	Parent ID	Type	AGE	CHAS	CRIM	DIS	INDUS	LSTAT	NOX	RAD	TAX	ZN
0	-1	Class										
1	0	Class						≥ 9.726				
2	0	Class						1.73 to 9.726				
3	1	Class						≥ 14.983				
4	1	Class						9.726 to 14.983				
5	2	Class						4.654 to 9.726				
6	2	Class						1.73 to 4.654				
7	3	Class			≥ 6.385			≥ 14.983				
8	3	Class			0.006 to 6.385			≥ 14.983				

Node rules up to depth three are shown above.

13. To close the Details table and exit maximize mode, click ↘↖ (**Restore**) on the object toolbar.
14. Click ⋮ (**More**) on the object toolbar (in the upper right-hand corner of the central canvas) and then click **Derive predicted**.

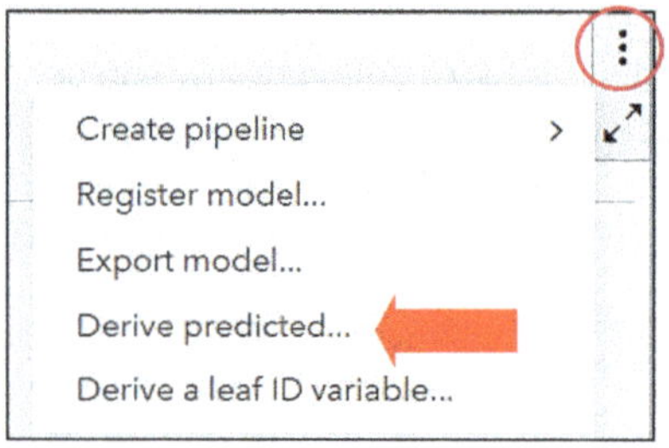

15. Click **OK**.

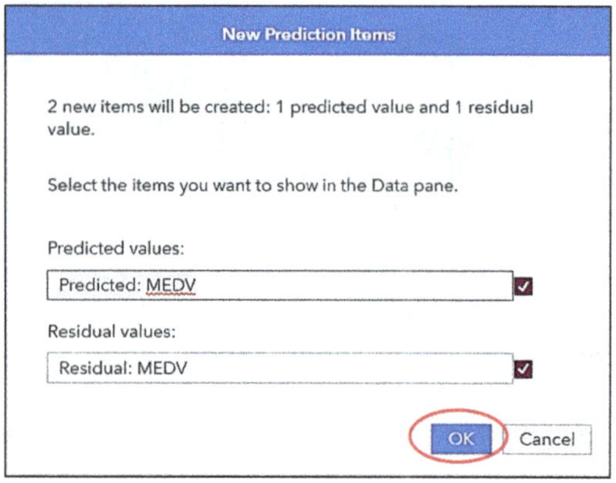

16. Select the current page (Page 1) and click the three vertical dots ⋮ (**Options**) ⇨ **Rename Page**. Rename Page 1 as **RegTree**.
17. Start a new page to explore the residuals. Click **+** (**New page**).

18. Click the **Data** icon in the left pane. Under Derived, select **Predicted: MEDV** and **Residual: MEDV** by holding down the Ctrl key and dragging and dropping them on the canvas. A scatter plot is created automatically.

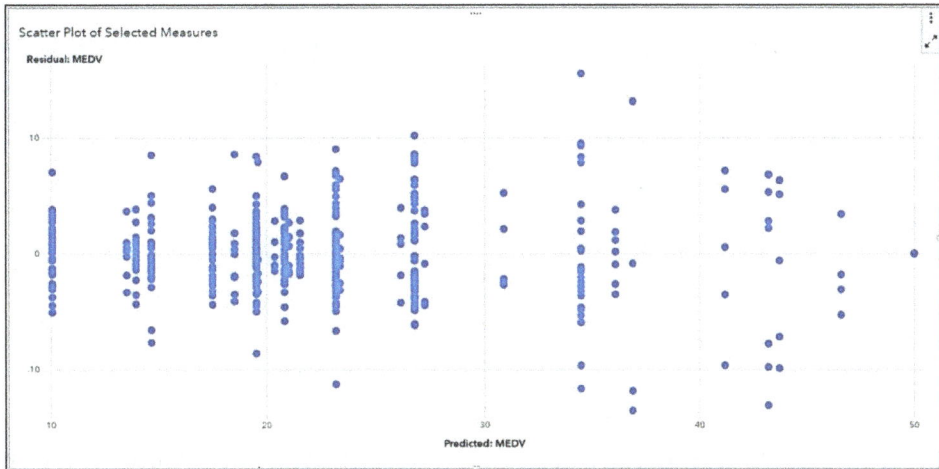

Scatter Plot of Selected Measures

The residuals versus predicted plot is a standard regression diagnostic commonly known as a *residual plot*. The absolute values of the residuals represent variability. The obvious flaring pattern indicates that the variance increases with the mean. Decision trees do not assume constant variance. However, stabilizing the variance might improve the predictive accuracy of the model.

19. Click ⋮ (Options) ⇨ **Rename Page**. Rename Page 2 as **ResPlot**.
20. Click the **Data** icon in the left pane again.
21. Create a log-transformed target.
    a. Click **+ New data item** ⇨ **Calculated item**.
    b. Enter **Log_MEDV** in the **Name** field.
    c. Click the **Operators** tab ⇨ **Numeric (advanced)** ⇨ **Ln** (double-click).
    d. Click the **Data Items** tab. Select the **MEDV** variable. You can either double-click or drag and drop it in the number box.
    e. Click **OK**.
    f. Observe that the **Log_MEDV** variable is created and now appearing under Measures in the Data pane.
22. Return to the RegTree page. To access the alternative menu, in the upper right corner of the canvas, click ⋮ (**More**) and then hold down the Alt key ⇨ **Duplicate on new page**.

23. Select the new decision tree on the canvas of Page 3 to make it the active object. In the Roles menu, select **MEDV** under Response and replace it with **Log_MEDV**.
24. Switch back to the Variable Importance chart if what you see is the Leaf Statistics plot.

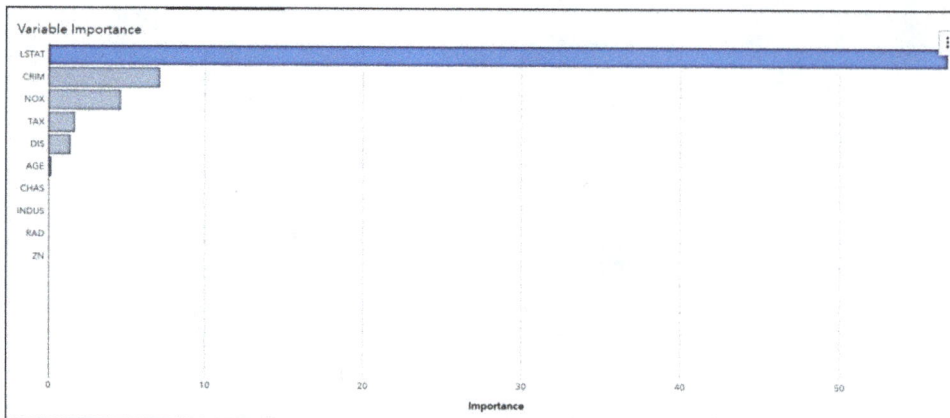

Observe that the relative importance of variables changed as compared to the previous model, using the non-transformed target. The variable **CRIM** (crime rate) is now second-most important.

25. Click ⋮ **(Options)** ⇨ **Rename Page**. Rename Page 3 as **RegTreeLn**.
26. Click ⋮ **(More)** on the object toolbar and then click **Derive predicted**. Click **OK**.
27. Start a new page to explore the residuals of the regression tree with the transformed target. Click **+ (New page)**.
28. Click the **Data** icon in the left pane. Under Derived, select **Predicted: Log_MEDV** and **Residual: Log_MEDV** by holding down the Ctrl key and dragging and dropping them on the canvas.

    A residual plot is created.

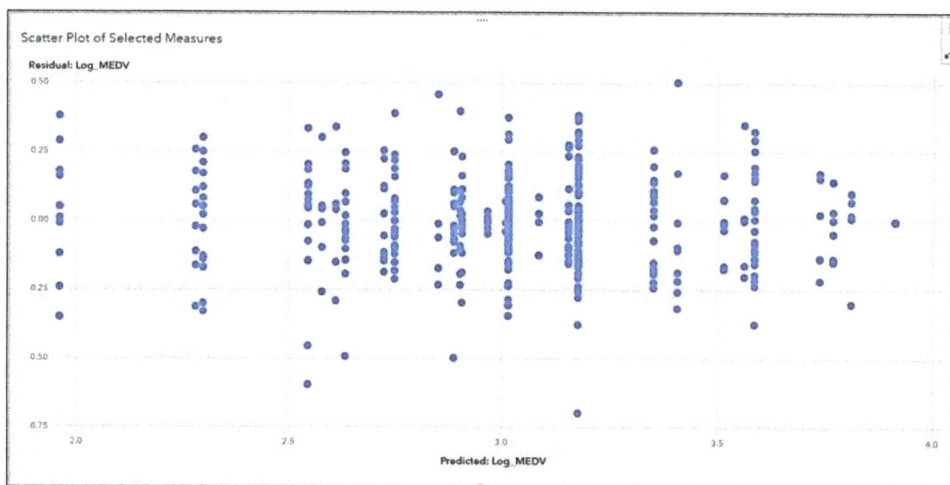

The target transformation somewhat stabilized the variance.

29. Click ⠿ (**Options**) ⇨ **Rename Page**. Rename Page 4 as **ResPlotLn**.
30. Explore the relationship between predicted values and the two most important variables in the tree model, **CRIM** and **LSTAT**.
    a. Click + (**New page**).
    b. Click the **Data** icon in the left pane. Select **LSTAT** and **CRIM** under Measures and **Predicted: Log_MEDV** under Derived. Drag and drop them on the canvas.

    A scatter plot matrix is created.

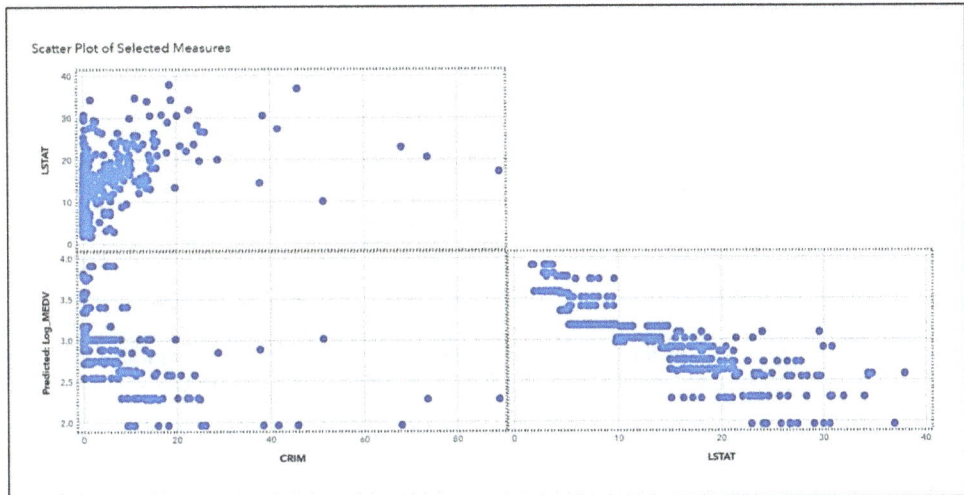

Scatter Plot of Selected Measures

The relationship between crime rates and predicted **log(MEDV)** appears to be nonlinear.

Compared with other regression and classification methods, tree models have the advantage that they are easy to interpret and visualize, especially when the tree is small. Tree-based methods scale well to large data, and they offer various methods of handling missing values, including surrogate splits. However, tree models have limitations. Regression tree models fit response surfaces that are constant over rectangular regions of the predictor space, so they often lack the flexibility needed to capture smooth relationships between the predictor variables and the response. Another limitation of tree model is that small changes in the data can lead to very different splits, and this undermines the interpretability of the model.

31. Click ⠿ (**Menu**) ⇨ **Save as** in the upper right corner of Report 1. Save this report as **Housing_RegTree** in the My Folder location.

32. Return to the **RegTree** page. Click **Create pipeline** ⇨ **Add to new project**.

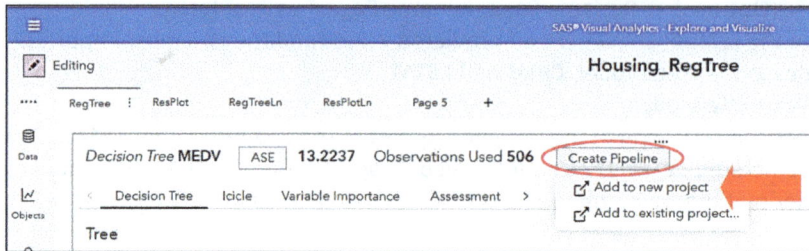

Observe that the ASE is 13.2237.

It was interesting to explore the two decision trees (one with the **MEDV** target and another with **Log_MEDV** target) in line with regression diagnostics. You can push either of them into Model Studio for further refinement.

In general, the purpose of a decision tree is to determine a set of *if-then* rules that deliver accurate prediction or classification of cases. Therefore, there is no implicit assumption that the underlying relationships between the target and the inputs are linear, follow some specific non-linear link function, or that they are even monotonic in nature. This encourages us to push the decision tree with the original target.

A new pipeline is created in the **Housing_RegTree** project with the same name that you entered for the SAS Visual Analytics report, that is, Interactive-Model Pipeline.

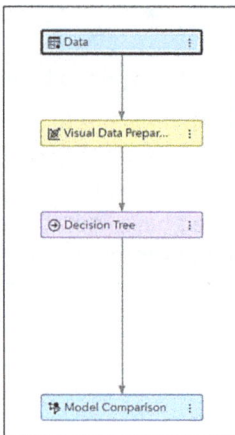

The Data node defines all the information about the data set.

The Visual Data Preparation node represents the interactive calculations and transformations from SAS Visual Analytics.

The Decision Tree node fits a model based on the interactive model decision tree.

33. Click the **Data** tab to see the project metadata before running this pipeline. Change the role of **RM** and **PTRATIO** variables to **Rejected**.

34. Go back to the **Pipelines** tab.
35. Click the **Run Pipeline** button and open the **Results** of the Decision Tree node.

Observe that the results represent the interactive model score code from SAS Visual Analytics. Model score code is applied to data to generate assessment results.
The properties of this node are not editable by default.

36. Close the **Results** of the Decision Tree node and now open the **Results** of the **Model Comparison** node.

	Champion	Name	Algorithm Name	Average Squared Error	Root Average Squared Error
Model Comparison					
	⊡	Decision Tree	Decision Tree	13.2237	3.6364

Since it was the same tree created in SAS Visual Statistics, an ASE of 13.2237 is matching.

37. Close the **Results** of the Model Comparison node.
38. Right-click the **Decision Tree** node and select **Enable Properties**.

39. Select **Yes** for the Enable Properties pop-up box. By enabling the properties of this node, the model will be retrained, which might yield new results. At this moment, no change is required in the properties on your right.
40. Click the **Run Pipeline** button to rerun the pipeline and open the **Results** of the Decision Tree node.

Observe that along with model score code outputs, usual trained outputs such as tree diagrams and treemaps are now available.

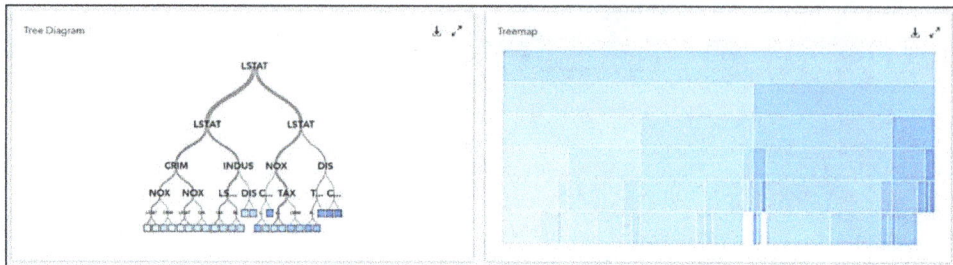

Explore the results as deemed fit.

41. Close the **Results**.
42. On the Interactive-Model Pipeline tab, click ⋮ (**Options**) ⇨ **Rename**.
43. Enter **VA Pipeline** in the **Name** field and click **OK**.

End of Demonstration

Quiz

1. Which of the following statements is true regarding a decision tree?
   a. represents a segmentation of the data that is created by applying a series of simple rules and therefore it's an unsupervised technique
   b. can be used to classify the observations but can't be used to estimate responses
   c. a flowchart-like structure in which internal node represents test on an input, each branch represents the outcome of test, and each leaf node represents class label or interval estimated
   d. all of the above

Answers

1. c

Chapter 3: Growing a Decision Tree

Recursive Partitioning

Recursive partitioning repeatedly splits parent nodes into child nodes to create a decision tree. It's a simple three-step procedure shown in Figure 3.1.

An easy algorithm first splits the entire data set, the root node, into at least two nodes. Each node is then split into more nodes, and so on. Each node in its turn is considered in isolation. The search for a split is done without any concern as to how another node is to be split. This process is called recursive partitioning and the majority of tree algorithms use it.

The algorithm splits the parent node into two or more child nodes in such a way that the values (or levels) of the response variable within each child region are as similar as possible. The splitting process is then repeated for each of the child nodes, and the recursion continues until a stopping criterion is satisfied. At that point, the tree is fully built (grown). At each step, the split is determined by finding the best predictor variable and the best split value that optimize a specified criterion in the response variable across the child nodes to which the parent node is split.

Recursive partitioning is myopic (Neville 1999). It focuses on optimizing individual splits and pays no attention to the quality of the entire tree. Alternatives to recursive partitioning have not become popular though. They are difficult to implement and take much longer to run. Moreover, recursive partitioning generally gives adequate results.

Figure 3.1: Recursive Partitioning Procedure

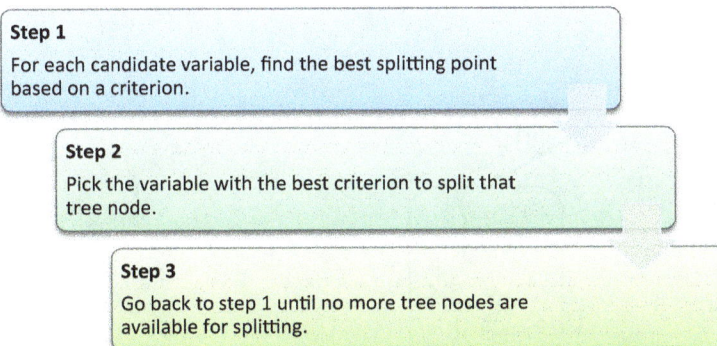

Step 1
For each candidate variable, find the best splitting point based on a criterion.

Step 2
Pick the variable with the best criterion to split that tree node.

Step 3
Go back to step 1 until no more tree nodes are available for splitting.

Recursive partitioning has three main components:

- a way to select a split test (split search)
- a rule to determine when a tree node is terminal (stopping rule)
- a rule for assigning a value to each terminal node (splitting rule)

We discuss these components later in this chapter.

Root-Node Split and One-Deep Space

Recursive partitioning is a top-down, greedy algorithm. Starting at the root node, several splits that involve a single input are examined. For interval inputs, the splits are disjoint ranges of the input values. For nominal inputs, the splits are disjoint subsets of the input categories. Various split-search strategies can be used to determine the set of candidate splits. A splitting criterion is used to choose the split. The splitting criterion measures the reduction in variability of the target distribution in the child nodes. The goal is to reduce variability and thus increase purity in the child nodes. The cases in the root node are then partitioned according to the selected split.

A greedy algorithm makes locally optimal choices at each step. Figure 3.2 shows the root node of a classification tree that was shown in the previous chapter. The analysis goal is to predict whether the handwritten digit is 1, 7, or 9.

The root node represents the number of cases for each digit. Starting at the root node, recursive partitioning uses an iterative process to select the best split for the node. This process is called a split search. First, recursive partitioning identifies the best of all possible splits for each input variable based on the value of a splitting criterion. These are the candidate splits. The process

Figure 3.2: Root-Node Split in Handwriting Recognition Example

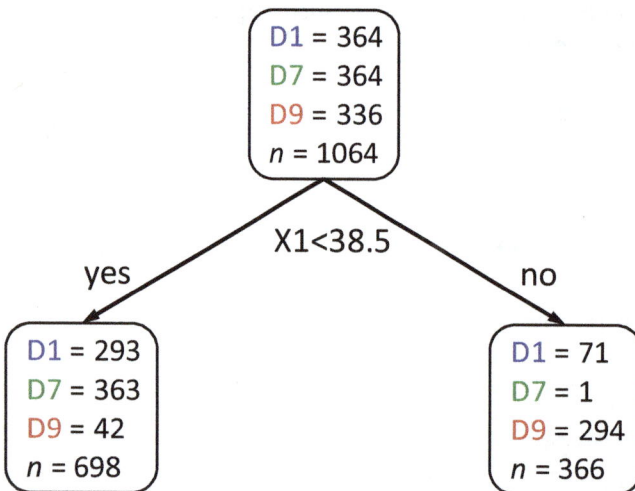

Figure 3.3: One-Deep Space in Handwriting Recognition Example

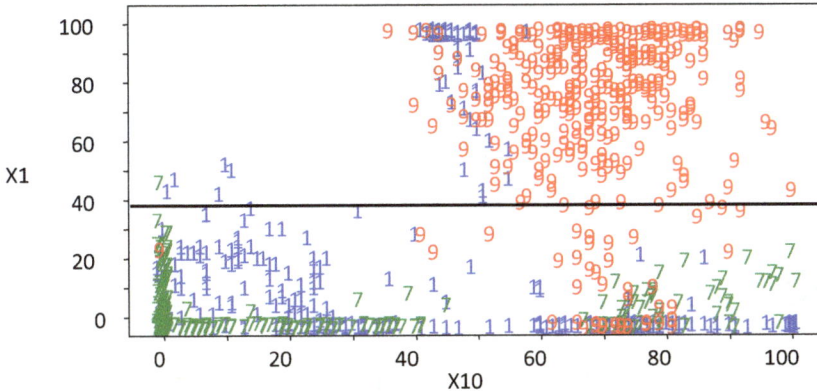

then selects the overall best split (that is, a split that is based on a single input). This split is expressed as an IF-THEN-ELSE rule. The cases in the root node are then partitioned according to the selected split. In this tree, the optimal split is to branch left when X1 is less than 38.5 and branch right when X1 is greater than or equal to 38.5. The 1064 observations in the root node have been partitioned such that 698 follow the rule to the left branch, and 366 follow the rule to the right branch.

The root-node split corresponds to a partition of the input space where the boundary is perpendicular to one input dimension shown in Figure 3.3.

Depth Two and Two-Deep Space

The partitioning is now repeated in each child node as if it were the root node of a new tree. The split selection at a node depends entirely on the cases in that local region of the input space. As the split search continues deeper into the tree, the data become more fragmented. The splitting criterion and other stopping rules determine the end of the process and the final depth of the tree.

After the second iteration of the split search, the resulting tree has a depth of two. It is now a two-deep space. The resulting splits are shown in Figure 3.4.

This node split corresponds to a partition of the input space where the boundary is perpendicular to the other input dimension shown in Figure 3.5.

The process is repeated. The depth is governed by stopping rules, which are discussed in Chapter 5.

Recursive partitioning is essentially a simple nonparametric technique for classification and regression tasks. The prediction technique is nonparametric in that it does not rely on any particular assumption about the type of dependence of the target variable on the inputs. When

Figure 3.4: Depth Two Split in Handwriting Recognition Example

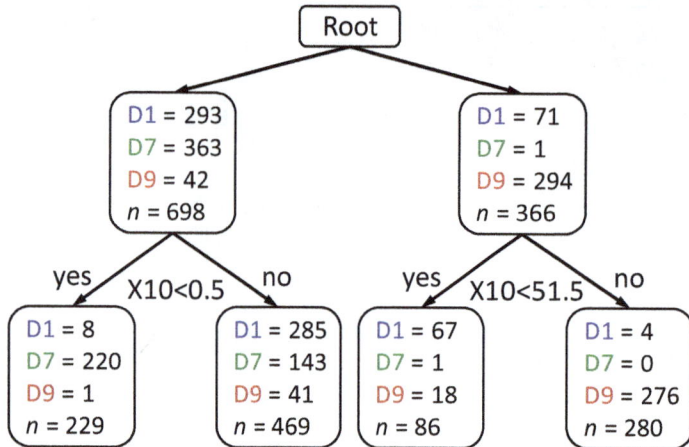

used in the standard way, recursive partitioning provides trees that display the succession of rules that need to be followed to derive prediction. This simple visual display explains the appeal to decision makers from many disciplines.

Alternatives to Recursive Partitioning

It seems reasonable that this greedy algorithm could be improved by incorporating some type of *look-ahead* or backup. Aside from the computational burden, trees built using limited look-ahead are not shown to be an improvement. In many cases, they give inferior trees (Murthy and Salzberg 1995).

Another variation is *oblique splits*. Standard decision trees partition the input space using boundaries that are parallel to the input coordinates as shown in Figure 3.6 that shows all the

Figure 3.5: Two-Deep Space in Handwriting Recognition Example

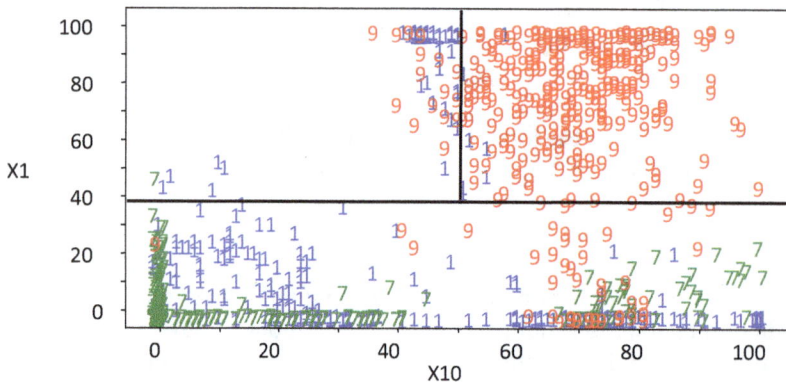

Figure 3.6: Decision Boundaries Parallel to the Input Coordinates

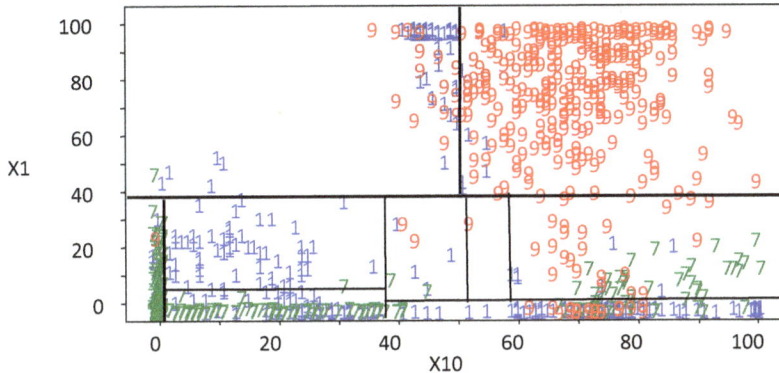

splits that occurred before the recursive partitioning process stopped. These coordinate-axis splits make a fitted decision tree easy to interpret and provide resistance to the curse of dimensionality.

Splits on linear combinations of inputs give oblique boundaries. Several algorithms were developed for inducing oblique decision trees (BFOS 1984, Murthy et al 1994, Loh and Shih 1997). Oblique decision trees are not supported by SAS Viya, and there is not strong evidence that this approach produces consistently stronger models.

Constructing Decision Trees Interactively

The Decision Tree object in SAS Visual Statistics enables you to perform multiway splitting of your database based on nominal, ordinal, and continuous variables. The object supports both automatic and interactive training. When you run the Decision Tree object in automatic mode, it automatically ranks the input variables based on the strength of their contributions to the tree. This ranking can be used to select variables for use in subsequent modeling. You can override any automatic step and prune explicit subtrees. Interactive training enables you to explore and evaluate a large set of trees as you develop them.

Demo 3.1: Interactively Building a Regression Tree in SAS Visual Statistics

This demonstration uses the Boston Housing data to interactively create a regression tree model using SAS Visual Analytics interface and builds on the **Housing_RegTree** report from the previous demonstration.

1. Go to the **Applications** menu (☰) in the upper left side.
2. Click **Explore and Visualize** in the application shortcut area. The SAS Visual Analytics window appears.

3. Select **My Folder** from the left pane.
4. Double-click the **Housing_RegTree** report tile to open it.

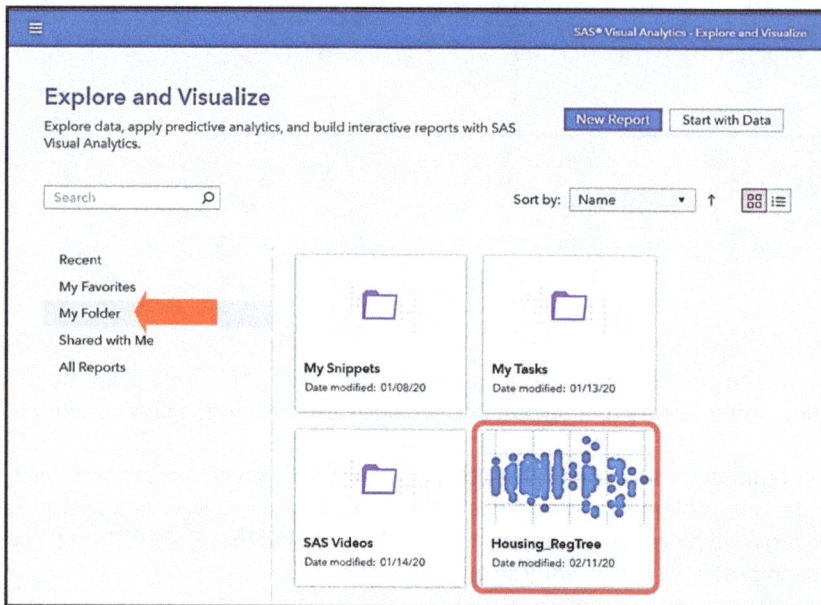

5. Ensure that the RegTree page is opened. Right-click the canvas and select **Enter inter-active mode**.

   Note: You can directly start making changes to the decision tree in the Tree window by right-clicking the node.

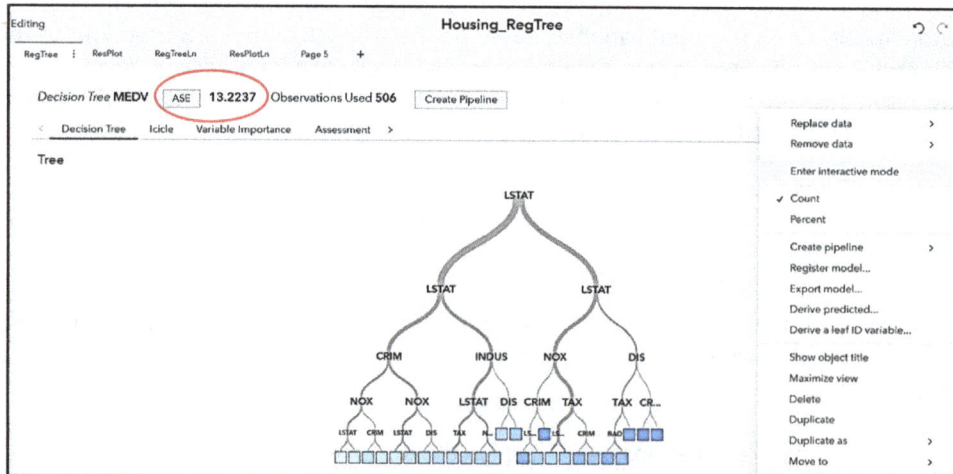

Observe that the ASE was 13.2237. We will try to improve this. You can manually train and prune nodes by entering interactive mode. In interactive mode, growth properties are locked. Certain modifications to predictors are allowed, such as converting a measure to a category. When you are in interactive mode and modify a predictor, the decision tree remains in interactive mode, but attempts to rebuild the splits and prunes using the same rules.

6. Recall that the default leaf size is 5, which means that a minimum of five observations are allowed in a leaf (terminal) node. Therefore, select a large enough leaf, such the first leaf from the right amongst the leaves in the maximum depth. Right-click and select **Split best from node**.

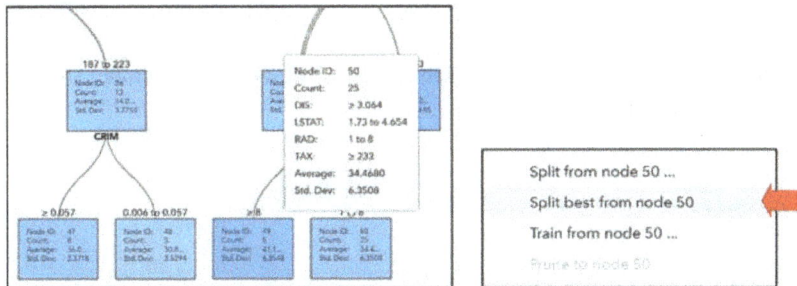

The Split best from node option splits the node based on the variable with the best gain ratio.

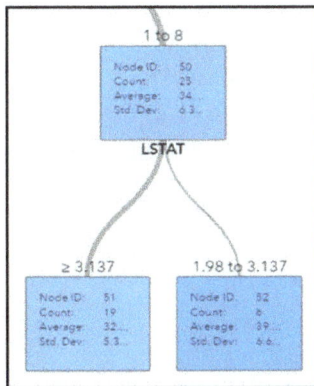

7. Now select another larger leaf, say the first leaf from left amongst the leaves in the maximum but one depth. Right-click and select **Split from node**.

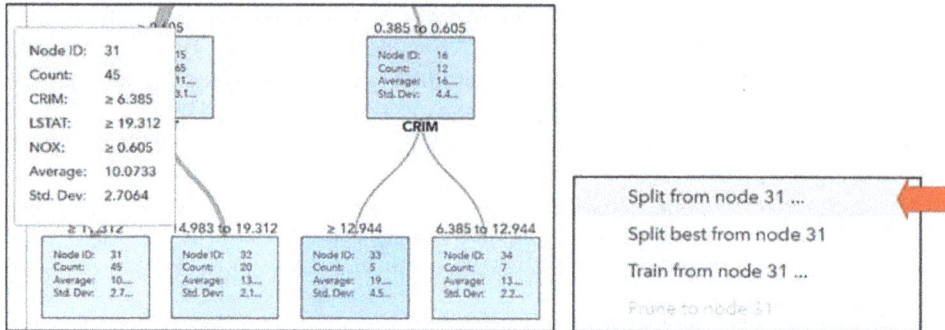

The Split from node option opens the **Split node** window. You can select the variable that will be used to split the node.

8. Select **Age** and then click **OK**.

Variables are sorted in descending order by their logworths. Logworth measures how well a variable divides the data. **NOX** has the highest logworth. However, you might have compelling business reasons to select **AGE**.

The manually trained decision tree is constructed and available on the canvas.

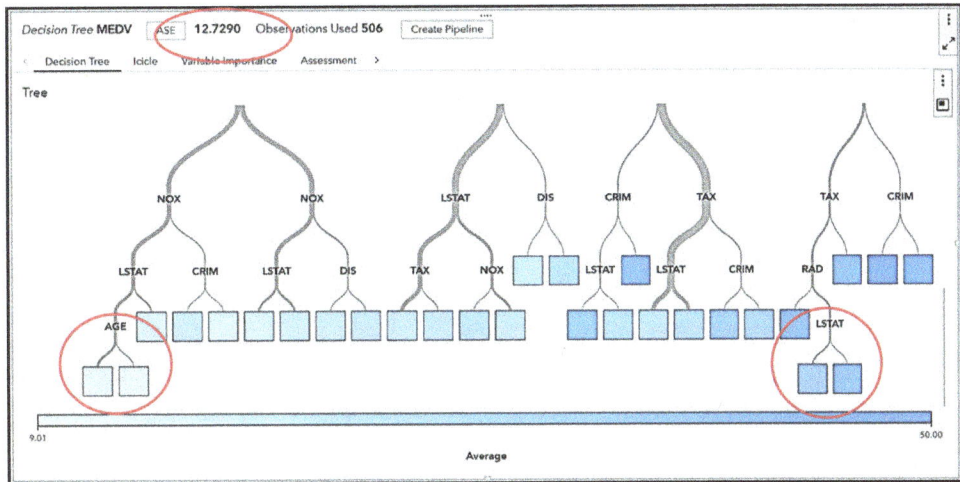

Observe the manually altered splits on the two extreme sides of the decision tree. The ASE has reduced from 13.2237 to 12.7290. It is a good sign! You can further try modifying this tree.

But evaluating the model on the same data the model was fit on usually leads to an optimistically biased assessment. The simplest strategy for correcting the optimism bias is using a holdout sample for empirical validation or performing the cross validation. Cross validation on this example is demonstrated in chapter 5.

9. Although we discuss pruning in chapter 5, select the root node, right-click, and then select **Prune to node**.

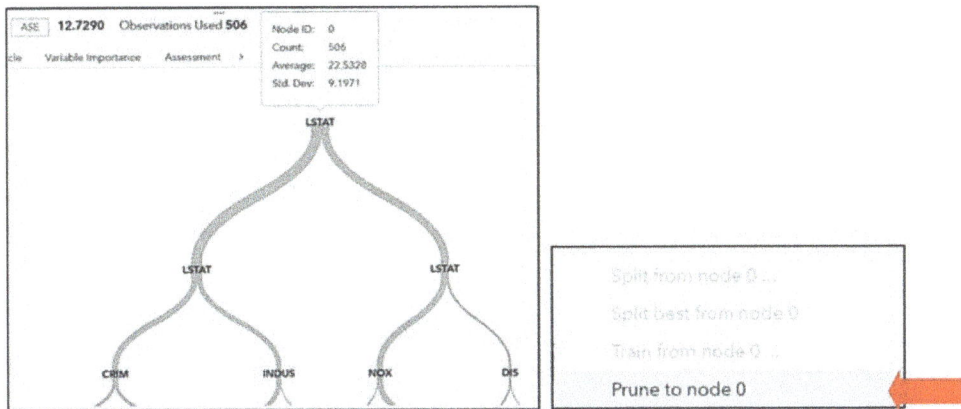

The Prune to node option removes all nodes beneath the selected node and turns that node into a leaf node.

10. Select the root node, right-click, and then select **Train from node**.

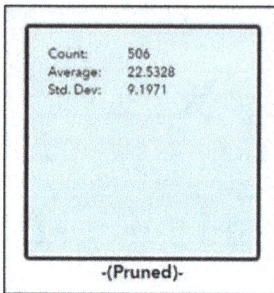

Count: 506
Average: 22.5328
Std. Dev: 9.1971

-(Pruned)-

Split from node 0 ...
Split best from node 0
Train from node 0 ...
Prune to node 0

The Train from node option trains more than one level beyond the leaf node (here it is the root node, because the tree was pruned up to the root node). Variables are still sorted in descending order by their logworths at that node.

11. The Train node window appears with all the variables selected by default. Deselect the last four variables in the list so that only six variables with the highest logworths are considered for splitting. Increase **Maximum depth of subtree** from 6 to **10** and then click **OK**.

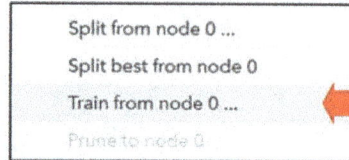

Train node

Select variables to use to train the node.

☐	Variable	Log Worth	‼
☑	LSTAT	18,818.167197	
☑	INDUS	11,083.225468	
☑	NOX	9,118.899734	
☑	TAX	8,618.084277	
☑	CRIM	7,862.807131	
☑	RAD	6,708.643325	
☐	ZN	6,669.062507	
☐	AGE	5,427.837963	
☐	DIS	4,801.003302	

Maximum depth of subtree: ⌄ 10

Log Worth - Measures how well a variable divides the data. This value is the negative log of the p-value from the chi-square test.

OK   Cancel

Obviously, a deeper tree is created.

The ASE has further reduced to 11.6002.

The tree was created by splitting the source data, constituting the root node of the tree, into subsets, which constitute the child nodes. The splitting was based on a set of splitting rules based on classification features (Shai and Shai 2014). This process is repeated on each derived subset in a recursive manner. The recursion is completed when the subset at a node has all the same values of the target variable, or when splitting no longer adds value to the predictions. This process of ***top-down induction of decision trees*** (Quinlan 1986) is an example of a greedy algorithm, which is based on the concept of heuristic problem solving by making an optimal local choice at each node. By making these local optimal choices, you reached the approximate optimal solution globally.

12. Without exiting the interactive mode, click **Create Pipeline** and select **Add to existing project**.

Note: To leave interactive mode, right-click in the Tree window and select **Exit interactive mode**. When you leave interactive mode, you lose all your changes.

13. Select the **Housing_RegTree** project in Model Studio. Select **Add pipeline**.
14. A new pipeline is added with an auto-populated name **Interactive-Model Pipeline**. Click the **Run Pipeline** button.

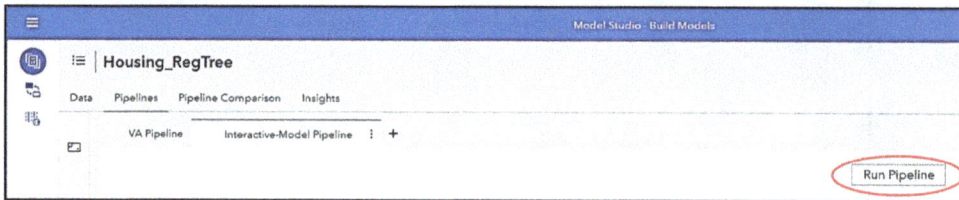

15. Click the **Pipeline Comparison** tab to compare the two regression trees. Click check boxes for both the trees and then click the **Compare** button.

The regression tree created interactively seems to be superior to the other regression tree. However, it might be an overfitted model. You can change the properties and retrain this tree to further improve it.

16. Close the Compare window.
17. Return to the **Pipelines** tab.

End of Demonstration

Feature Selection Using Split Search

To select useful inputs, decision trees use a split-search algorithm. Decision trees confront the curse of dimensionality by ignoring irrelevant inputs.

> Curiously, trees have no built-in method for ignoring redundant inputs. Because trees can be trained quickly and have a simple structure, this is usually not an issue for model creation. However, it can be an issue for model deployment, in that trees might arbitrarily select from a set of correlated inputs. To avoid this problem, you should use an algorithm that is external to the tree to manage input redundancy.

Consider a simple prediction problem with two inputs and a binary target as shown in Figure 3.7.

Figure 3.7: Simple Prediction Illustration

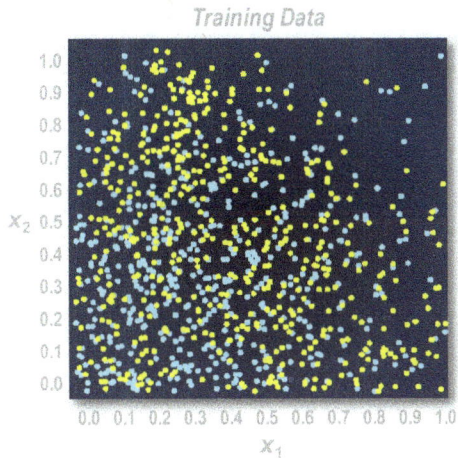

The inputs, x1 and x2, locate the case in the unit square. The target outcome is represented by a color; yellow is primary and blue is secondary. The analysis goal is to predict the outcome based on the location in the unit square.

To predict cases, decision trees use rules that involve the values of the input variables as shown in Figure 3.8.

The rules are arranged hierarchically in a tree-like structure with nodes connected by lines. The nodes represent decision rules, and the lines order the rules. The first rule, at the base (top) of the tree, is named the *root node*. Subsequent rules are named *interior nodes*. Nodes with only one connection are *leaf nodes*.

Figure 3.8: Decision Tree Prediction Rules

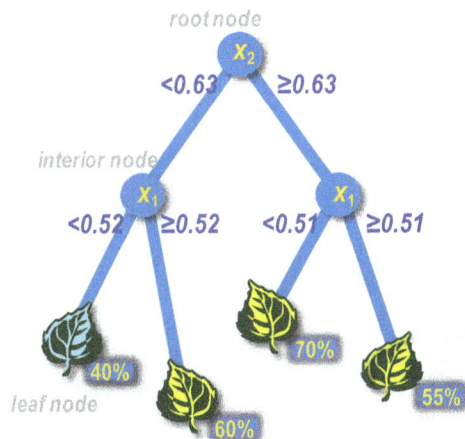

To score a new case, examine the input values and apply the rules defined by the decision tree.

The input values of a new case eventually lead to a single leaf in the tree as shown in Figure 3.9. A tree leaf provides a decision (for example, classify as yellow) and an estimate (for example, the primary-target proportion).

Understanding a split-search algorithm for building trees enables you to better use the decision tree in SAS Viya and interpret your results. The description presented here assumes a binary target, but the algorithm for interval targets is similar.

The split search starts by selecting an input for splitting the training data. If the measurement scale of the selected input is interval, binned or clustered values serve as a potential split point for the data. If the input is categorical, the average value of the target is taken within each categorical input level. The averages serve the same role as the binned or clustered interval input values in the discussion that follows.

For a selected input and fixed split point, two groups are generated. Cases with input values less than the split point are said to branch left. Cases with input values greater than the split point are said to branch right. The groups, combined with the target outcomes, form a 2x2 contingency table as shown in Figure 3.10. The columns specify branch direction (left or right) and rows specify target value (0 or 1).

An information gain statistic that is based on the entropy of the root node and the entropy of the data in each partition of the split can be used to quantify the separation of counts in the table's columns. Large values for the gain statistic suggest that the proportion of zeros and ones in the left branch is different from the proportion in the right branch. A major difference in outcome proportions indicates a good split.

The best split for a predictor is the split that yields the highest information gain. Assume that there are 100 total observations and a 50/50 split of yellow/blue in the training data. Also, there

Figure 3.9: Scoring A New Case Using Decision Tree

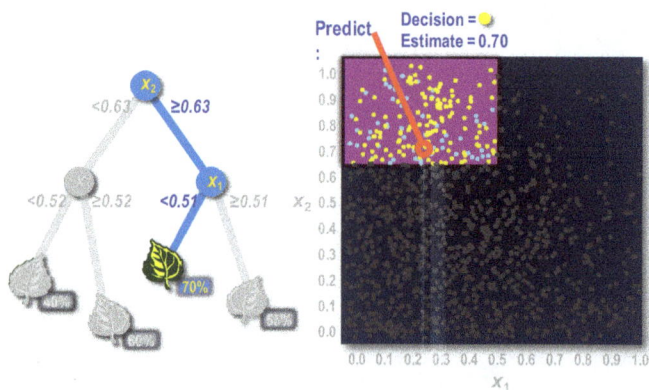

Figure 3.10: Assessing Every Partition on Input X$_1$

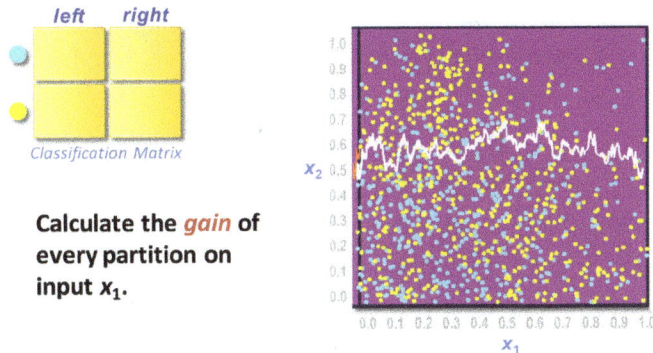

Calculate the *gain* of every partition on input x_1.

are 52 observations to the left and 48 observations to the right of the split. Based on this and the numbers given in the table in Figure 3.11, gain can be computed as shown below:

Entropy Total   : $-0.5 \times log_2(0.5) - 0.5 \times log_2(0.5) = 1$

Entropy Left   : $-0.53 \times log_2(0.53) - 0.47 \times log_2(0.47) = 0.997$

Entropy Right   : $-0.42 \times log_2(0.42) - 0.58 \times log_2(0.58) = 0.98$

Gain   : $1 - (52/100) \times 0.997 - (48/100) \times 0.98 = 0.0112$

Entropy is discussed in detail later in this chapter.

The partitioning process is repeated for every input in the training data. Again, the optimal split for the input is the one that maximizes the gain function as shown in Figure 3.12.

Figure 3.11: Selecting Partition with the Maximum Gain

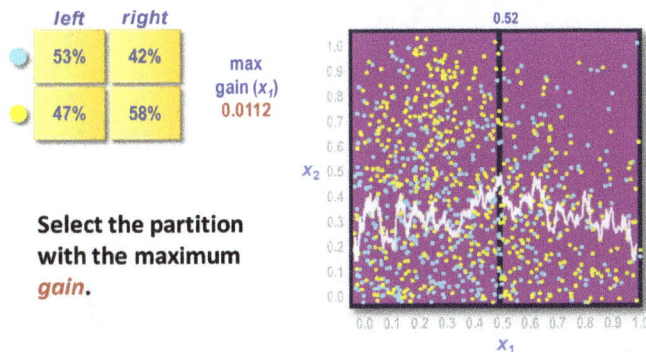

Select the partition with the maximum *gain*.

Figure 3.12: Selecting Partition with the Maximum Gain on Input X_2

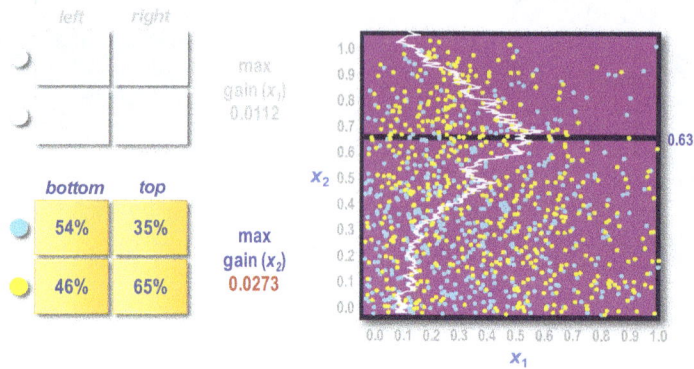

After you determine the best split for every input, the tree algorithm compares each best split's corresponding gain. The split with the highest gain is regarded as best. In this example, the max gain of input x_2 (0.0273) is larger than the max gain of input x_1 (0.0112); therefore, x_2 will be used for splitting as shown in Figure 3.13.

The best split rule is used to partition the data. The process is repeated in each subset down the tree as shown in Figures 3.14 through 3.18.

The split search is repeated within each new leaf. Gain statistics are compared as before.

The resulting partition of the predictor variable space is known as the *maximal tree*. Development of the maximal tree is based exclusively on splitting criteria of gain on the data. There are various splitting criteria available in SAS Visual Data Mining and Machine Learning.

Figure 3.13: Creating a Partition Rule from the Best Partition Across All Inputs

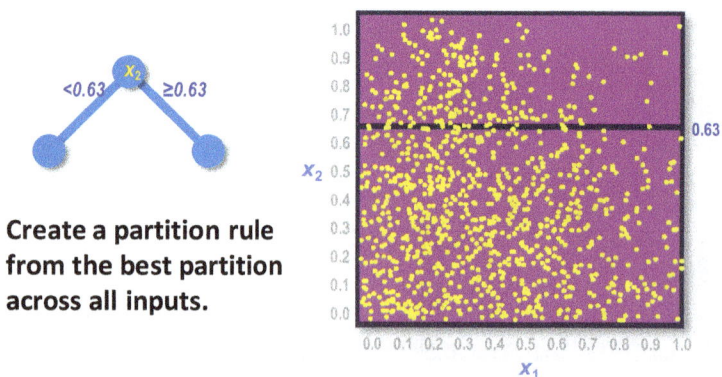

Figure 3.14: Repeating the Process in Each Subset

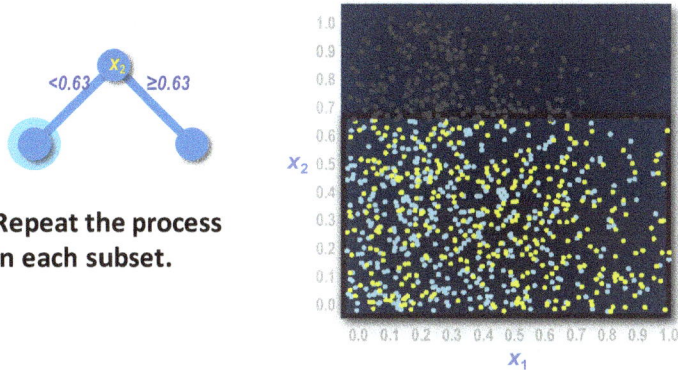

Repeat the process
in each subset.

Figure 3.15: Assessing Every Partition on Input X₁ in the Subset

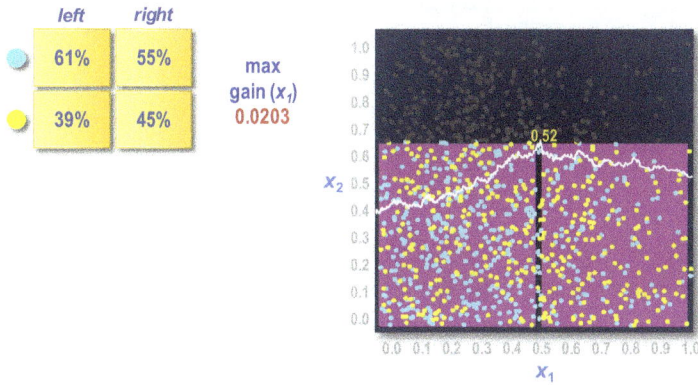

	left	right
●	61%	55%
●	39%	45%

max
gain (x_1)
0.0203

Figure 3.16: Assessing Every Partition on Input X₂ in the Subset

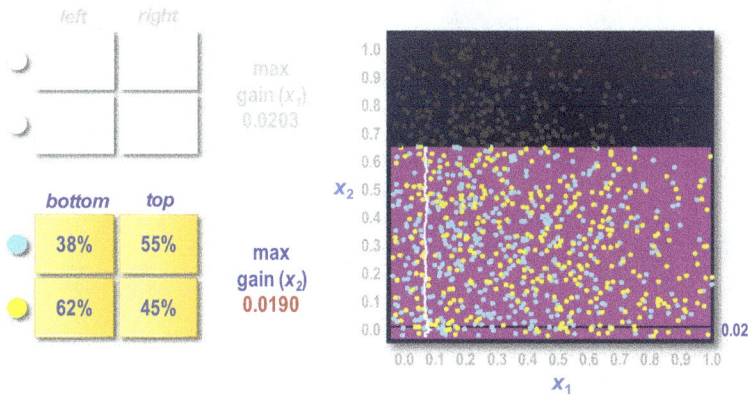

	left	right
●		
●		

max
gain (x_1)
0.0203

	bottom	top
●	38%	55%
●	62%	45%

max
gain (x_2)
0.0190

Figure 3.17: Selecting the Partition with the Maximum Gain on Input X₂

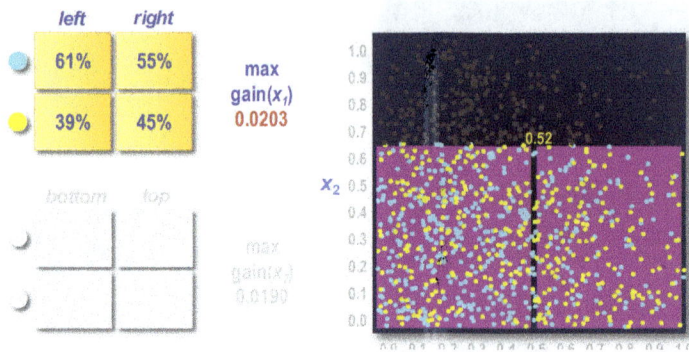

Figure 3.18: Creating a Second Partition Rule

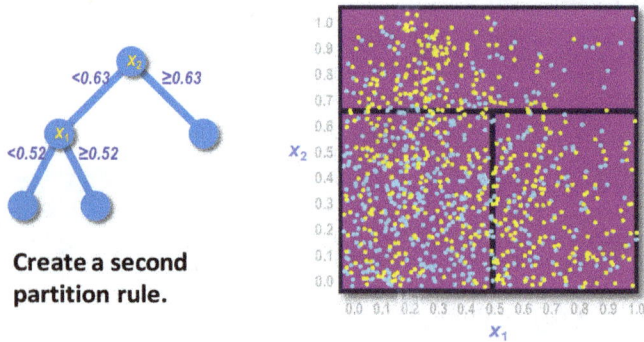

Create a second
partition rule.

Figure 3.19: Creating a Maximal Tree

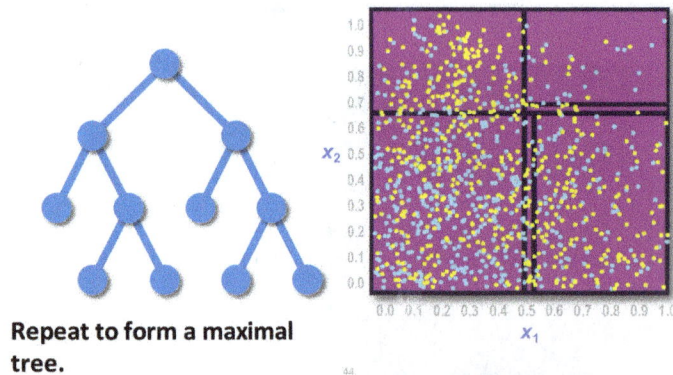

Repeat to form a maximal
tree.

Split Points

Split search is the first part of the decision tree algorithm. The split search starts by selecting an input for partitioning the available training data. Potential split points for the data depend on the measurement scale of the selected input.

Decision trees treat splits on inputs with nominal and ordinal measurement scales differently.

Splitting on a Nominal Input

Splits on a nominal input are not restricted. For a nominal input with *L* distinct levels, there are $S(L,B)$ partitions into *B* branches, where $S(L,B)$ is a Stirling number of the second kind. The total number of partitions is one less than the Bell number for *L* levels,

$$B_L = \sum_{i=0}^{L} S(L,i).$$

> From combinatorics, the Stirling number of the second kind is the number of ways to partition L objects into B non-empty subsets.
>
> $$S(L,B) = \frac{1}{B!} \sum_{j=0}^{B} (-1)^j \binom{B}{j} (B-j)^2$$

The snowballing number of partitions is illustrated with the increasing number of levels of a nominal input in Figure 3.20. The number of partitions are calculated as $S(L,B) = B \cdot S(L-1,B) + S(L-1,B-1)$.

An example of a nominal variable with four levels is also shown with probable partitions in the right side of Figure 3.20. In all, there would be 14 probable partitions: seven two-way splits, six three-way splits, and one four-way split.

How Are the Nominal Variables Encoded?

Behind the scenes, the nominal variables are encoded. For a given nominal variable, a unique continuous index mapping is performed first using the formatted values. This process is often called levelization (for example, variable C contains C1, C2, C3, and C4, and it is levelized to four levels). By default, a hash-based order is assigned, not based on the alphabetic order. So, the generated mapping could be like this:

Formatted Value	Index
C1	1
C2	0
C3	3
C4	2

Figure 3.20: Number of Partitions Illustrated with Increasing Number of Nominal Levels

Number of Levels (L)							Number of Partitions	
2					1		1	1 | 2,3,4
3				3	1		4	2 | 1,3,4
4			7	6	1		14	3 | 1,2,4
5		15	25	10	1		51	4 | 1,2,3
6	31	90	65	15	1		202	1,2 | 3,4
7	63	301	350	140	21	1	876	1,3 | 2,4

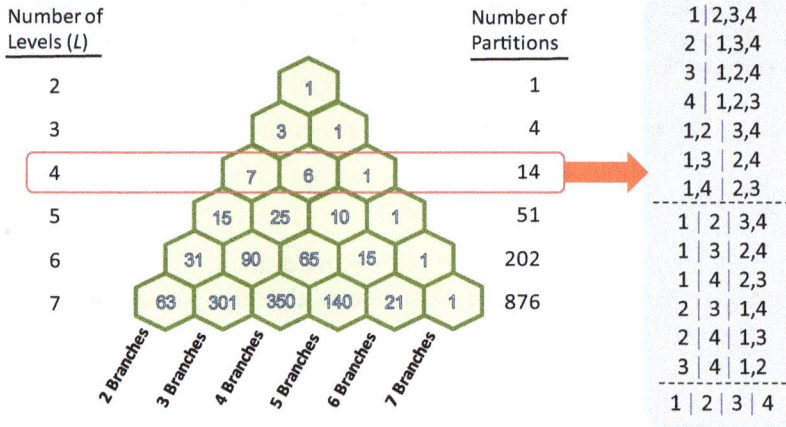

The values are not sorted because SAS Viya often uses decision trees to partition data with high cardinality variables. Sorting would be expensive. You should not care about the ordering, and all these mappings are internal. The ordering might cause different splitting points if the variable has too many levels. In terms of building a good model, it would be better to preprocess (collapse the number of levels) the high cardinality variables.

If one binary target variable Y contains only either 0 or 1, it must be added into the nominal list to train a binary classification model. Because the levels are not sorted, the internal index could be like this:

Formatted Value of Y	Index
0	1
1	0

You should not rely on any internal index and should use only the formatted values (for example, to assess the model performance).

Splitting on an Ordinal Input

Splits on ordinal inputs are restricted to preserve the ordering. Only adjacent values are grouped. Swelling number of partitions are illustrated with increasing number of levels of an ordinal input in Figure 3.21. For an ordinal input with L distinct levels, there are $\binom{L-1}{B-1} = \frac{(L-1)!}{(B-1)!\,(L-B)!}$ partitions into B branches. There are $\sum_{l=2}^{L} \binom{L-1}{l-1} = 2^{L-1} - 1$ possible splits on a single ordinal input.

Figure 3.21: Number of Partitions Illustrated with Increasing Number of Ordinal Levels

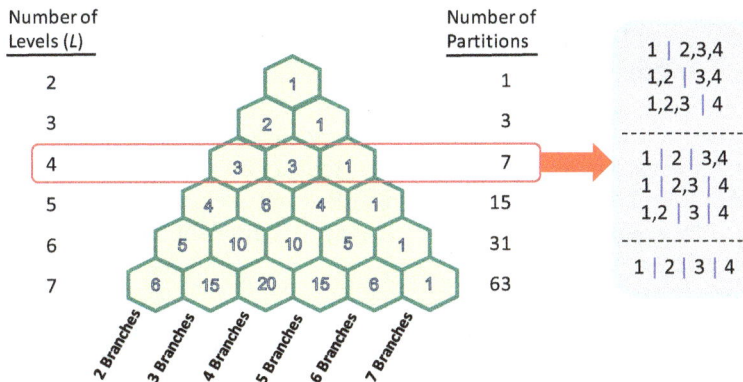

An example of an ordinal variable with four levels is also shown with probable partitions in the right side of Figure 3.21. In all, there would be seven probable partitions: three two-way splits, three three-way splits, and one four-way split.

Splitting on an Interval Input

If the measurement scale of the selected input is interval, each unique value serves as a potential split point for the data. This is called the exhaustive search approach of splitting, which is illustrated in Figure 3.22. The first table shows a small data set. Assume that X is an interval variable and the target Y is binary with Yes and No values. The sufficient statistics that need be collected are the crosstab counts for each unique pair of X and Y shown in the second table. Assuming binary splitting (only two branches/children will be generated from variable X), the exact approach is to enumerate all groupings of X values and for each combination, and the splitting gain is computed. The best combination wins (discussed later in this chapter).

Figure 3.22: Exhaustive Search Illustration

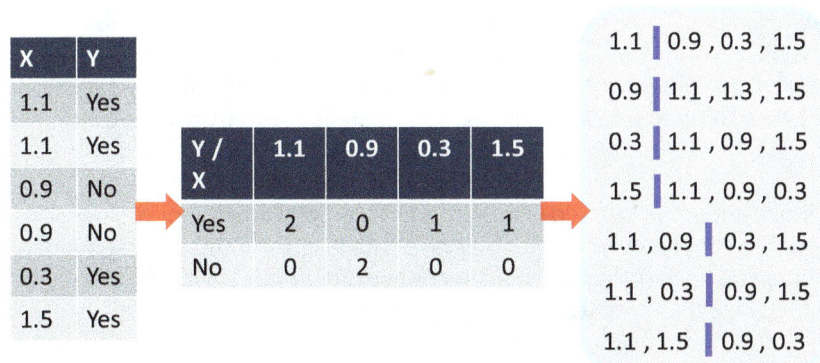

For this example, all the possible groupings are shown on the right-hand side. For binary splitting, the total number of possible combinations is ($2^{n-1} - 1$), where n is the number of distinct values of X. In the above example, there would be seven combinations. Exhaustive search considers an interval variable as if it is nominal and searches for the best split point.

In SAS Viya, if you do not want to perform the greedy search, the k-means clustering-based split search (discussed later) is used. If the cardinality of an input is less than 16 an exhaustive search is used. For larger cardinality, the levels are ordered. In boosting, the levels are ordered by the average gradient. In other cases, the levels are ordered by frequency. The gradient and frequency depend only on the data in the node being split.

An exhaustive search algorithm considers all possible partitions of all inputs at every node in the tree. When the distinct values of an interval input grow rapidly, the possible combinations are enormous. The combinatorial explosion usually makes an exhaustive search prohibitively expensive. There is no way to do the exhaustive search and find the global best splitting point that produces the largest splitting gain. The common *approximation* is to assume the certain order of the variable values that includes:

- ranking
- binning
- k-means clustering
- multi-way splits

Decision trees in SAS Viya use an intricate strategy of split search based on how many branches are required and how many distinct values of variables. Whenever the computational cost is affordable, the greedy and exhaustive search will be applied. It also has one unique k-means clustering-based method to cluster the values into distinct groups when the possible combination is huge. Another unique feature for SAS Viya trees is to support not only binary splitting but also any arbitrary number of branches up to 10.

Ranking

To reduce the possible split points, the common approximation is to assume a certain order of the variable values. For example, if X is numerical, it can simply be sorted in the ascending order, and only (n-1) possible partitions will be performed as shown in Figure 3.23. This reduces the combination number to $O(n)$. To see more visible patterns and reduce variability of data, natural logarithms can also be computed before ranking. This analysis is also true if X is a nominal variable.

For Gini and entropy splitting criteria (discussed later in this chapter) and interval or ordinal inputs, splits in a decision tree depend only on the ordering of the levels. This makes tree models robust to outliers in input space. The application of a rank or any monotonic transformation to an interval variable does not change the fitted tree.

Figure 3.23: Ranking Approximation

X (sorted)	0.9	1.1	1.3	1.5
ln(X)	-0.11	0.10	0.26	0.41
Rank (X)	1	2	3	4

0.9 | 1.1, 1.3 , 1.5

0.9 , 1.1 | 1.3 , 1.5

0.9 , 1.1, 1.3 | 1.5

However, for the splitting criteria that depend on *p*-values (that is, ProbChisq and ProbF discussed below), the invariance to monotonic transformations might not always hold. To avoid overly severe Bonferroni corrections, interval variable values can be binned. Applying transformations can affect this binning and the resulting split choices. The Bonferroni correction, discussed later in the chapter, is a multiple-comparison correction used when several dependent or independent statistical tests are being performed simultaneously.

Binning

Numeric variables with millions of values can be computationally expensive regarding the number of split points. Collecting all the statistics could be expensive, particularly when the data is distributed among threads and workers. So, approximations must be made. The most popular one is to bin the numerical variables and collect the sufficient statistics only for each bin. Binning could be done using either quantile-based or equal-width based methods as shown in Figure 3.24. When the data is highly skewed, the quantile-based binning could be helpful. The number of bins is also a critical parameter.

Figure 3.24: Binning Approximation

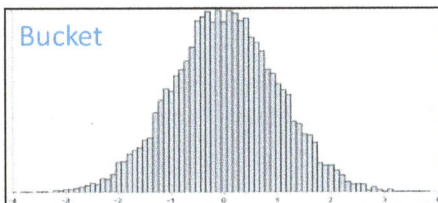

Bucket

Equal interval width of each bin

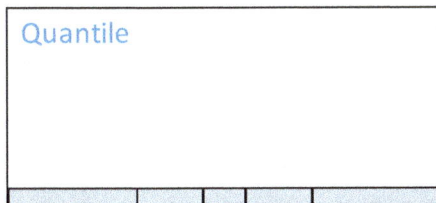

Quantile

Equal number of observations in each bin

Accuracy in terms of finding the optimal splitting points usually improves as the number of bins increases, but at the cost of an increase in computation time. However, more bins need more memory to store the crosstab counts and need more time to aggregate the final statistics across threads and workers. The memory cost could be easily increased to hundreds of gigabytes if the number of numerical variables and bins are huge. Therefore, the number of bins incorporates a trade-off between speed versus accuracy. If one numerical variable contains only few distinct values, it would be better to treat it as the nominal variable since binning would waste too much memory.

> The decision tree in SAS Viya bins the numerical variables only once using the entire training data at the beginning. The binning information such as min and max of variables will not be updated when the data is further partitioned into different tree nodes. This could cause some accuracy issues for decision tree or random forest. However, gradient boosting can mitigate this issue because the target information is updated whenever a new tree is built. Increasing the number of bins can help improve the accuracy. Forest and gradient boosting models are ensemble of trees discussed in subsequent chapters.

Related binning approximation properties of the Decision Tree node in Model Studio are shown in Figure 3.25.

The **Number of interval bins** property specifies the number of bins used for interval inputs. Bin size is (maximum value – minimum value) / interval bins. The default value is 50.

The **Interval bin method** property specifies the method used to bin the interval input variables. Select **Bucket** to divide inputs into evenly spaced intervals based on the difference between maximum and minimum values. Select **Quantile** to divide inputs into approximately equal sized groups. The default value is Bucket.

k-Means Clustering

Once an input variable is binned, you can also determine the splits at each node through clustering each input variable, and then choose the splitting variable based on which variable and

Figure 3.25: Binning Approximation Properties

Number of interval bins:

2

Interval bin method:

Bucket ▼

Bucket

Quantile

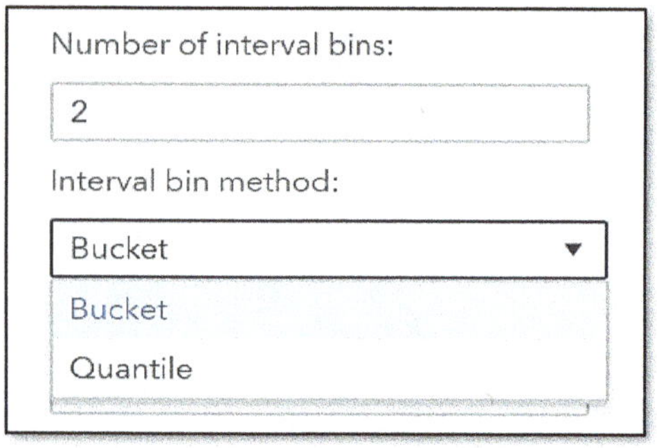

Figure 3.26: Two-Means Clustering

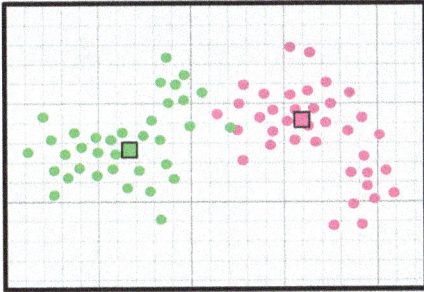

split optimize the chosen criterion. One of the most used methods for clustering is the k-means algorithm. It is a straightforward algorithm that scales well to large data sets. If you believe that the interval input variable is divisible (can be broken down) further, you can use the **Perform clustering-based split search** property, which specifies that a clustering-based search algorithm, instead of an exhaustive search, be used for determining the best split for each input for each tree node. When this option is selected, the order of bins is ignored for interval inputs. By default, this is deselected.

A cluster split writes k number of clusters ($2 \leq k \leq$ max branches), which minimizes the sum (over clusters) of within-cluster-sum-of-square-distance-from-average of:

- y for y interval
- proportion in class for a categorical target
- average gradient for boosting.

Figure 3.26 is an example of a two-means clustering. The within-cluster sum of squares is a measure of the variability of the observations within each cluster. For a categorical target, the within-cluster-sum-of-squares for cluster C is sum of observations in C over bins and classes of [(proportion in bin and class) − (proportion in class)]^2.

Multi-Way Splits

In theory, multi-way splits are not more flexible than binary splits if sufficient binary splitting is allowed. So, the tree could be shallow with multiple branches (>=2) and the model size could be small as shown in Figure 3.27. Trees with multi-way splits tend to be wider and shorter than trees with binary splits. However, research has shown that trees with multi-way splits do not necessarily outperform trees with binary splits.

The Maximum number of branches property specifies the maximum number of branches that a splitting rule produces. The default value of 2 results in binary splits. Possible values range from 2 to 10.

Figure 3.27: Binary Split Versus Multi-Way Split

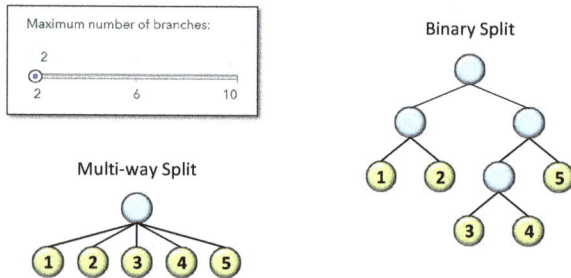

You are often interested in ranking the attributes according to how much they contribute to the classification of instances. Such a ranking can be read from a multiway tree simply by tracking the classification of an instance, starting at the root, and collecting the attributes being tested. Decision trees in which numeric attributes (or in general, any attribute whose values form a totally ordered set) are split several ways are more comprehensible than the usual binary trees, because attributes rarely appear more than once in any path from root to leaf. Multi-way splits, therefore, often give more interpretable trees because split variables tend to be used fewer times. They are favored over repeated binary splits when only one input exists. Many prefer binary splits because there are generally fewer splits to consider and an exhaustive search is more feasible. Further, multi-way splits deplete data quickly, which starves subsequent searches on other variables.

Demo 3.2: Exploring Split Search Tree Growth Options

In this demonstration, you return to the classification tree created earlier on the **insurance_part** data set and experiment with some of the split search options.

   1. Click the **View all projects** icon to return to the **Insurance_ClassTree** project.

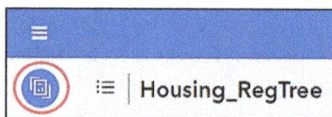

   2. Double-click the **Insurance_ClassTree** tile to open the project.

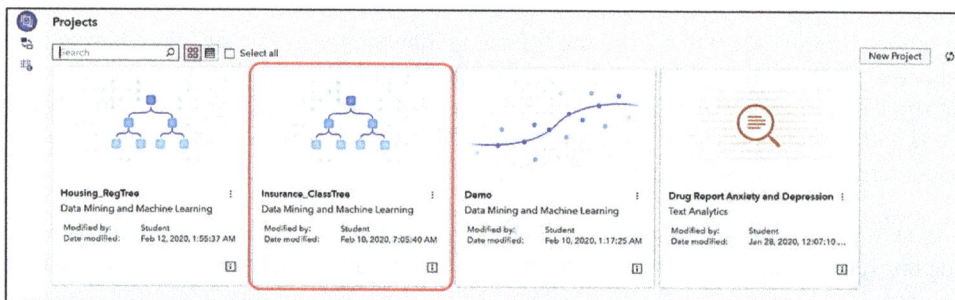

3. Click the **Pipelines** tab to access the pipeline created earlier.
4. Right-click the **Data** node and select **Add child node** ⇨ **Supervised Learning** ⇨ **Decision Tree**.
5. Right-click the newly added Decision Tree node, select **Rename**, and enter **DT Multiway Splits** in the **Name** field to give a more explicit name.
6. Under Splitting Options, set the maximum number of branches to **4** (scroll the slider or type in).

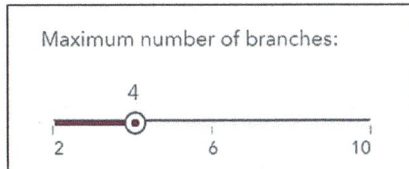

You can always split a node into two branches to avoid having to decide what an appropriate number of branches would be. But this easy approach might poorly communicate structure in the data if the data more naturally splits into more branches.

A multiway split can always be accomplished with a sequence of binary splits on the same input. An algorithm that proceeds in binary steps can split with more than one input and thus will consider more multistep partitions than an algorithm can consider in a single-step multiway split. The extra branches reduce the data available deeper in the tree, degrading the statistics and splits in deeper nodes.

7. Click the **Run Pipeline** button.

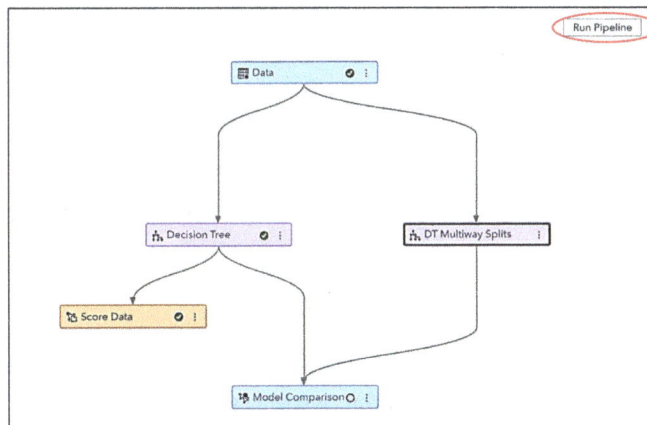

8. Right-click the **Decision Tree** node after the pipeline runs successfully. Click **Log** to see the processing time.

```
NOTE: PROCEDURE TREESPLIT used (Total process time):
      real time               0.72 seconds
      cpu time                0.12 seconds
```

The TREESPLIT procedure took 0.72 seconds for a binary split search.
(Use find functionality (Ctrl+F) to search for "treesplit" for speedy tracing.)

9. Close the log and repeat the same for the **DT Multiway Splits** node.

```
NOTE: PROCEDURE TREESPLIT used (Total process time):
      real time              1.73 seconds
      cpu time               0.10 seconds
```

The TREESPLIT procedure took more time (1.73 seconds in this case) for a multi-way split decision tree.

Tree algorithms usually take a shortcut to reduce the split search by restricting searches to binary splits. Multi-way splits request more evaluations for the candidate splits because all inputs in all *n*-way splits must be considered (for example, in four-way splits, all candidates for two-way splits, three-way splits, and four-way splits were evaluated).

10. Close the log and open the Results of the **DT Multiway Splits** node to see whether multi-way splits bring us some gain.
11. Maximize the tree diagram.

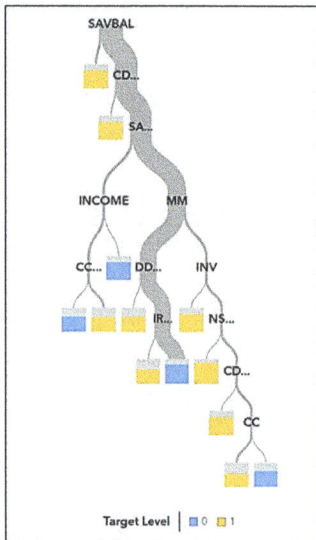

Surprisingly, none of the splits are more than the two-way splits. Why?

12. Close the tree diagram and observe the Output window.

Model Information	
Split Criterion	IGR
Pruning Method	Cost Complexity
Max Branches per Node	4
Max Tree Depth	10
Tree Depth Before Pruning	10
Tree Depth After Pruning	8
Number of Leaves Before Pruning	137
Number of Leaves After Pruning	13

In fact, this is a complex tree with 137 leaves, but pruned to 13 leaves.

13. Close the Results of the DT Multiway Splits node.
14. Under Pruning Options, change the **Selection method** from Automatic to **Largest** to see the maximal tree.

> ∨ Pruning Options
>
> Subtree method:
>
> [Cost complexity ▼]
>
> Selection method:
>
> [Largest ▼]

Note: Pruning is discussed in detail in the next chapter.

15. Rerun the pipeline by clicking the **Run Pipeline** button.
16. Open the Results of the DT Multiway Splits node and maximize the tree diagram. Adjust the zoomed-in view to clearly see the decision tree with multiple branching allowed.

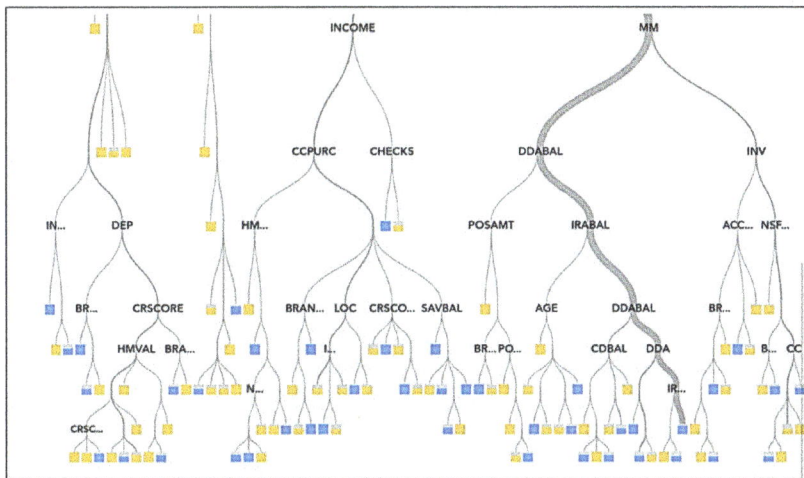

There are many two-way splits, and quite a few nodes resulted in three-way and four-way splits (for example, ACCTAGE, HMVAL, CDBAL and PHONE amongst others have three-way splits whereas BRANCH, ATMAMT, and so on, have four-way splits).

17. Close the tree diagram and close the Results of the DT Multiway Splits node.
18. Open the Results of the Model Comparison node.

Model Comparison								
Champi...	Name	Algorith...	KS (You...	Misclas...	Misclas...	Root Av...	Averag...	Sum of ...
⊡	Decision Tree	Decision Tree	0.4204	0.2678	0.2678	0.4303	0.1852	5,807
	DT Multiway Splits	Decision Tree	0.4041	0.2752	0.2752	0.4401	0.1937	5,807

The average square error of multi-way splits decision tree is 0.1937 and misclassification rate is 0.2752 on the validation partition. Even though the multi-way splits' decision tree is abundantly more complex, there is hardly any gain over the decision tree with binary splits that stood at an ASE of 0.1852 and MISC of 0.2678.

19. Close the results of the Model Comparison node.
20. Return to the results of the decision tree with two-way splits. Right-click the **Decision Tree** node and open the results.

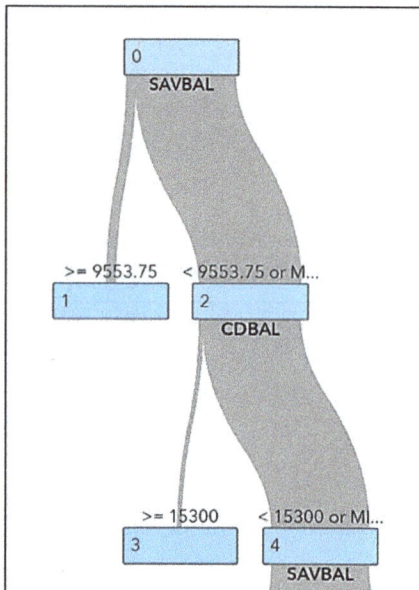

The tree with the default setting of two-way splits has grown up to 103 leaves. **SAVBAL** is the first variable (with a split point of 9553.75), followed by **CDBAL** (with a split point of 15300) that was used in splitting.

21. Maximize the **Variable importance** table.

Variable Label	Role	Variable Name	Validation Import...	Importance Sta...	Relative Importance	Count
Saving Balance	INPUT	SAVBAL	222.0828	0	1	2
Money Market	INPUT	MM	91.0402	0	0.4099	1
CD Balance	INPUT	CDBAL	72.2555	0	0.3254	2
Checking Balance	INPUT	DDABAL	65.7168	0	0.2959	2
Checking Account	INPUT	DDA	36.4405	0	0.1641	1
IRA Balance	INPUT	IRABAL	5.5151	0	0.0248	1
Credit Card Purchases	INPUT	CCPURC	2.9355	0	0.0132	1
Age	INPUT	AGE	1.9427	0	0.0087	2
Branch of Bank	INPUT	BRANCH	0.9591	0	0.0043	2
Amount NSF	INPUT	NSFAMT	0.6847	0	0.0031	1
Credit Card	INPUT	CC	0.5572	0	0.0025	1
Number of Checks	INPUT	CHECKS	0.2705	0	0.0012	1

Decision tree algorithm search for a good split on all the inputs, one at a time, and then select the input with the best split. Splitting each variable significantly biases the selection toward nominal variables with many categories. You have two nominal variables in this data: branch of bank (**BRANCH**) and area classification (**RES**), each has 19 and 3 levels, respectively. **BRANCH** is used once for splitting and **RES** not even once.

Note: You can consider the possibility of searching for a split on each input and then penalizing those splits that are prone to bias by adjusting the *p*-values. This will be discussed later in this chapter.

Now focus on the most important variables.

The top four variables are saving balance (**SAVBAL**), money market (**MM**), certificate of deposit balance (**CDBAL**), and checking account balance (**DDABAL**). Money market is a binary input and the remaining three are interval – each have 8552, 554, and 15269 distinct values, respectively. (You can explore them in SAS Visual Analytics.) Why is **DDABAL** less important although it has most distinct values?

There are two problems with the exhaustive search approach. One, the computational complexity and two, bias in variable selection, which could be a more fundamental problem from the standpoint of tree interpretation. Unrestrained search tends to select variables that have more splits. This makes it hard to draw reliable conclusions from the tree structures. Interval variables restrict this by binning. Recall that the number of interval bins were 50; therefore, each interval input, in fact, has 49 plausible split points. It is interesting to note that their importance is almost in line with their correlation with the target.

22. Close the Variable Importance table and close the Results of the Decision Tree node.

23. Keep the Decision Tree node selected, and under Splitting Options, change the **Minimum leaf size** from 5 to **20** and the **Number of interval bins** from 50 to **100**.

Maximum depth:

10

Minimum leaf size:

20

Missing values:

Use in search ▾

Minimum missing use in search:

1

Number of interval bins:

100

Interval bin method:

Quantile ▾

The greater the number of bins, the better accuracy in terms of finding the optimal split points. However, more bins need more processing time. Increasing the minimum leaf size avoids creating smaller leaves and might compensate a bit on tree complexity.

24. Run the **Decision Tree** node and open the **Results**.

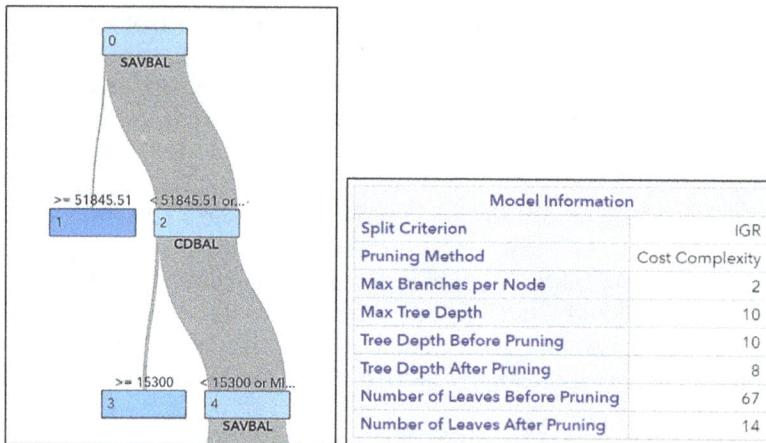

Model Information	
Split Criterion	IGR
Pruning Method	Cost Complexity
Max Branches per Node	2
Max Tree Depth	10
Tree Depth Before Pruning	10
Tree Depth After Pruning	8
Number of Leaves Before Pruning	67
Number of Leaves After Pruning	14

This tree has grown up to 67 leaves.

The final tree is a simpler tree with 14 leaves (as against 31 leaves earlier with default settings). There is no change in the order of the variable importance of the top few variables. However, split points have changed for many of them (for example, SAVBAL has a new split point at 51845.51 instead of 9553.75 in the default tree).

Variable Label	Role	Variable Name	Validation Imp...	Importance Sta...	Relative Import...	Count
Saving Balance	INPUT	SAVBAL	219.6877	0	1	3
Money Market	INPUT	MM	93.6858	0	0.4265	1
CD Balance	INPUT	CDBAL	75.5051	0	0.3437	2
Checking Balance	INPUT	DDABAL	27.0782	0	0.1233	2
IRA Balance	INPUT	IRABAL	6.0739	0	0.0276	1
Credit Card	INPUT	CC	3.3834	0	0.0154	1
Number Point of Sale	INPUT	POS	1.1803	0	0.0054	1
Investment	INPUT	INV	-0.1806	0	-0.0008	1
Branch of Bank	INPUT	BRANCH	-0.2230	0	-0.0010	1

25. Click on the **Assessment** tab. Scroll down and maximize the **Fit Statistics** table.

Target ...	Data Role	Partitio...	Formatt...	Sum of ...	Averag...	Divisor ...	Root Av...	Misclas...	Multi-Cl...	KS (You...
INS	TRAIN	1	1	13,550	0.1865	13,550	0.4318	0.2613	0.5584	0.3888
INS	VALIDATE	0	0	5,807	0.1897	5,807	0.4355	0.2676	0.5660	0.3818

There is barely any improvement in model performance. ASE is 0.1897 as against 0.1852 of the default tree and MISC is 0.2676 as against 0.2678 of the default tree.

A clustering algorithm to find the splits for interval inputs generally leads to an improvement of accuracy and reduced tree size.

26. Close the **Results** of the Decision Tree node and return to the node options on your right.

27. Change the **Number of interval bins** to **10** and check the **Perform clustering-based split search** box.

Number of interval bins:

10

Interval bin method:

Quantile ▼

Surrogate rules:

0

☐ Use input once

☑ Perform clustering-based split search

Because you have a binary split decision tree, the cluster split performs two-means clustering that minimizes the sum (over these two clusters) of within-cluster-sum-of-square-distance-from-average in the class (that is, proportion) of INS.

Binning of interval inputs operates the same way with clustering-based split search as without it. For a numerical input with clustering-based split search performed, the number of interval bins is the initial number of clusters of the values. Many bins might lead to a large number of plausible split points, which might not be very practical. Hence, you can control this by reducing the number of bins appropriately.

Note: Enabling the *k*-means fast search algorithm ignores bin ordering. Disable this option if you want to use the greedy search method and respect bin ordering.

28. Run the **Decision Tree** node again.
29. Right-click the **Decision Tree** node and open the **Results**.
30. Maximize the **Tree Diagram** and concentrate on the root node and one-depth below that.

Model Information	
Split Criterion	IGR
Pruning Method	Cost Complexity
Max Branches per Node	2
Max Tree Depth	10
Tree Depth Before Pruning	10
Tree Depth After Pruning	8
Number of Leaves Before Pruning	172
Number of Leaves After Pruning	18

This tree with clustering-based split search performed has grown up to 172 leaves (as against 67 leaves earlier with the number of bins set to 100). The final tree is a slightly complex tree with 18 leaves (as against 14 leaves earlier with the number of bins set to 100). **SAVBAL** and **DDABAL** are still being used early in the tree, however, with altogether different split points. **CDBAL** becomes an unimportant variable. However, nominal input **BRANCH** is now used early in the tree.

31. Close the Tree Diagram and maximize the **Variable Importance** table.

Variable Label	Role	Variable Name	Validation Importa...	Impor...	Relative Importance	Count
Saving Balance	INPUT	SAVBAL	218.9077	0	1	1
Checking Balance	INPUT	DDABAL	137.1400	0	0.6265	2
Money Market	INPUT	MM	115.2705	0	0.5266	1
Certificate of Deposit	INPUT	CD	35.7783	0	0.1634	2
Branch of Bank	INPUT	BRANCH	8.4289	0	0.0385	3
Age of Oldest Account	INPUT	ACCTAGE	6.3551	0	0.0290	1
Number of Checks	INPUT	CHECKS	2.4767	0	0.0113	2
Line of Credit	INPUT	LOC	2.3284	0	0.0106	1
Investment	INPUT	INV	2.1738	0	0.0099	2
Installment Loan	INPUT	ILS	0.8901	0	0.0041	1
Retirement Account	INPUT	IRA	-0.3899	0	-0.0018	1

There is some change in the variable importance of first few variables also.

32. Close the **Variable Importance** table and maximize the **Fit Statistics** table under the **Assessment** tab.

Target...	Data Role	Par...	Fo...	Sum of ...	Averag...	Divisor ...	Root Av...	Misclas...	Multi-Cl...	KS (You...
INS	TRAIN	1	1	13,550	0.1792	13,550	0.4233	0.2551	0.5393	0.4288
INS	VALIDATE	0	0	5,807	0.1808	5,807	0.4252	0.2616	0.5422	0.4259

The ASE and MISC have marginally improved from 0.1897 to 0.1808 and 0.2676 to 0.2616 respectively.

33. Close the Fit Statistics table and close the results.

End of Demonstration

Splitting Criteria

The predictor variables for tree models can be categorical or continuous. The model is based on partitioning the predictor space (the set of all possible combinations of the predictor variables) into nonoverlapping segments, which correspond to the terminal nodes (called leaves) of the tree. Partitioning is done repeatedly, starting with the root node, which contains all the data, and continuing until a stopping criterion is met. At each step, the parent node is split into child nodes by selecting a predictor variable and a split value for that variable that minimize the variability, according to a specified measure, in the response variable across the child nodes. Whether it is an exhaustive search or approximations for possible split points, we enumerate all possible groupings of X values, and for each combination, the splitting gain is computed. The best combination wins. Figure 3.28 shows splitting gain represented in star ratings are easy to comprehend. Various measures, such as the Gini index, entropy, and residual sum of squares, can be used to assess candidate splits for each node.

Figure 3.28: Splitting Gain

Figure 3.29: Splitting Criteria and Type of Decision Tree

	Criteria Based on Impurity	Criteria Based on Statistical Test
Classification Tree	• Gini • Entropy • IGR	• Chi-Square • CHAID
Regression Tree	• Variance	• F-Test • CHAID

There are two types of criteria for splitting a parent node: criteria that maximize a decrease in node impurity, as defined by an impurity function, and criteria that are defined by a statistical test. The choice of these criteria is also dependent on the type of decision tree you want to build. Splitting criteria are shown in Figure 3.29 with a distinctive connection between type of criteria and type of decision tree.

The **Grow criterion** property in Model Studio specifies the criterion for splitting a parent node into child nodes. The following options are available:

- The **Class target criterion** property specifies the splitting criterion to use for determining best splits on inputs that are given a class target. Possible values are shown in Figure 3.30. The default value is **Information gain ratio**.
- The **Interval target criterion** property specifies the splitting criterion to use for determining the best splits on inputs that are given an interval target. Possible values are shown in Figure 3.30. The default value is **Variance**.

You can change how the splits are evaluated in the split-search phase of the tree algorithm. For categorical targets, changing from the chi-square criterion typically yields similar splits

Figure 3.30: Splitting Criteria in Model Studio

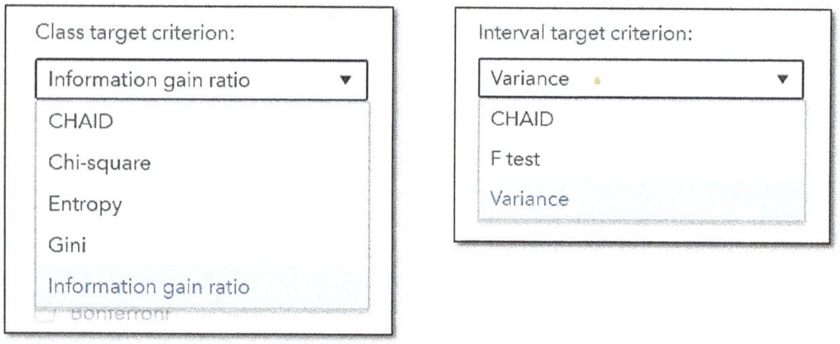

Class target criterion:

Information gain ratio ▼
CHAID
Chi-square
Entropy
Gini
Information gain ratio
Bonferroni

Interval target criterion:

Variance ▼
CHAID
F test
Variance

Figure 3.31: Impurity Reduction

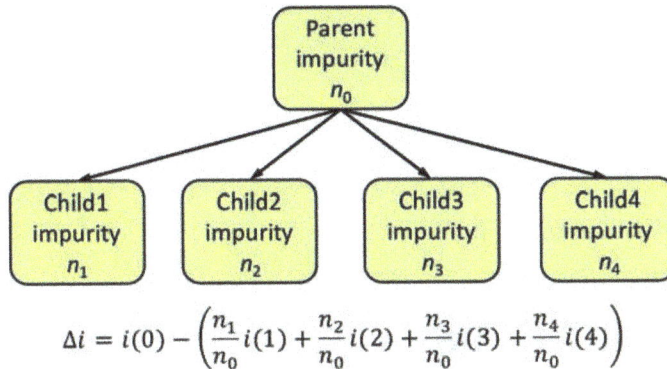

$$\Delta i = i(0) - \left(\frac{n_1}{n_0}i(1) + \frac{n_2}{n_0}i(2) + \frac{n_3}{n_0}i(3) + \frac{n_4}{n_0}i(4)\right)$$

if the number of distinct levels in each input is similar. If not, the other split methods tend to favor inputs with more levels due to the multiple comparison problem, which will be discussed later. You can also cause the chi-square method to favor inputs with more levels by turning off the Bonferroni adjustments. Because Gini reduction and entropy reduction criteria lack the significance threshold feature of the chi-square criterion, they tend to grow enormous trees.

Splitting Criteria Based on Impurity

The impurity of a parent node is a nonnegative number that is equal to zero for a pure node (a node in which all the observations have the same value of the response variable). Nodes in which the observations have very different values of the response variable have a large impurity.

Let $i(.)$ be some measure of within-node impurity and let Δi represent the overall reduction in impurity for the tree. Figure 3.31 illustrates impurity reduction from a parent node to several child nodes. Many splitting criteria (including Gini and entropy) are based on the reduction in node impurity (that is, the reduction of within-node variability) induced by the split.

The entropy, Gini index, and variance (RSS – residual sum of squares) criteria decrease impurity. The impurity of a parent node τ is defined as $i(\tau)$. The algorithm selects the best split variable and the best split value to produce the highest reduction in impurity,

$$\Delta i(s, \tau) = i(\tau) - \sum_{b=1}^{B} p(\tau_b|\tau)\ i(\tau_b)$$

where τ_b denotes the bth child node, $p(\tau_b|\tau)$ is the proportion of observations in τ that are assigned to τ_b, and B is the number of branches after splitting τ.

Figure 3.32: Gini Impurity Example

$$1 - \sum_{j=1}^{J} p_j^2 = 2 \sum_{j<k} p_j p_k$$

high diversity, low purity

Pr(interspecific encounter) = 1 - 2(3/8)^2 + 2(1/8)^2 = **0.69**

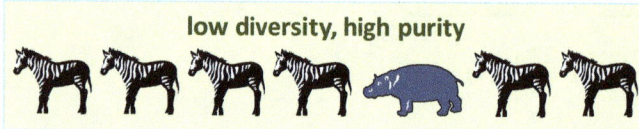

low diversity, high purity

Pr(interspecific encounter) = 1 - (6/7)^2 + (1/7)^2 = **0.24**

Impurity Reduction Criteria for Classification Trees

Gini Index Criterion

This criterion defines $i(\tau)$ as the Gini index that corresponds to the average squared error of a class response and is given by

$$i(\tau) = -\sum_{j=1}^{J} p_j^2$$

where p_j is the proportion of observations that have the jth response value.

This can also be represented as Gini Impurity or Gini Diversity Index: Pr(interspecific encounter) = 1-Gini Index.

The Gini index is a measure of variability for categorical data (developed by the eminent Italian statistician Corrado Gini in 1912). The Gini index can be used as a measure of node impurity where p_1, p_2, ..., p_r are the proportions of each target class in a node. The ΔGini splitting criteria was proposed by Breiman et al. (BFOS 1984).

The Gini index can be interpreted as the probability that any two elements of a multi-set, chosen at random (with replacement), are different. Figure 3.32 illustrates an example wherein different types of animals represent the extent of diversity and consequent purity in a node. A pure node has a Gini index of 0. As the number of evenly distributed classes increases, the Gini index approaches 1. For more information, see Hastie, Tibshirani, and Friedman (2009).

In mathematical ecology, the Gini index is known as *Simpson's diversity index*. In cryptanalysis, it is 1 minus the *repeat rate* (Patil and Taillie 1982).

Figure 3.33: Entropy Example

$$i(\tau) = -\sum_{j=1}^{J} p_j \ log_2(p_j)$$

high diversity, low purity

$$2[(3/8) \log_2(3/8)^2] + 2[(1/8) \log_2(1/8)^2] = \textbf{1.81}$$

low diversity, high purity

$$[(6/7) \log_2(6/7)^2] + [(1/7) \log_2(1/7)^2] = \textbf{0.59}$$

Entropy Criterion

The entropy impurity of node **t** is defined as

$$i(\tau) = -\sum_{j=1}^{J} p_j \ log_2(p_j)$$

Entropy uses the gain in information or the decrease in entropy to split each variable and then to determine the split. A minimum of decrease in entropy or increase in information gain ratio can be specified.

Figure 3.33 extends the different type of animals example to entropy criteria. Entropy can be thought of as a measure of purity. Higher values of entropy mean low levels of purity.

Entropy, as with the Gini index, is a measure of variability for categorical data. Consider r mutually exclusive events with probabilities $p_1, p_2, ..., p_r$. The rarity of a particular outcome can be measured as $-\log_2 (p_j)$. Entropy is the average rarity and thus measures the uncertainty of the outcome. In communication (information) theory, entropy was developed to measure the uncertainty of a transmitted message, measured in bits (Shannon 1948). Entropy has some desirable properties.

$$0 \leq H(p_1, p_2, ..., p_r) \leq H\left(\tfrac{1}{r}, \tfrac{1}{r}, ..., \tfrac{1}{r}\right)$$
$$H\left(\tfrac{1}{r}, \tfrac{1}{r}, ..., \tfrac{1}{r}\right) = \log_2(r)$$
$$H(1, 0, ..., 0) = 0$$

The Δentropy splitting criterion was proposed by Quinlan (1993). The Δentropy splitting criterion is equivalent to using the likelihood ratio chi-square test statistic for association between the branches and the target categories (Ripley 1996). Quinlan (1993) prefers the use of the gain ratio, a modification of Δentropy that reduces its bias toward a larger number of branches and attributes with many values.

Figure 3.34: Relationship of Entropy and Gini Functions

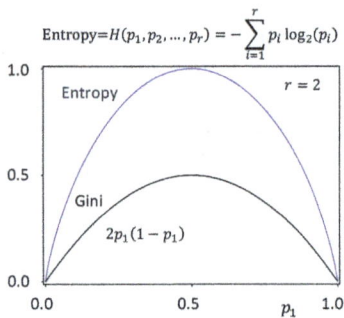

Both impurity measures usually yield similar results. The graph in Figure 3.34 shows that the Gini index and entropy are similar impurity criterion. Entropy might be a little slower to compute because it uses a logarithm.

> For classification trees with binary splits, Breiman (1996) showed that the ΔGini criterion tends to favor isolating the largest target class in one branch, whereas the Δentropy criterion tends to favor split balance.

Information Gain Criterion

The information gain method chooses a split based on which attribute provides the greatest information gain. The gain is measured in bits. Although this method provides good results, it favors splitting on variables that have many attributes. The information gain ratio method incorporates the value of a split to determine what proportion of the information gain is valuable for that split. The split with the greatest information gain ratio is chosen.

Shown in Figure 3.35 is information gain if you split the node based on color of animals that has two levels: brownish and grayish.

The information gain calculation starts by determining the information of the training data. The information in a response value, *r*, is calculated in the following expression:

$$-log_2\left(\frac{freq(r,T)}{|T|}\right)$$

T represents the training data and $|T|$ is the number of observations. To determine the expected information of the training data, sum this expression for every possible response value:

$$I(T) = -\sum_{i=1}^{n}\frac{freq(r_i,T)}{|T|} \times log_2\left(\frac{freq(r_i,T)}{|T|}\right)$$

Figure 3.35: Information Gain per Color of Animals

$$g(S) = i(\tau) - \sum_{i=1}^{M} \frac{\tau_i}{\tau} i(\tau_i)$$

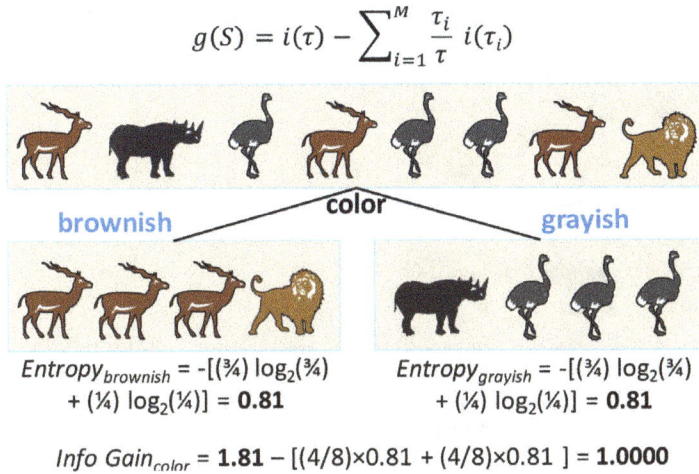

brownish   color   grayish

$Entropy_{brownish}$ = -[(¾) $\log_2$(¾)
+ (¼) $\log_2$(¼)] = **0.81**

$Entropy_{grayish}$ = -[(¾) $\log_2$(¾)
+ (¼) $\log_2$(¼)] = **0.81**

$Info\ Gain_{color}$ = **1.81** − [(4/8)×0.81 + (4/8)×0.81] = **1.0000**

Here, *n* is the total number of response values. This value is also referred to as the *entropy* of the training data.

Next, consider a split *S* on a variable *X* with *m* possible attributes. The expected information provided by that split is calculated by the following equation:

$$I_S(T) = -\sum_{j=1}^{m} \frac{|T_j|}{|T|} \times I(T_j)$$

In this equation, T_j represents the observations that contain the *j*th attribute.

The information gain of split *S* is calculated by the following equation:

$$G(S) = I(T) - I_S(T)$$

Information gain ratio attempts to correct the information gain calculation by introducing a split information value. The split information is calculated by the following equation:

$$SI(S) = -\sum_{j=1}^{m} \frac{|T_j|}{|T|} \times \log_2 \left(\frac{|T_j|}{|T|} \right)$$

As its name suggests, the information gain ratio is the ratio of the information gain to the split information:

$$GR(S) = \frac{G(S)}{SI(S)}$$

Shown in Figure 3.36 is information gain if you split the node based on size of animals of three levels: small, big, and huge. Because information gain on size is more than information gain on color, the first split would be at size. Variables that have more levels generally dominate.

Figure 3.36: Information Gain per Size of Animals

$$g(S) = i(\tau) - \sum_{i=1}^{M} \frac{\tau_i}{\tau} i(\tau_i)$$

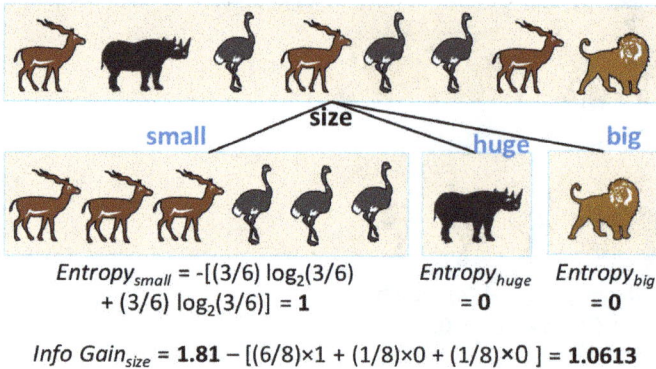

small   size   huge   big

$Entropy_{small}$ = -[(3/6) log$_2$(3/6) + (3/6) log$_2$(3/6)] = **1** $Entropy_{huge}$ = **0** $Entropy_{big}$ = **0**

Info Gain$_{size}$ = **1.81** − [(6/8)×1 + (1/8)×0 + (1/8)×0] = **1.0613**

Now, reconsider the pen-based recognition of handwritten digits example referenced in the previous chapter where the cases were digits written on a pressure-sensitive tablet. Figure 3.37 demonstrates values of Gini and entropy criteria for that classification tree.

After a set of candidate splits is determined, a splitting criterion is used to determine the best one. In some situations, the worth of a split is obvious. If target distributions are the same in the child nodes as they are in the parent node, then no improvement was made, and the split is worthless. In contrast, if a split results in pure children, then the split is the best.

In classification trees, the three most well-known splitting criteria are based on the Gini index (BFOS 1984), entropy (Quinlan 1993), and the chi-square test (Kass 1980). Well-known algorithms and software products associated with these three splitting criteria are CART (classification and

Figure 3.37: Classification Tree – Gini and Entropy Criteria

X1: <38.5 ≥38.5

	<38.5	≥38.5	ΔGini	Δentropy
1	293	71		
7	363	1	.197	.504
9	42	294		

X10: <0.5 1-41 42-51 ≥51.5

	<0.5	1-41	42-51	≥51.5	ΔGini	Δentropy
1	9	143	65	147		
7	221	88	1	54	.255	.600
9	1	4	16	315		

(higher is better)

Figure 3.38: Variance Example

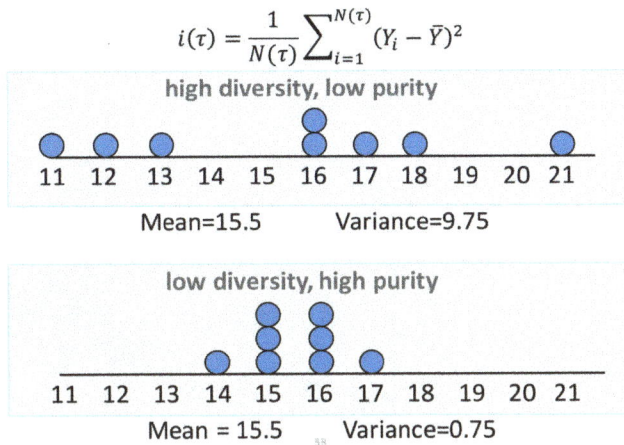

$$i(\tau) = \frac{1}{N(\tau)} \sum_{i=1}^{N(\tau)} (Y_i - \bar{Y})^2$$

high diversity, low purity

| | | | | | | | | | | |
|11|12|13|14|15|16|17|18|19|20|21|

Mean=15.5   Variance=9.75

low diversity, high purity

| | | | | | | | | | | |
|11|12|13|14|15|16|17|18|19|20|21|

Mean = 15.5   Variance=0.75

regression tree); C5.0 (developed by the machine learning researcher Quinlin); and the CHAID algorithm (chi-squared automatic interaction detection).

Impurity Reduction Criterion for Regression Trees

Variance

Only one impurity reduction criterion, the variance, is available for regression trees. This criterion, also called the ANOVA criterion or the RSS criterion, defines $i(\tau)$ as the residual sum of squares,

$$i(\tau) = \frac{1}{N(\tau)} \sum_{i=1}^{N(\tau)} (Y_i - \bar{Y})^2$$

where $N(\tau)$ is the number of observations in τ, Y_i is the response value of observation i, and $\bar{Y}$ is the average response of the observations in τ.

The variance reduction of a node is defined as the total reduction of the variance of the target variable due to the split at this node. This method uses the standard formula of variance to choose the best split. Variance calculation is a two-step process. First, calculate variance for each node, and second, calculate variance for each split as the weighted average of each node variance. The split with the lower variance is selected as the criteria to split the data. Figure 3.38 illustrates dispersion (diversity) and consequent purity along with the variance values.

Recall the Boston housing data example in the previous lesson, where the cases were 506 census tracts in Boston. The target was the median home value (MEDV) and one of the 13 inputs was the nitrogen oxide concentration in the air (NOX).

Figure 3.39: Regression Tree – Variance Reduction Criterion

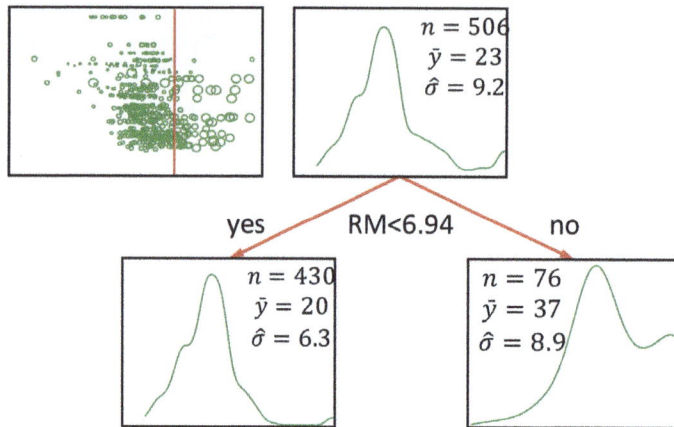

Regression trees endeavor to partition the input space into segments where the target values are alike, that is, each segment or node has low variability as shown in Figure 3.39. All target values would be equal in a pure node. In other words, the variance of the target would be zero within a pure node.

Other more robust measures of spread such as the least absolute deviation (LAD) were proposed (BFOS 1984).

Splitting Criteria Based on Statistical Tests

The chi-square, F test, and CHAID criteria are defined by statistical tests. These criteria calculate the worth of a split by testing for a significant difference in the response variable across the branches that are defined by a split.

This is a three-step process:
    Step 1: Calculate the worth of every partition on an input variable.
    Step 2: Select the partition with the maximum worth.
    Step 3: Repeat for all other input variables.

The worth is defined as -log(p), where p is the p-value of the test. In Model Studio, the **Significance level** property specifies the significance level for the splitting criteria CHAID, chi-square, and F test. The default value is 0.2.

Statistical Test Criterion for Classification Trees

Any split in a classification tree can be arranged in a contingency table. The rows represent the child nodes, and the columns represent the classes. Each cell contains the frequency of cases of that class in that child node.

Figure 3.40: Classification Tree – Chi-Square Criterion

	Observed X1: <38.5 ≥38.5			Expected		$\frac{(O-E)^2}{E}$	
1	293	71	.342	239	125	12	23
7	363	1	.342	239	125	64	123
9	42	294	.316	225	116	149	273
	.656	.344	n=1064				

Expected = $row\ proportion \times column\ proportion \times n$

$$\chi_v^2 = \sum \frac{(O-E)^2}{E} = 644 \qquad v = (3-1)(2-1) = 2$$

Chi-Square Test

The Pearson chi-square test can be used to judge the worth of the split. It tests whether the column distributions (class proportions) are the same in each row (child node). The test statistic measures the difference between the observed cell counts and what would be expected if the branches and target classes (rows and columns) were independent. Observed frequencies, expected frequencies, and chi-square statistic calculations for the pen-based recognition of handwritten digits example is shown in Figure 3.40.

The statistical significance of the test is not monotonically related to the size of the chi-square test statistic. The degrees of freedom of the test is $(r-1)(B-1)$, where r (target levels) and B (branches) are the dimensions of the table. The expected value of a chi-square test statistic with v degrees of freedom equals v. Consequently, larger tables (more branches) naturally have larger chi-square statistics. The p-value of the test is the probability that the chi-square statistic has a value at least as large as the one that was observed, given that there is no association between rows and columns. (The target distribution is the same across the branches.)

In the chi-square criterion for categorical response variables, the worth is based on the p-value for the Pearson chi-square test, which compares the frequencies of the levels of the response across the child nodes.

> The chi-square splitting criterion uses the p-value of the chi-square test (Kass 1980). When the p-values are very small, it is more convenient to use *logworth* = $-\log_{10}(p\text{-value})$, which increases as p-value decreases.
>
> The ΔGini and Δentropy splitting criteria also tend to increase as the number of branches increase. However, they do not have an analogous degree of freedom adjustment. Consequently, they favor multi-way splits with large B.

Figure 3.41: Null Hypothesis and F Statistic

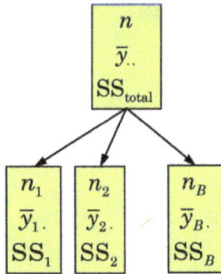

H_0: Means of the response values are identical across the child nodes

$$F = \left(\frac{\text{SS}_{\text{between}}}{\text{SS}_{\text{within}}} \right) \left(\frac{n - B}{B - 1} \right) \sim F_{B-1, n-B}$$

$$\text{SS}_{\text{between}} = \sum_{i=1}^{B} n_i \left(\bar{y}_{i.} - \bar{y}_{..} \right)^2$$

$$\text{SS}_{\text{within}} = \sum_{i=1}^{B} \text{SS}_i = \sum_{i=1}^{B} \sum_{j=1}^{n_i} \left(y_{ij} - \bar{y}_{i.} \right)^2$$

$$\text{SS}_{\text{total}} = \sum_{i=1}^{B} \sum_{j=1}^{n_i} \left(y_{ij} - \bar{y}_{..} \right)^2$$

Distribution of F assumes $(Y_i | x_i) \sim \text{iid } N(\mu, \sigma^2)$

Statistical Test Criterion for Regression Trees

Basic regression trees partition a data set into smaller subgroups and then fit a simple constant for each observation in the subgroup. The F test can be used analogously to the chi-square test for regression trees.

F Test

A split at a node can be thought of as a one-way analysis of variance where the B branches are the B treatments. Let $\bar{y}_{i.} = \frac{1}{n_i} \sum_{j=1}^{n_i} y_{ij}$ be the mean of the target in each node and $\bar{y}_{xx}$ be the mean in the root node (the overall mean).

As shown in Figure 3.41, the between-node sum of squares ($\text{SS}_{\text{between}}$) is a measure of the distance between the node means and the overall mean. The within-node sum of squares (SS_{wthin}) measures the variability within a node. Large values of the F statistic indicate departures from the null hypothesis that all the node means are equal. When the target values, conditional on the inputs, are independently, normally distributed with constant variance, then the F statistic follows an F distribution with $B - 1$ and $n - B$ degrees of freedom. The p-value of the test is used in the same way as the p-value for a chi-square test for classification trees.

The total sum of squares (SS_{total}) can be considered fixed with regard to comparing splits at a particular node. Thus, it follows from the ANOVA identity

$$\text{SS}_{\text{total}} = \text{SS}_{\text{between}} + \text{SS}_{\text{within}}$$

Figure 3.42: Heteroscedasticity

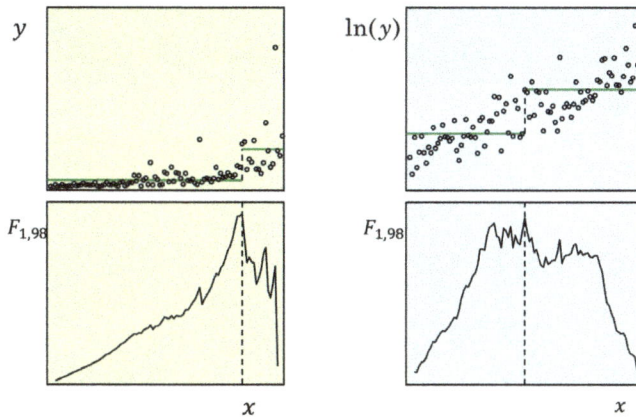

that the *F* test statistic can be thought of as either maximizing the differences between the node means or reducing the within-node variance. This latter interpretation indicates the equivalency between the *F* statistic and the reduction in impurity (variance) splitting criterion.

$$\Delta\mathrm{var} = \frac{\mathrm{SS}_{\mathrm{total}}}{n} - \sum_{i=1}^{B}\left(\frac{n_i}{n}\right)\left(\frac{\mathrm{SS}_i}{n_i}\right) = \frac{1}{n}\left(\mathrm{SS}_{\mathrm{total}} - \mathrm{SS}_{\mathrm{within}}\right) = \frac{\mathrm{SS}_{\mathrm{between}}}{n}$$

Thus, using Δvariance is equivalent to **not** adjusting the *F* test for degrees of freedom (number of branches).

The *F* test has many optimal properties when the distribution of the target (conditional on the inputs) is independently and normally distributed with constant variance. The *F* test is relatively robust to departures from the normality assumption. However, as shown in Figure 3.42, variance heterogeneity (heteroscedasticity) can have disastrous effects (Scheffe 1959).

For example, the *F* test is too liberal (overstates the significance of the effect) when small nodes have larger variance. Consider the common case of a nonnegative target with variance increasing with the mean. Using the *F* test as the splitting criterion tends to favor small splits of the largest values. This means that large split values are favored, which results in few observations in the leaves.

Decision trees are usually regarded as robust and nonparametric. This means that decision trees are generally robust to outliers in the input space. However, regression trees are not robust to outliers or heteroscedasticity in an interval target. As in classical regression models, finding a suitable variance stabilizing transformation for the target can improve the model.

Figure 3.43: Determining the Number of Branches in CHAID

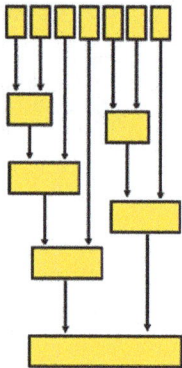

> To summarize, although decision trees are generally robust to outliers in the input space, regression trees are not robust to outliers in the target.

Statistical Criterion for Both Classification and Regression Trees

Available for both categorical and continuous response variables, CHAID is an approach that regards every possible split as representing a test.

CHAID

CHAID tests the hypothesis of no association between the values of the response (target) variable and the branches of a node.

CHAID recursively partitions the data (for example, with a nominal target using multiway splits on nominal and ordinal inputs). A split must achieve a threshold level of significance in a chi-square test of independence between the nominal target values and the branches, or else the node is not split. CHAID generally uses a Bonferroni adjustment (discussed in the next section) for the number of categorical values of the input variable, thereby mitigating the bias toward inputs with many values.

CHAID also uses significance testing to determine the number of branches as shown in Figure 3.43. The search for a split on a given input proceeds as follows. First, assign a different branch to each value of the input. For each pair of branches, form a two-way table of counts of cases in each branch by target value. Find the pair of branches with the smallest chi-square measure of independence. If the significance level is below a threshold, merge the branches and repeat. Otherwise, consider re-splitting branches containing three or more input values. If a binary split that exceeds a threshold significance level is found, split the branch in two and go back to merging branches.

The search ends when no more merges or re-splits are significant. The last split on the input is chosen as the candidate split for the input. Notice that it is generally not the most significant split examined.

For categorical predictor variables, CHAID uses the value of a chi-square statistic (for a classification tree) or an F statistic (for a regression tree) to merge similar levels of the predictor variable until the number of children in the proposed split reaches the number that you specify in the **Maximum number of branches** option. The *p*-values for the final split determine the variable on which to split.

For continuous predictor variables, CHAID chooses the best single split until the number of children in the proposed split reaches the value that you specify in the **Maximum number of branches** option.

A general issue that arises when applying CHAID is that the final trees can become very large. In practice, when the input data are complex and contain many different categories for classification problems and many possible predictors for performing the classification, for example, then the resulting trees can become very large. You learn pruning strategies as a remedy in the next section.

Bonferroni Adjustment for Splitting Criteria Based on Statistical Tests

When you test for the independence of column categories in a contingency table, it is possible to obtain significant (large) values of the chi-squared statistic even when there are no differences in the true, underlying proportions between split branches. In other words, if there are many ways to split the variable that labels the rows of the table (and thus many Chi-square tables and tests), then you are likely to get at least one with a very small *p*-value even when the variable has no true effect. As the number of possible split points increases, the likelihood of obtaining significant values also increases. In this way, an input with a multitude of unique input values has a greater chance of accidentally having a large worth than an input with only a few distinct input values.

Statisticians face a similar problem when they combine the results from multiple statistical tests. As the number of tests increases, the chance of a false positive result likewise increases. To maintain overall confidence in the statistical findings, statisticians inflate the *p*-values of each test by a factor equal to the number of tests being conducted. If an inflated *p*-value shows a significant result, then the significance of the overall results is assured. This type of *p*-value adjustment is known as a *Bonferroni correction*.

When you control the comparisonwise error rate (CER), you fix the level of alpha for a single comparison without taking into consideration all the pairwise comparisons that you are making. The experimentwise error rate (EER) uses an alpha that takes into consideration all the pairwise comparisons that you are making. Presuming no differences exist, the chance that you falsely

Figure 3.44: CER, EER, and Bonferroni Adjustment

Significance level:

0.2

☐ Bonferroni

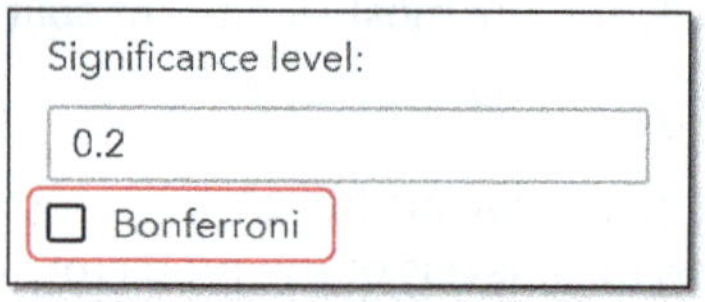

Controls EER at the top *p*-values for the splitting criteria CHAID, chi-square, and F-test.

Number of Groups Compared	Number of Comparisons	Experimentwise Error Rate (α=0.05)
2	1	.05
3	3	.14
4	6	.26
5	10	.40

Comparisonwise Error Rate = α = 0.05

$EER \leq 1 - (1 - \alpha)^{nc}$
where *nc*=number of comparisons

conclude that *at least one* difference exists is much higher when you consider all possible comparisons as illustrated in Figure 3.44. If you want to make sure that the error rate is 0.05 for the entire set of comparisons, use a method like Bonferroni adjustment that controls the experimentwise error rate at 0.05.

Because each split point corresponds to a statistical test, Bonferroni corrections are automatically applied to the worth calculations for an input. These corrections penalize inputs with many split points by reducing the worth of a split by an amount equal to the log of the number of distinct input values. This is equivalent to the Bonferroni correction because subtracting this constant from worth is equivalent to multiplying the corresponding chi-squared *p*-value by the number of split points. The adjustment enables a fairer comparison of inputs with many levels and few levels later in the split-search algorithm.

> There is some disagreement among statisticians about whether and how to control the experimentwise error rate.

The Bonferroni property specifies whether to apply a Bonferroni adjustment to the top p-values for the splitting criteria CHAID, Chi-Square, and F Test. By default, this option is deselected.

The use of a splitting criterion can be thought of as a two-step process. First, select the best split on each input variable, and then select the best of these. Both steps might require adjustments to help control for possibly spurious associations discovered simply because so many splits were examined.

- *Comparing splits on the same input variable:* The chi-square test statistic (as well as Gini and entropy) favors splits into greater numbers of branches. The *p*-value (or worth) adjusts for this bias through the degrees of freedom. For binary splits, no adjustment is necessary.

Figure 3.45: Classification Tree – *p*-Value Adjustments

X1:   38.5

				χ_ν^2	ν	$-\log_{10}(P)$	m	$-\log_{10}(mP)$
1	293	71						
7	363	1		644	2	140	96	138
9	42	294						

X1:   17.5   36.5

1	249	42	73					
7	338	25	1	660	4	141	4560	137
9	26	16	294					

X10:   0.5   41.5   51.5

1	9	143	65	147				
7	221	88	1	54	814	6	172	156849	167
9	1	4	16	315				

- *Comparing splits on different input variables:* The selected split on each input results from choosing the best from a multiplicity of possible splits. There are more splits to consider on input variables with more levels.
  For example, as shown in Figure 3.45, there is only one possible split on a binary input. Degrees of freedom for a χ^2 is represented by ν and given by (#rows-1)×(#columns-1). The maximum worth tends to become larger as the number of splits (*m*) increases. Consequently, input variables with a larger *m* are favored. Nominal inputs are favored over ordinal inputs with the same number of levels. Among inputs with the same measurement scale, those with more levels are favored.

Bonferroni adjustments of the *p*-values account for this bias. Let *a* be the probability of a Type I error on each test (that is, discovering an erroneous association). For a set of *m* tests, a conservative upper bound on the probability of at least one Type I error is *m a* (Bonferroni inequality). Consequently, the Bonferroni adjustment multiplies the *p*-values by *m* (equivalently, subtract $\log_{10}(m)$ from the worth). The multiplier *m* is $\binom{L-1}{B-1}$ for ordinal inputs and $S(L,B)$ for nominal inputs.

Several peripheral factors make the split search somewhat more complicated. The tree algorithm settings disallow certain partitions of the data. Settings, such as the minimum number of observations required for a split search force a minimum number of cases in a split partition. This minimum number of cases reduces the number of potential partitions for each input in the split search. Other properties, as represented in Figure 3.46, such as the maximum depth and using input once also affect tree growth.

The family of adjustments that you modify most often when building trees are the rules that limit the growth of the tree. Changing the minimum number of observations in a leaf prevents the creation of leaves with only one or a handful of cases. Changing the maximum depth allows for larger trees that can be more sensitive to complex input and target associations.

Figure 3.46: Other Tree Growth Options

Maximum depth:

10

Minimum leaf size:

1

☐ Use input once

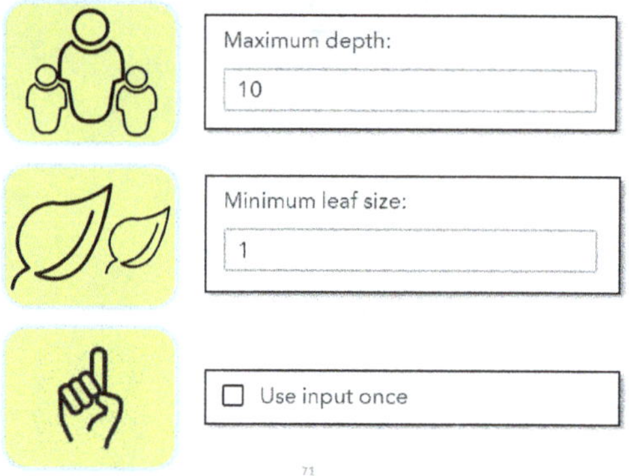

The **Maximum depth** property specifies the maximum number of generations in nodes. The original node, generation 0, is called the root node. The children of the root node are the first generation. Possible values range from 1 to 50. The default value is 10.

The **Minimum leaf size** property specifies the smallest number of training observations that a leaf can have. The default value is 5.

The **Use input once** property specifies that no splitting rule will be based on an input variable that has already been used in a splitting rule of an ancestor node. By default, this is deselected.

Demo 3.3: Experimenting with the Splitting Criteria

This demonstration builds further on the **Insurance_ClassTree** project on the **insurance_part** data. You experiment by changing the splitting criteria and other tree growth options.

1. Ensure that the **Insurance_ClassTree** project is open in Model Studio and that you are on the Pipelines tab.
2. Select the **DT Multiway Split** node.
3. Under **Splitting Options** ⇨ **Grow Criterion**, change the **Class target criterion** from Information gain ratio to **Chi-square**, the **Minimum leaf size** from 5 to **20**, and the **Number of interval bins** from 50 to **100**.

∨ Grow Criterion

Class target criterion:

Chi-square

Interval target criterion:

Variance ▼

Significance level:

0.20

☐ Bonferroni

Maximum depth:

10

Minimum leaf size:

20

Missing values:

Use in search ▼

Minimum missing use in search:

1

Number of interval bins:

100

In most of the cases, the choice of splitting criteria does not make much difference on the tree performance. Each criterion is superior in some cases and inferior in others, as the "no free lunch" theorem suggests. However, you must experiment.

Note: Information gain and Gini index are largely biased toward multivalued attributes, whereas information gain ratio tends to prefer unbalanced splits in which one partition is much smaller than the other. The Gini index also has difficulties when the number of classes is large and tends to favor tests that result in equal-sized partitions and purity in both partitions.

Increasing the minimum leaf size avoids orphan nodes. For large data sets, you might want to increase the number of leaves to obtain additional modeling resolution.

4. Run the **DT Multiway Split** node and open the **Results**.

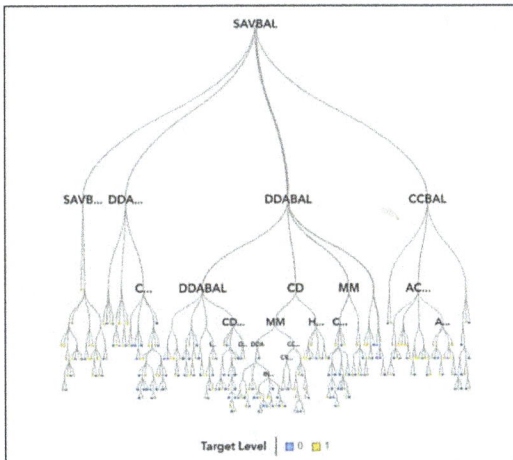

Model Information	
Split Criterion	Chi Square
Pruning Method	Cost Complexity
Max Branches per Node	4
Max Tree Depth	10
Tree Depth Before Pruning	10
Tree Depth After Pruning	10
Number of Leaves Before Pruning	415
Number of Leaves After Pruning	244

This tree has grown up to 415 leaves and the chosen tree is also considerably larger with 244 leaves with the following important variables.

Variable Label	Role	Variable Name	Validation Importa...	Import...	Relative Importance	Count
Saving Balance	INPUT	SAVBAL	246.3824	0	1	3
Checking Balance	INPUT	DDABAL	213.1465	0	0.8651	3
Money Market	INPUT	MM	39.6320	0	0.1609	2
CD Balance	INPUT	CDBAL	31.8027	0	0.1291	5
Certificate of Deposit	INPUT	CD	21.6419	0	0.0878	1
Money Market Balance	INPUT	MMBAL	15.9925	0	0.0649	1
Checking Account	INPUT	DDA	6.8244	0	0.0277	1
Number Insufficient Fund	INPUT	NSF	1.2006	0	0.0049	1

5. Close the **Results** and get back to the properties panel on your right.
6. Continue with the DT Multiway Split node. Under **Splitting Options** ⇨ **Grow Criterion**, change the **Significance level** from 0.20 to **0.01**.

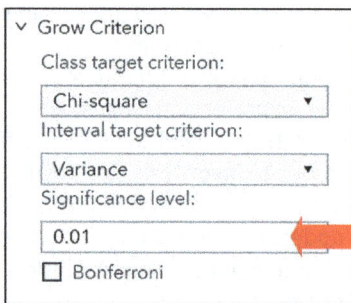

Lower values of significance level (alpha) tend to produce trees with fewer nodes.
7. Rerun the **DT Multiway Split** node and open the **Results**.

This tree (with $\alpha=0.01$) has lesser number of leaves (207 as compared to 415 leaves in the previous tree with $\alpha=0.20$).

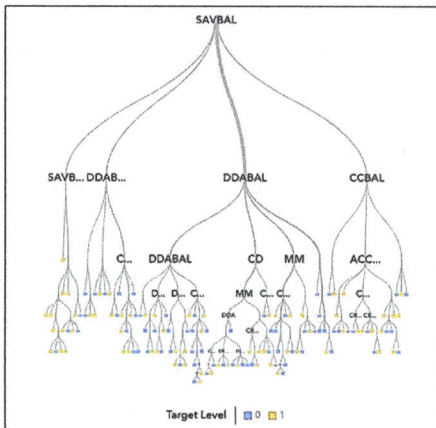

Model Information	
Split Criterion	Chi Square
Pruning Method	Cost Complexity
Max Branches per Node	4
Max Tree Depth	10
Tree Depth Before Pruning	10
Tree Depth After Pruning	9
Number of Leaves Before Pruning	207
Number of Leaves After Pruning	138

Changing the logworth threshold changes the minimum logworth required for a split to be considered by the tree algorithm. The higher the value of chi-square, the higher the statistical significance of differences between child and parent node.

Variable Label	Role	Variable Name	Validation Importa...	Im...	Relative Importance	Count
Saving Balance	INPUT	SAVBAL	251.2035	0	1	2
Checking Balance	INPUT	DDABAL	213.1465	0	0.8485	3
Money Market	INPUT	MM	39.6320	0	0.1578	2
CD Balance	INPUT	CDBAL	31.8027	0	0.1266	5
Certificate of Deposit	INPUT	CD	21.6419	0	0.0862	1
Money Market Balance	INPUT	MMBAL	15.9925	0	0.0637	1
Checking Account	INPUT	DDA	6.8244	0	0.0272	1
Number Insufficient Fund	INPUT	NSF	1.2006	0	0.0048 .	1
Investment	INPUT	INV	0.8242	0	0.0033	1

8. Close the Results and once again get back to the properties panel.
9. Keep the DT Multiway Split node selected. Under **Splitting Options** ⇨ **Grow Criterion**, check the **Bonferroni** box.

```
∨ Grow Criterion
   Class target criterion:
   [ Chi-square          ▼ ]
   Interval target criterion:
   [ Variance            ▼ ]
   Significance level:
   [ 0.01 ]
   ☑ Bonferroni
```

The Bonferroni method will adjust significance values based on the number of tests, which directly relates to the number of categories and measurement level of a predictor. This is generally desirable because it better controls the false-positive error rate. Disabling this option will increase the power of your analysis to find true differences, but at the cost of an increased false-positive rate. Disabling this option can be recommended for small samples. If you want big trees and insist on using the chi-squared split-worth criterion, deactivate the Bonferroni adjustment.

10. Click the **Run Pipeline** button.
11. Open the **Results** of the **DT Multiway Split** node.

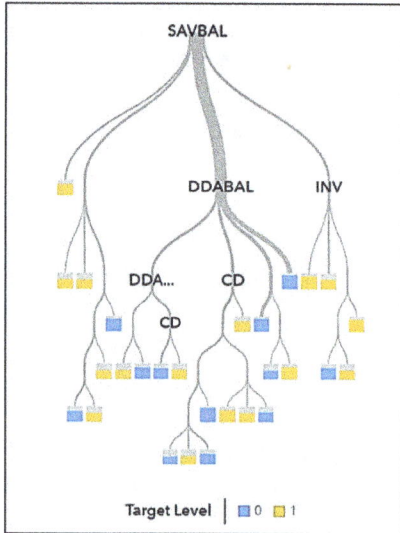

Model Information	
Split Criterion	Chi Square
Pruning Method	Cost Complexity
Max Branches per Node	4
Max Tree Depth	10
Tree Depth Before Pruning	9
Tree Depth After Pruning	6
Number of Leaves Before Pruning	49
Number of Leaves After Pruning	28

The tree with p-values adjusted is a much simpler tree with 28 leaves, originally grown up to 49 leaves.

Careful examination of the variable importance table below reveals some finer observations. For example, **CD** becomes a more important input displacing the **CDBAL**. There was only one possible split on this binary input. Because you had adjusted the p-values, the maximum worth of **CDBAL** did not tend to become large enough. Similarly, the **MMBAL** variable is now turned out to be an unimportant variable. The bottom line is other input variables with a larger number of plausible splits were not favored after adjusting the p-values.

Variable Label	Role	Variable Name	Validation Import...	Impor...	Relative Importance	Count
Saving Balance	INPUT	SAVBAL	252.9172	0	1	1
Checking Balance	INPUT	DDABAL	207.4270	0	0.8201	3
Money Market	INPUT	MM	39.3148	0	0.1554	3
Certificate of Deposit	INPUT	CD	38.7504	0	0.1532	5
Checking Account	INPUT	DDA	6.8244	0	0.0270	1
Investment	INPUT	INV	3.9949	0	0.0158	2
Credit Card	INPUT	CC	0.3424	0	0.0014	1
Mortgage	INPUT	MTG	-2.4343	0	-0.0096	1
Age of Oldest Account	INPUT	ACCTAGE	-4.6809	0	-0.0185	2

12. Close the **Results** of the **DT Multiway Split** node.

13. Open the **Results** of the **Model Comparison** node. Maximize the **Model Comparison** table.

Champi...	Name	Algorith...	KS (You...	Misclas...	Misclas...	Root Av...	Averag...
▣	Decision Tree	Decision Tree	0.4259	0.2616	0.2616	0.4252	0.1808
	DT Multiway Splits	Decision Tree	0.4139	0.2654	0.2654	0.4239	0.1797

The Average Squared Error (ASE) of the multi-way split tree has improved from 0.1882 (without **Bonferroni**) to 0.1797 and the misclassification rate has improved from 0.2903 (without **Bonferroni**) to 0.2654. Consequently, this model is certainly better than the two-way split decision tree and stands as champion so far.

14. Close the **Model Comparison** table.
15. Close the **Results** of the **Model Comparison** node.

End of Demonstration

Quiz

1. Decision trees in SAS Viya consider linear combinations of inputs in the split search.

   a. True
   b. False

2. When you evaluate the worth of a split of a parent node, decision trees in SAS Viya use a look-ahead algorithm that assesses the distribution of the target levels in both child and grandchild nodes.

   a. True
   b. False

3. You have created a decision tree for a binary target (events and nonevents). A leaf can be said to be purer if most of the data instances belong to which of the following?

   a. events
   b. nonevents
   c. either events or nonevents
   d. both events and nonevents

4. Which of the following statements is true regarding split worthiness in decision trees?

   a. The CHAID algorithm uses the chi-squared test to assess split worthiness.
   b. The CART algorithm uses the Entropy index to assess split worthiness.
   c. Machine learning algorithms use the Gini index to assess split worthiness.
   d. Multi-way (>2 branch) splits using Gini and Entropy in the Decision Tree node favor larger way splits.
   e. answers a. and d.

5. Which of the following statements is true regarding robustness in decision trees?

   a. Decision trees are robust to outliers in input space.
   b. Regression trees are robust to outliers in the target.
   c. Neither of the above

6. SAS Viya supports many splitting criteria in decision trees that are based on the reduction in node impurity induced by the split. Which of the following statements is true?

   a. Chi-square, Entropy, Gini, and Information Gain Ratio are splitting criterion for interval targets.
   b. CHAID can be used for both categorical and interval targets.
   c. F test and Variance are splitting criterion for categorical targets.
   d. A pure node has a Gini index of 1.

Answers

1. False	2. False	3. c	4. e	5. a	6. b

Chapter 4: Decision Trees: Strengths, Weaknesses, and Uses

Missing Values in Decision Trees

You might have to contend with missing values among the inputs as shown by question marks in Figure 4.1. Decision trees accommodate missing values very well compared to other modeling methods. Decision trees that split on one input at a time are more tolerant to missing data than models such as regression that combine several inputs. In regression models, an observation missing any input value is discarded (complete case analysis).

For the simplest of tree algorithms, only observations that need to be excluded are those missing the input currently being considered to split on. They can be included when considering splitting on a different input (for example, tree algorithms that treat missing observations as a special value use all the observations). Trees, therefore, might be the best modeling tool for imputing missing values because of their tolerance to missing data, their acceptance of different data types, and their robustness to assumptions about the input distributions.

Some of the important points to note about missing values in decision trees include the following:

- missing values do not prevent splitting the data for building a decision tree
- missing values can affect the choice of the splitting criteria
- missingness can also be present in the scoring data along with the training data

Figure 4.1: Missing Values

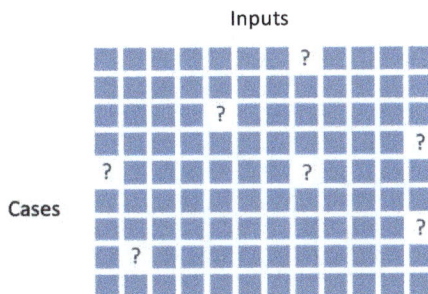

Compared with other regression and classification methods, decision trees have the advantage that they are easy to interpret and visualize, especially when the tree is small. Tree-based methods scale well to large data, and they offer various methods of handling missing values, including surrogate splits. Surrogate splitting rules enable you to use the values of other input variables to perform a split for observations with missing values. You can specify how to handle missing values of predictor variables during training and how to handle missing values and unknown levels of predictor variables (after all surrogate rules have been applied) during scoring. An unknown level of a categorical predictor variable is a level that does not exist in the training data but is encountered during scoring. During scoring, unknown levels are treated as missing values.

There are four separate ways of handling missing values in decision trees in SAS Visual Data Mining and Machine Learning shown in Figure 4.2.

1. The most trivial way is to exclude observations with a missing value from training.
2. Another approach is to somehow distribute observations with missing values. It first creates a rule without the observations with missing values, and then the program uses the missing values to decide which branch the rule should put them in. That rule is then used during scoring.
3. A more computationally intensive approach includes using data points with missing values in the evaluation of a split.
4. Use surrogate rules as backup splitting rules before assigning the observation to the branch for missing values. This is a more practical approach.

A Simple Example

Consider a simple example of data collected on people in a city park in the vicinity of a hotdog and ice cream stand. The owner of the concession stand wants to know what predisposes people to buy ice cream with the help of attributes like maximum temperature of the day, whether they have some extra money to spend, and whether they really crave the ice cream. The data is shown in Table 4.1.

Figure 4.2: Handling Missing Values in Decision Trees

- Exclude
  - Ignore
- Assign
  - Largest branch
  - Most correlated branch
  - Separate branch
- Use
  - Use as machine smallest
  - Use in search
- Surrogate Rules

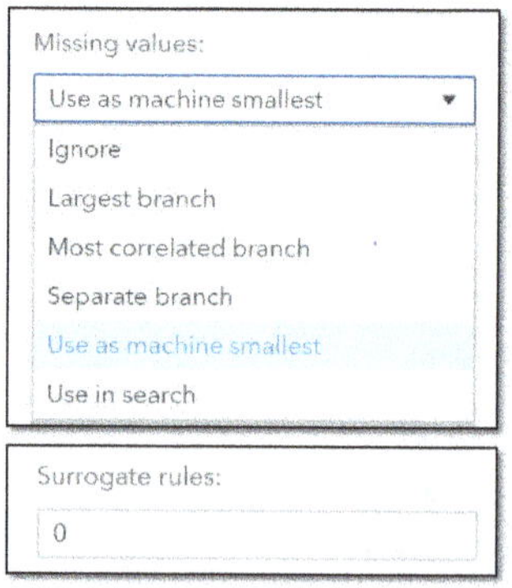

Table 4.1: Ice Cream Data

Case	Day Temp	Crave	Extra Money	Buy Ice cream
1	25		No	Yes
2	28	Yes	Yes	Yes
3	.	No	No	No
4	20	Yes	Yes	Yes
5	18		Yes	No
6	10	Yes	No	No
7	12	No	Yes	No
8	.	Yes		Yes

Observe that there are five missing values. In general, there is no best, universal method of handling missing values. It would be interesting to explore these methods using this example data.

Ignore

During the training phase, the **Ignore** option excludes any observation that has a missing value for any predictor variable. In the scoring phase, if an observation has a missing value or an unknown level for a predictor variable, then the observation is assigned to the child node that contains the most training observations.

A decision tree on the ice cream data with ignored cases is shown in Figure 4.3. Note that the root node has only 4 observations.

In both the training and the scoring phases, missing values in continuous predictor variables are treated as the smallest possible value.

Figure 4.3: Ignore Option

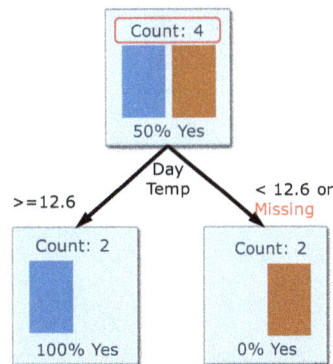

Case	Day Temp	Crave	Extra Money	Buy Ice cream
1	25		No	Yes
2	28	Yes	Yes	Yes
3	.	No	No	No
4	20	Yes	Yes	Yes
5	18		Yes	No
6	10	Yes	No	No
7	12	No	Yes	No
8	.	Yes		Yes

The Ignore option can significantly reduce your training data!

Count: 4
50% Yes

Day Temp
>=12.6   < 12.6 or Missing

Count: 2   Count: 2
100% Yes   0% Yes

All cases with missing attribute values are deleted from the data set. It is also called *listwise deletion* (or *casewise deletion*, or *complete case analysis*) in statistics. Obviously, a lot of information is missing in the constructed tree (only four observations). However, there might be some reasons to consider it a good method, which is shown in Allison (2002) and Little and Rubin (2002).

> The **Ignore** option in Model Studio is equivalent to the ASSIGNMISSING=NONE option in the TREESPLIT procedure.

Because maximum day temperature was the only important variable in the decision tree, for simplicity let us ignore other predictors and re-create the decision tree with **DayTemp** as the sole input variable. Observe in Figure 4.4 that the count of the root node has changed from 4 to 6.

For brevity and triviality, we will consider DayTemp as the only input in the decision tree to explore remaining options for handling missing values.

Largest Branch

During the training phase, the **Largest branch** option assigns any observation that has a missing value in the predictor variable to the child node that has the most training observations. In the scoring phase, if an observation has a missing value or an unknown level for a predictor variable, then the observation is assigned to the child node that contains the most training observations.

Before allocation, both the branches have an equal number of training observations, that is, three each. Therefore, missing values can be assigned on either branch. Figure 4.5 shows that the missing values are assigned to the left branch, so the leaf size becomes 5 as against the right branch with leaf size of 3.

Figure 4.4: Ignore Option with Reduced Columns

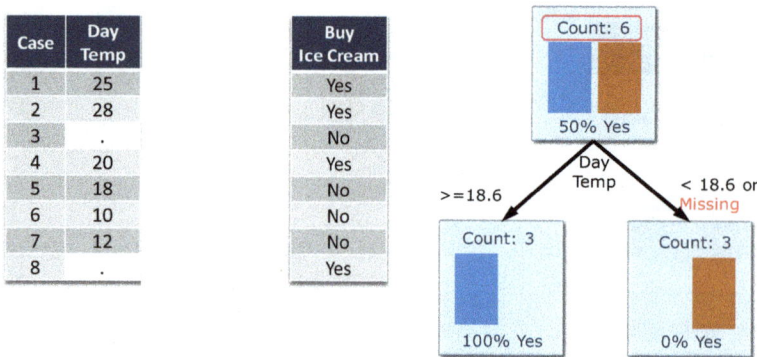

Case	Day Temp
1	25
2	28
3	.
4	20
5	18
6	10
7	12
8	.

Buy Ice Cream
Yes
Yes
No
Yes
No
No
No
Yes

Figure 4.5: Largest Branch Option

Case	Day Temp	Buy Ice cream
1	25	Yes
2	28	Yes
3	.	No
4	20	Yes
5	18	No
6	10	No
7	12	No
8	.	Yes

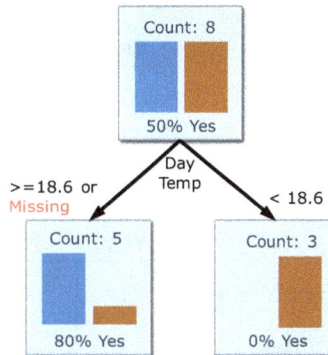

The **Largest branch** option in Model Studio is the equivalent of the ASSIGNMISSING=POPULAR option in the TREESPLIT procedure.

Most Correlated Branch

During the training phase, the **Most correlated branch** option assigns any observation that has a missing value in the predictor variable to the child node whose observations are most like it. This similarity is determined using the chi-square criterion for categorical responses or the F test criterion for continuous responses. If all observations have nonmissing values for a predictor variable, then no branch is selected to contain observations with missing values. In the scoring phase, if an observation has a missing value for a predictor variable and no branch is selected to contain observations with missing values, or if an observation has an unknown level for a predictor variable, then the observation is assigned to the child node that contains the most training observations.

Figure 4.6: Most Correlated Branch Option

Case	Day Temp	Buy Ice cream
1	25	Yes
2	28	Yes
3	.	No
4	20	Yes
5	18	No
6	10	No
7	12	No
8	.	Yes

Similarity is determined using a chi-square test for categorical target or an F-test for interval target.

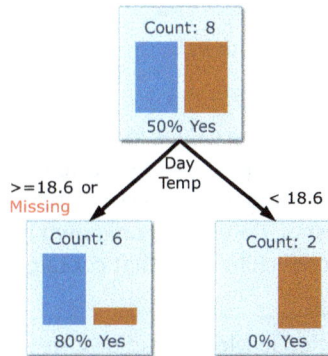

Figure 4.7: Competing Tests for Missing Values Assignment

	BuyIC	
	Yes	No
<18.6	0	3
≥18.6 or Missing	4	1

Branches

	BuyIC	
	Yes	No
<18.6 or Missing	1	4
≥18.6	3	0

Branches

In the ice cream data example shown in Figure 4.6, why are the missing values assigned to the left branch? In fact, chi-square tests were used to evaluate, as shown in Figure 4.7.

The top and bottom 2×2 contingency tables both have equal *p*-values of 0.1429 (Fisher's exact χ^2 test). So, there is a tie. Otherwise, missing values are assigned to the branch that has smaller *p*-value. Here, missing values are assigned to the left branch.

> The **Most correlated branch** option in Model Studio is the equivalent of the ASSIGNMISSING=-SIMILAR option in the TREESPLIT procedure.

Separate Branch

During the training phase, the **Separate branch** option assigns any observation that has a missing value for the predictor variable to a specially created child node (branch), provided you have set up enough maximum number of branches, or else there might be no splits at all.

In the ice cream data example, note in Figure 4.8 that all the missing values are assigned in a separate branch.

If all observations have nonmissing values for a predictor variable, then no branch is created to contain observations with missing values. In the scoring phase, if an observation has a missing value for a predictor variable and no special branch was created to contain observations with missing values, or if an observation has an unknown level for a predictor variable, then the observation is assigned to the child node that contains the most training observations.

> The **Separate branch** option in Model Studio is the equivalent of the ASSIGNMISS-ING=BRANCH option in the TREESPLIT procedure.

Figure 4.8: Separate Branch Option

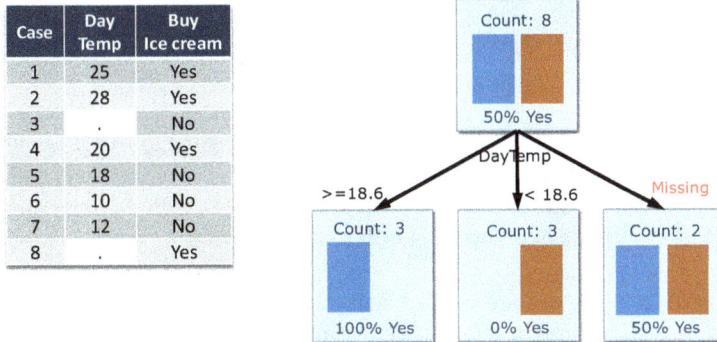

Case	Day Temp	Buy Ice cream
1	25	Yes
2	28	Yes
3	.	No
4	20	Yes
5	18	No
6	10	No
7	12	No
8	.	Yes

Count: 8
50% Yes
DayTemp
>=18.6 | < 18.6 | Missing
Count: 3 — 100% Yes
Count: 3 — 0% Yes
Count: 2 — 50% Yes

Use as Machine Smallest

During the training phase, the **Use as machine smallest** option treats a missing value in a categorical predictor variable as a separate, legitimate value. If all observations have nonmissing values for a categorical predictor variable, then no branch is selected to contain observations with missing values. In the scoring phase, if an observation has a missing value for a categorical predictor variable and no branch is selected to contain observations with missing values, or if an observation has an unknown level for a predictor variable, then the observation is assigned to the child node that contains the most training observations.

In both the training and the scoring phases, missing values in continuous predictor variables are treated as the smallest possible value.

In the ice cream data example as shown in Figure 4.9, the two missing values were included in the split search evaluation by considering them as the smallest values of **DayTemp** and therefore assigned to the right branch (<18.6).

Figure 4.9: Use as Machine Smallest Option

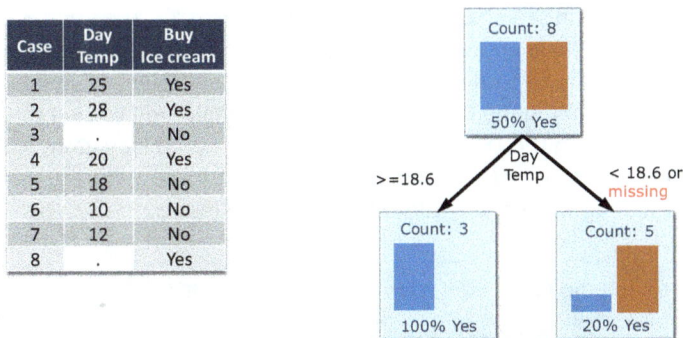

Case	Day Temp	Buy Ice cream
1	25	Yes
2	28	Yes
3	.	No
4	20	Yes
5	18	No
6	10	No
7	12	No
8	.	Yes

Count: 8
50% Yes
Day Temp
>=18.6 | < 18.6 or missing
Count: 3 — 100% Yes
Count: 5 — 20% Yes

> The **Use as machine smallest** option in Model Studio is the equivalent of the ASSIGNMISS-ING=MACSMALL option in the TREESPLIT procedure.

Use in Search

Another simple strategy is to regard a missing value as a special nonmissing value. For a nominal input, a missing value simply constitutes a new categorical value. For an input whose values are ordered, each missing value constitutes a special value that is assigned a place in the ordering that yields the best split. That place is usually different in different nodes of the tree.

This strategy is beneficial when missing values are predictive of certain target values. For example, people who have large incomes might be more reluctant to disclose their income than people who have ordinary incomes. If income were predictive of a target, then a missing income value would be predictive of the target and the missing values would be regarded as a special large-income value. The strategy seems harmless when the distribution of missing values is uncorrelated with the target because no choice of branch for the missing values would help predict the target. A linear regression could use the same strategy by adding binary indicator variables to designate whether a value is missing.

During the training phase, the **Use in search** option treats a missing value in a predictor variable as a separate, legitimate value. If all observations have nonmissing values for a predictor variable, then no branch is selected to contain observations with missing values. In the scoring phase, if an observation has a missing value for a predictor variable and no branch is selected to contain observations with missing values, or if an observation has an unknown level for a predictor variable, then the observation is assigned to the child node that contains the most training observations.

You can specify the minimum number of observations to use with the use in search policy for handling missing values. An additional option of **Minimum missing use in search** specifies a threshold for using missing values in the split search. If the number of observations that have missing values for the splitting variable is greater than or equal to the number that you specify, then only the Decision Tree node uses the use in search policy for missing values. By default, the minimum missing use in search=1.

The ice cream data example using the Use in search option is shown in Figure 4.10. Note that a different tree than the use as machine smallest option is built. The difference between the two is in the data table to which the decision tree model is to be saved. The *MissingOnNode* column exists in the USEINSERACH outmodel table, but not in the MACSMALL outmodel table. You can get the outmodel table using OUTMODEL= option in the TREESPLIT procedure. The two observations with missing **DayTemp** are included in the split search; the resulting rule assigns missing values to the left branch (>=18.6) on these data. During scoring, the rule will be the same (assign to the left).

> The **Use in search** option in Model Studio is the equivalent of the ASSIGNMISSING=USEIN-SEARCH option in the TREESPLIT procedure.

Figure 4.10: Use in Search Option

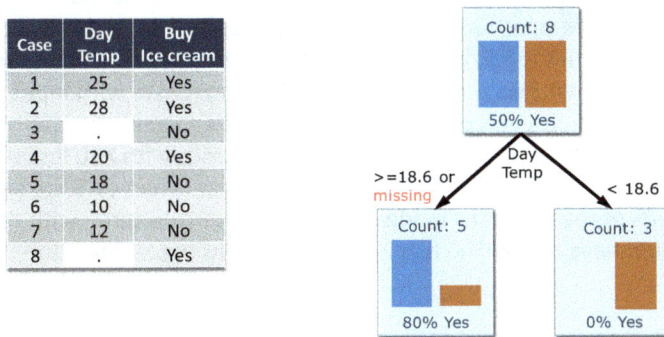

Surrogate Splits

Surrogate splits can be used to handle missing values (BFOS 1984). A surrogate split is a partition using a different input that mimics the selected split. A perfect surrogate maps all the cases that are in the same node of the primary split to the same node of the surrogate split. The agreement between two splits can be measured as the proportion of cases that are sent to the same branch. The split with the greatest agreement is taken as the best surrogate.

A surrogate splitting rule is a backup to the primary splitting rule. For example, the primary splitting rule might use **COUNTY** as input, and the surrogate might use **REGION**. If the **COUNTY** is unknown and the **REGION** is known, the surrogate is used. If several surrogate rules exist, each surrogate is considered in sequence until one can be applied to the observation. If none can be applied, the primary rule assigns the observation to the branch that is designated for missing values.

Figure 4.11: Surrogate Rules on Handwritten Digits Example

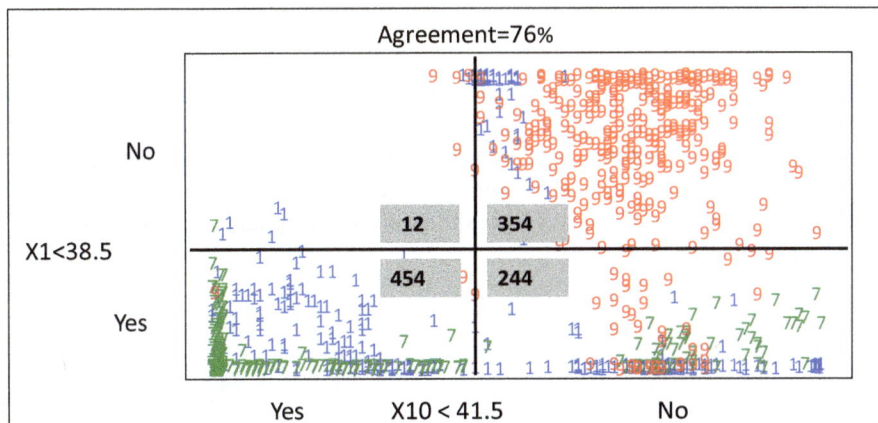

Consider the handwriting example that was discussed earlier, with inputs X1 (on the Y axis) and X10 (on the X axis). In this example (illustrated in Figure 4.11), let us say that the main splitting rule is based on X1, and the surrogate is based on X10. The plot above shows both splits. For the X1 input and a fixed split point (38.5, as shown in the plot above), two groups are generated. Cases with input values less than 38.5 are said to branch bottom. Cases with input values greater than 38.5 are said to branch top. However, if X1 is an unknown value and X10 is known, then the surrogate rule is used. When the surrogate rule is used, cases with X10 less than the split point 41.5 are said to branch left, and cases with input values greater than 41.5 are said to branch right. In this example, 76% of the decisions made by the surrogate rule agree with the decisions that the original rule would have made. The higher the agreement percentage, the better the surrogate rule is.

The surrogates are considered in the order of their agreement with the primary splitting rule. The agreement is measured as the proportion of training observations that the surrogate rule and the primary rule assign to the same branch. The measure excludes the observations to which the primary rule cannot be applied. Among the remaining observations, those on which the surrogate rule cannot be applied count as observations that are not assigned to the same branch. Thus, an observation that has used a missing value on the input in the surrogate rule but not the input in the primary rule counts against the surrogate.

When surrogate rules are requested and if a new case has a missing value on the splitting variable, then the best surrogate is used to classify the case. If the surrogate variable is missing as well, then the second-best surrogate is used. If the new case has a missing value on all the surrogates, it is sent to the branch that contains the missing values of the training data.

The **Surrogate rules** property specifies the number of surrogate rules. The default value is 0. You can determine the number of surrogates that are sought. A surrogate is discarded if its agreement is less than or equal to the largest proportion of observations in any branch. Consequently, a node might have fewer surrogates specified than the number in the Surrogate rules property.

Missing Values in Ordinal Inputs

An ordinal input with missing values requires modifications of the split search strategy and the Bonferroni adjustment. The missing values cannot usually be placed in order among the input levels. The missing value acts as a nominal level. Consequently, the split search should not place any restrictions on the branch that contains the missing level. It might be in a branch by itself or mixed in with the adjacent ordinal levels. This increases the number of splits to consider from

$$\binom{L-1}{B-1} \text{ to } B\binom{L-1}{B-1} + \binom{L-1}{B-2}.$$

Figure 4.12: Splits to Consider for a Three-Level Input with Missing Values

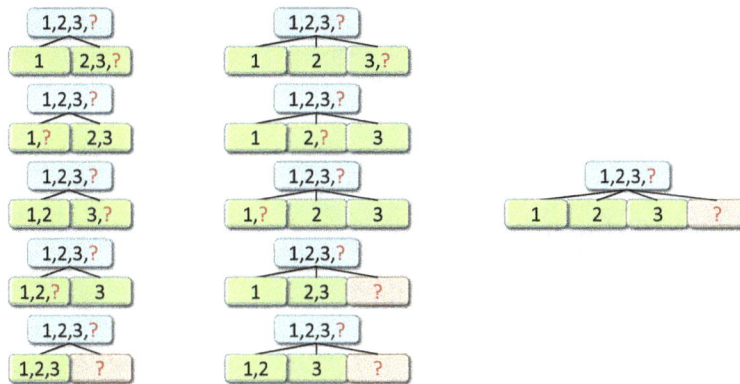

For example, as shown in Figure 4.12, a three-level input with missing values has five binary splits, five three-way splits, and one four-way split. The total number of splits increased from 2^{L-1}

$$-1 \text{ to } \sum_{i=2}^{L}\left(i\left(\binom{L-1}{i-1}+\binom{L-1}{i-2}\right)\right)=2^{L-2}(L+3)-1.$$

Variable Importance

The importance of a variable is the contribution that it makes to the success of the model. For a decision tree, or for that matter any predictive model, success means good prediction. Often the prediction relies on a few variables. A good measure of importance reveals those variables. The better the prediction, the more closely the model represents reality and the more plausible it is that the important variables represent the true cause of prediction. Some people prefer a simple model so that they can understand it. However, a simple model usually relinquishes details of reality. Sometimes it is better to first find a good model and then ask which variables are important than to first ask which model is good for variable importance and then train that model.

Analysis data usually contains many predictor variables, some of which are useful for predicting the target and others of which are not. Variable importance is an indication of which predictor variables are the most useful for predicting the response variable. The most important variables might not be the ones near the top of the tree. The Decision Tree node implements several methods for computing variable importance.

BFOS (1984) devised a measure of variable importance for trees. It can be particularly useful for tree interpretation.

Figure 4.13: Variable Importance in Decision Trees

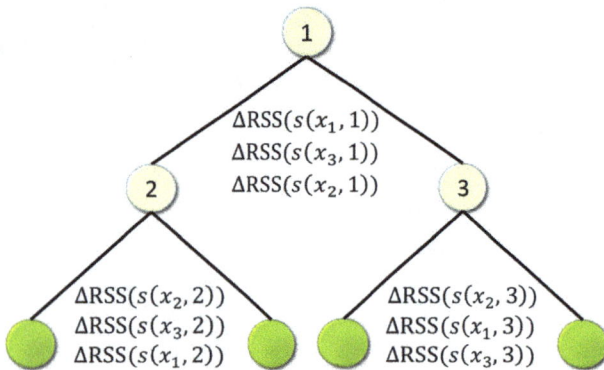

Let $s(x_j, t)$ be a surrogate split (including the primary split) at the tth internal node using the jth input. Importance is a weighted average of the reduction in impurity for the surrogate splits using the jth input across all the internal nodes in the tree. The weights are the node sizes.

$$\text{Importance } (x_j) = \sum_{t=1}^{T} \frac{n_t}{n} \Delta(i)(s(x_j), t),$$

In this definition, $\Delta(i)$ represents impurity reduction that is based on residual sum of squares (RSS). As shown in Figure 4.13, for instance, input x_1 is used as a primary splitting variable in node 1 (the root node). However, down the tree it is used as the first and second surrogate splitting variable in nodes 3 and 2, respectively.

Methods for Computing Variable Importance

SAS Viya implements several methods for computing variable importance in decision trees that includes:

- Count-Based Importance
- Residual Sum of Squares (RSS) Importance
- Relative Variable Importance
- Random Branch Assignment (RBA) Importance

By default, a SAS Viya decision tree model calculates the variable importance using three methods (count based, change in the residual sum of square errors, and relative importance), and writes the results in one Variable Importance table that includes columns named Valid Importance, Importance Standard Deviation, Relative Importance, and Count.

You can request that the TREESPLIT procedure also calculate the variable importance by random branch assignment (RBA) by specifying the RBAIMP option. However, this option is not available in the Decision Tree node in Model Studio. See the *SAS Visual Statistics 8.5: Procedures* documentation for more details.

Count-Based Variable Importance

Count-based variable importance simply counts the number of times in the tree that a variable is used in a split. If the **Surrogate rules** property is specified with the value other than 0, then surrogate-count-based variable importance also counts the number of times that a variable is used in a surrogate splitting rule.

Relative Variable Importance

Relative variable importance measures variable importance based on the change of residual sum of squares (RSS) when a split is found at a node. It is a number between 0 and 1, which is calculated as the RSS-based importance of this variable divided by the maximum RSS-based importance among all the variables. The RSS and relative importance are calculated from the validation data. If no validation data exist, they are calculated instead from the training data.

So, it's important to discuss RSS-based variable importance.

Residual Sum of Squares (RSS) Importance

A residual sum of squares (RSS) is a statistical technique used to measure the amount of variance in a data set that is not explained by a (regression) model itself. Instead, it estimates the variance in the residuals, or error term. Precisely, it is a measure of the discrepancy between the data and an estimated model. A small RSS indicates a tight fit of the model to the data. The residual sum of squares for regression trees is defined in Figure 4.14. RSS is estimated from observations in a leaf by subdtracting the model predictions from the observed target values, squaring these differences, and averaging across all data points in the leaf.

> See the *SAS Visual Statistics 8.5: Procedures* documentation for the definition of the residual sum of squares for classification trees.

Figure 4.14: Residual Sum of Squares (RSS)

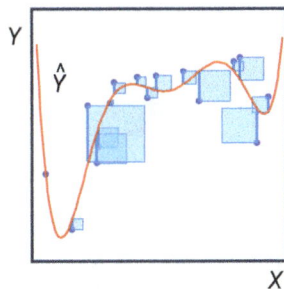

$$RSS = \sum_{\lambda} \sum_{i \in \lambda} \left(y_i - \hat{y}_\lambda^T \right)^2$$

i is an observation on leaf λ

y_i is the actual value of observation i on leaf λ

$\hat{y}_\lambda^T$ is the predicted value on leaf λ

Variable importance is measured as the decrease in residual sum of squares (RSS) at a node by splitting on the variable of interest, averaged over all nodes in which the variable is split. This is analogous to residual sum of squares-based methods for multiple linear regression and is applicable for all tree-based methods. This type of variable importance measure is also referred as *node purity* – see Liaw and Wiener (2002).

RSS importance measures variable importance based on the change of RSS when a split is found at a node. The change in RSS for a variable is the difference between the RSS if the node is treated as a leaf and after it has been split. If the change in RSS is negative (which is possible when you measure it on the validation set), then the change is set to 0.

The change for variable υ is:

$$\Delta_d^v = RSS_d - \sum_i RSS_i^d$$

where d denotes the node, i denotes the index of a child that this node includes, RSS_d is the RSS if the node is treated as a leaf, and RSS_i^d is the RSS of the node after it has been split. This is illustrated in Figure 4.15 with a simple example that includes a split of the right node in two possible binary splits.

> If surrogate rules are in effect, they are also credited with a portion of the change in RSS. The credit is proportional to the agreement between the primary and surrogate splitting rules at the node. See the *SAS Visual Statistics 8.3: Procedures* documentation for more details.

Random Branch Assignment (RBA) Importance

To compute the importance of a variable υ, randomize the branch assignment rules that involve υ and then apply the randomized model to the data and compute a goodness-of-fit measure. The randomized rule is one that randomly assigns an observation to a branch with a

Figure 4.15: Change in RSS

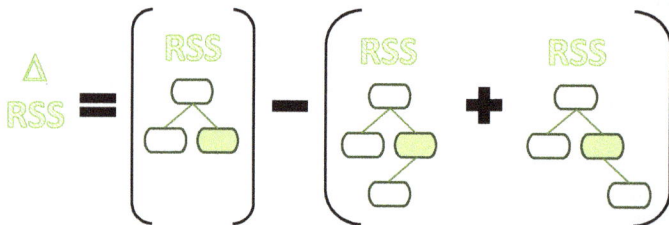

Figure 4.16: Random Branch Assignment Example

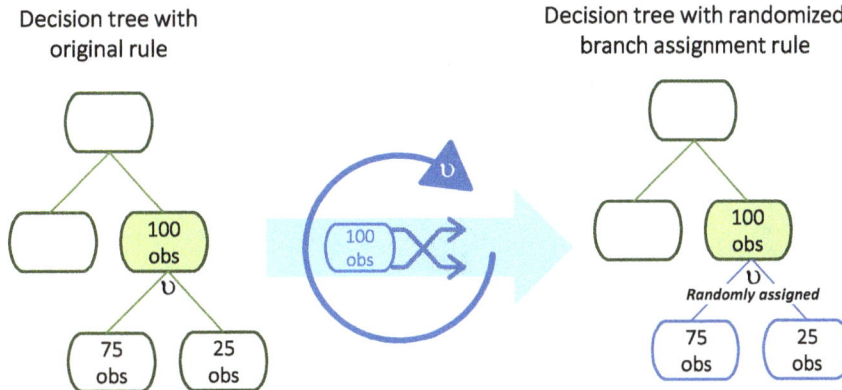

probability that is proportional to the number of observations in the branch. For example, as illustrated in Figure 4.16, suppose a node that contains 100 training observations is split by values of υ into two nodes: one contains 25 training observations and the other contains 75. When the importance of υ is evaluated, an observation that reaches the node is randomly assigned to the smaller branch with probability 0.25.

Neville and Tan (2014) claim that RBA satisfies the objectives of the methods of Breiman and of Strobl et al. The purpose of permuting the values is to break any relationship between the response variable and υ without changing the univariate distribution of υ.

The random branch assignment (RBA) method computes the importance of an input variable υ by comparing how well the data fit the predictions before and after they are modified. To modify the predictions, all splitting rules that use variable υ are replaced by a rule that randomly assigns an observation to a branch. The probability of assigning an observation to a branch is proportional to the number of observations that are assigned to the branch in the current data. The current data are the training data when RBA is computed during training. Otherwise, the current data are those being scored on an existing model.

The importance of a variable is proportional to the randomized fit minus the fit without randomization, illustrated in Figure 4.17. The RBA importance can be expressed mathematically as:

$$I_{\text{RBA}}(v) \propto \sum_{i=1}^{n} \text{Loss}(y_i, \ddot{y}_i) - \sum_{i=1}^{n} \text{Loss}(y_i, \hat{y}_i)$$

where $\ddot{y}_i$ is the modified prediction for observation i and $\hat{y}_i$ is the standard prediction.

Figure 4.17: RBA-Based Importance

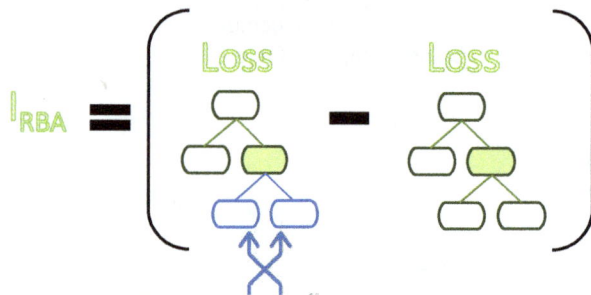

Introducing the TREESPLIT Procedure

The TREESPLIT procedure builds tree-based statistical models for classification and regression in SAS Viya. It is a programmatic equivalent of the Decision Tree node in Model Studio.

```
PROC TREESPLIT < options >;

CLASS variables;

GROW criterion < options > ;

MODEL response = variable. . .;

OUTPUT OUT=CAS-libref.data-table output-options;

PARTITION < partition-options>;

PRUNE prune-method < (prune-options) >;

WEIGHT variable;
```

The PROC TREESPLIT statement invokes the procedure.

The CLASS statement causes the specified variables to be treated as categorical variables in the analysis.

The GROW statement specifies the criterion by which to split a parent node into child nodes.

The MODEL statement causes PROC TREESPLIT to create a tree model by using response as the response variable and one or more variables as predictors.

The OUTPUT statement creates an output data table that contains the results of PROC TREESPLIT.

The PARTITION statement specifies how observations in the input data set are logically partitioned into disjoint subsets for model training, validation, and testing.

The PRUNE statement specifies the pruning method and related options.

The variable in the WEIGHT statement is used as a weight to perform a weighted analysis of the data.

Demo 4.1: Handling Missing Values and Determining Important Variables

This demonstration builds further on the previous demonstration on the **insurance_part** data. You will experiment by changing the missing value methods and determine the variable importance. A SAS Code node is used to compute RBA-based variable importance.

1. Ensure that the **Insurance_ClassTree** project is open in Model Studio and that you are on the **Pipelines** tab.
2. Select the **Decision Tree** node. Under Splitting Options, observe that **Missing values** is set to **Use in Search**, which is the default.
3. Open the **Results** of the **Decision Tree** node. Maximize the **Variable Importance** table.

Variable Label	Role	Variable Name	Validation Importa...	Impor...	Relative Importance	Count
Saving Balance	INPUT	SAVBAL	218.9077	0	1	1
Checking Balance	INPUT	DDABAL	137.1400	0	0.6265	2
Money Market	INPUT	MM	115.2705	0	0.5266	1
Certificate of Deposit	INPUT	CD	35.7783	0	0.1634	2
Branch of Bank	INPUT	BRANCH	8.4289	0	0.0385	3
Age of Oldest Account	INPUT	ACCTAGE	6.3551	0	0.0290	1
Number of Checks	INPUT	CHECKS	2.4767	0	0.0113	2
Line of Credit	INPUT	LOC	2.3284	0	0.0106	1
Investment	INPUT	INV	2.1738	0	0.0099	2
Installment Loan	INPUT	ILS	0.8901	0	0.0041	1
Retirement Account	INPUT	IRA	-0.3899	0	-0.0018	1

Recall that the Variable Importance table provides insight into the importance of inputs in the decision tree.

The Validation Importance column shows the maximum RSS-based variable importance. The Relative Importance column measures variable importance based on the change of residual sum of squares (RSS) when a split is found at a node. It is a number between 0 and 1, which is calculated as the RSS-based importance of this variable divided by the maximum RSS-based importance among all the variables (for example, 218.9077 / 218.9077 = 1; 137.1400 / 218.9077 = 0.6265 and so on). The magnitude of the relative importance statistic relates to the amount of variability in the target explained by the corresponding input relative to the input at the top of the table (for example, DDABAL explains 62.65% of the variability explained by SAVBAL). The RSS and relative importance are calculated from the validation data. If no validation data exist, they are calculated instead from the training data.

Notice that the input IRA has a negative validation importance. If the change in RSS is negative (which is possible when you measure it on the validation set), then the change is set to 0.

The Importance Standard Deviation column shows the dispersion of the importance taken over partially independent several trees. That is the reason it has all zero values for a single tree. The numbers would be nonzero for forest and gradient boosting, ensemble models discussed later in chapters 4 and 5. You can conveniently ignore this column here in a decision tree model.

The Count column shows how many times a variable is used in splitting in the decision tree. The BRANCH input seems to be the most important per count-based importance method.

4. Close the Variable Importance table and close the results of the Decision Tree node.
5. To identify the inputs with missing values, click the **Data** tab.
6. Click the **Missing** column heading twice.

	Variable Name	Label	Type	Role	Level	Order	Comment	Number of Levels	Missing ↓
☐	AGE	Age	Numeric	Input	Interval	Default		79	19.7241
☐	HMVAL	Home Value	Numeric	Input	Interval	Default		173	18.0142
☐	INCOME	Income	Numeric	Input	Interval	Default		192	18.0142
☐	LORES	Length of Residence	Numeric	Input	Interval	Default		39	18.0142
☐	HMOWN	Owns Home	Numeric	Input	Binary	Default		2	17.2444
☐	CC	Credit Card	Numeric	Input	Binary	Default		2	12.7344
☐	CCBAL	Credit Card Balance	Numeric	Input	Interval	Default		>254	12.7344
☐	CCPURC	Credit Card Purchases	Numeric	Input	Interval	Default		6	12.7344
☐	INV	Investment	Numeric	Input	Binary	Default		2	12.7344
☐	INVBAL	Investment Balance	Numeric	Input	Interval	Default		>254	12.7344
☐	PHONE	Number Telephone Banking	Numeric	Input	Interval	Default		19	12.7344
☐	POS	Number Point of Sale	Numeric	Input	Interval	Default		35	12.7344
☐	POSAMT	Amount Point of Sale	Numeric	Input	Interval	Default		>254	12.7344
☐	ACCTAGE	Age of Oldest Account	Numeric	Input	Interval	Default		>254	6.4111
☐	CRSCORE	Credit Score	Numeric	Input	Interval	Default		>254	2.0509

The variables have been sorted by percentage of missing values. Quite a few variables have missing values. Focus on those variables that have missing values as well, as they appear in the variable importance table. For example, two of these variables, ACCTAGE and INV, were also important variables in the Variable Importance table, each used once and twice respectively in splitting. They have around 6% and 13% missing values, respectively.

7. Return to the **Pipelines** tab and reopen the **Results** of the Decision Tree node.
8. Maximize the Tree Diagram and adjust the view (zoom-in) to focus on the splits containing missing observations of the two variables.

Why are missing values assigned to the left branch of ACCTAGE and not to the right branch?

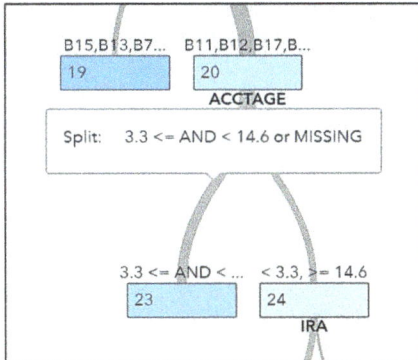

The use in search policy uses missing values in the calculation of the worth of a splitting rule. This consequently produces a splitting rule that assigns the missing values to the branch that maximized the worth of the split.

Note that the bin ordering is ignored because the *k*-means fast search algorithm was enabled.

Regarding missing values as a special value in search is sometimes inappropriate. If a substantial proportion of values is missing, then they can unduly influence the creation of the split. Evaluating a split using only known information improves credibility.

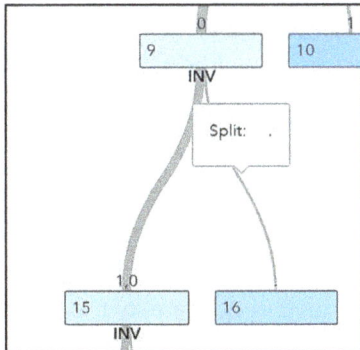

The simple strategy is to regard a missing value as a special nonmissing value. For a nominal input, like INV, a missing value simply constitutes a new categorical value.

Use in search is a desirable policy when the existence of a missing value is predictive of a target value. If not, you can want to exclude the missing values from the split search, and a new policy is needed for assigning missing observations to branches. One policy is to use missing values in a separate branch.

Note: Unique missing value indicators were created (using an Imputation node) for these three variables and then pushed in a logistic regression model (using a Logistic Regression node). None of them significantly predicted the target. Close the **Tree Diagram** and then close the **Results**.

9. Keep the **Decision Tree** node selected and change the **Missing values** from Use in Search to **Separate branch**.
10. Click the **Run Pipeline** button.
11. Open the **Results** of the **Decision Tree** node and maximize the **Variable Important** table.

Variable Label	Role	Variable Name	Validation Importa...	Impor...	Relative Importance	Count
Saving Balance	INPUT	SAVBAL	218.9077	0	1	1
Checking Balance	INPUT	DDABAL	137.1400	0	0.6265	2
Money Market	INPUT	MM	115.2705	0	0.5266	1
Certificate of Deposit	INPUT	CD	35.7783	0	0.1634	2
Branch of Bank	INPUT	BRANCH	7.2500	0	0.0331	2
Credit Card	INPUT	CC	2.4269	0	0.0111	1
Line of Credit	INPUT	LOC	2.3284	0	0.0106	1
Number of Checks	INPUT	CHECKS	1.0719	0	0.0049	1
Installment Loan	INPUT	ILS	0.8901	0	0.0041	1
Investment	INPUT	INV	-0.2531	0	-0.0012	1

The top few important variables are the same as what you had in a decision tree with the Use in search option. However, the two variables with missing values (**ACCTAGE** and **INV**) did not appear as important variables.

Assigning all the observations with missing values to a single branch is apt to reduce the purity of the branch, thereby degrading the split. If this is unavoidable, assigning the largest branch results in the least dilution of node purity.

12. Close the **Variable Importance** table and close the **Results** of the **Decision Tree** node.
13. Open the **Results** of the **Model Comparison** node and maximize the **Model Comparison** table.

Champi...	Name	Algorith...	KS (You...	Misclas...	Misclas...	Root Av...	Averag...
★	Decision Tree	Decision Tree	0.4192	0.2633	0.2633	0.4261	0.1815
	DT Multiway Splits	Decision Tree	0.4139	0.2654	0.2654	0.4239	0.1797

The ASE (Average Squared Error) and MISC (misclassification rate) of the binary-split tree has slightly deteriorated from 0.1808 to 0.1815 and from 0.2616 to 0.2633.

14. Close the **Model Comparison** table.
15. Close the **Results** of the **Model Comparison** node.
16. Go to step 27.
    Note: Steps 18-26 are optional for exploring surrogate rules.

17. Keep the Decision Tree node selected. Under Splitting options, change the **Surrogate rules** from 0 to **1**.

Missing values:

Separate branch ▾

Number of interval bins:

10

Interval bin method:

Quantile ▾

Surrogate rules:

1

☐ Use input once

☑ Perform clustering-based split search

Using surrogate splitting rules is another policy for assigning missing observations to branches when you want to exclude missing values from the split search. A surrogate splitting rule mimics a regular splitting rule. Using them to select more variables is an afterthought. A surrogate variable is typically correlated with the main splitting variable, so the selected variables will now have some redundancy. However, surrogate rules enable you to make better use of the data.

18. Run the **Decision Tree** node.
19. Open the **Results** and maximize the **Variable Importance** table.

Variable Label	Variable Name	Validation Import...	Importance Stand...	Relative Importance	Count	Times Used as a S...
Saving Balance	SAVBAL	395.0410	158.2083	1	2	0
Saving Account	SAV	280.0745	0	0.7090	0	1
Checking Balance	DDABAL	202.7041	77.1615	0.5131	1	1
Money Market	MM	164.7835	80.0664	0.4171	1	1
Money Market Balance	MMBAL	164.2039	79.7423	0.4157	1	1
CD Balance	CDBAL	158.0073	60.1177	0.4000	1	1
Age of Oldest Account	ACCTAGE	137.3895	63.8108	0.3478	0	2
Amount Deposited	DEPAMT	130.6266	0	0.3307	0	1
Certificate of Deposit	CD	19.3743	0	0.0490	1	0
Income	INCOME	17.1837	1.4241	0.0435	0	2
Home Value	HMVAL	13.0545	0	0.0330	1	0
Age	AGE	10.9537	0	0.0277	1	0
Branch of Bank	BRANCH	5.0282	0	0.0127	1	0

Observe that you have an additional column of Times Used as a Surrogate and the variable importance order has changed. Few variables appeared in the table due to surrogate rules and not because they were used in splitting. For example, the SAV variable (second in the list) was used once in the surrogate rules but not actually used in splitting in the decision tree. A surrogate rule substitutes for the main rule when the main rule cannot handle an observation. A good surrogate rule is one that mimics the main rule very well, even if it does not define a good split.

For example, the main rule might split on home value (**HMVAL**), and the surrogate might split on money market balance (**MMBAL**). The surrogate applies to observations with an unknown **HMVAL** and a known **MMBAL**. The surrogate might be less effective as a main splitting rule because **MMBAL** represents coarser information than **HMVAL**. The surrogate policy relies on redundant inputs. Another policy is needed when no good surrogate exists.

20. Close the **Variable Importance** table.
21. Click the **Assessment** tab and maximize the **Fit Statistics** table.

Target ...	Data Role	Partitio...	Formatt...	Sum of ...	Averag...	Divisor ...	Root Av...	Misclas...
INS	TRAIN	1	1	13,550	0.1794	13,550	0.4235	0.2661
INS	VALIDATE	0	0	5,807	0.1900	5,807	0.4359	0.2862

This is a smaller tree. However, the performance has deteriorated. The ASE and MISC have increased from 0.1841 to 0.1900 and 0.2636 to 0.2862, respectively. That's the price that you would pay to select more variables and to better handle the missing values.

22. Minimize the **Fit Statistics** table and close the **Results**.
23. Because the primary objective of this decision tree is prediction and not variable selection, change back the **Surrogate rules** property to **0**.
24. Run the **Decision Tree** node again.
25. Open the Results of the Decision Tree node if you have closed them. Scroll down and maximize the **Training Code** window.
26. Scroll down and copy (Ctrl+C) the Component Code on your clipboard.

```
64    *------------------------------------------------------*;
65    * Component Code;
66    *------------------------------------------------------*;
67  ⊖ proc treesplit data=&dm_datalib..'DM_2CSKLUS6H7NTJ428T07H1L9D1'n(&dm_data_caslib)
68         maxdepth=10 numbin=10 minleafsize=20
69         nsurrogates=0 maxbranch=2 assignmissing=BRANCH binmethod=QUANTILE
70         pruningtable
71         outmodel=&dm_datalib.._BFE9B3INOQN5V8X7TO9D2SE5W_model treeplot printtarget;
72    grow IGR
73    ;
74    target 'INS'n / level=nominal;
75    input %dm_interval_input / level=interval;
76    input %dm_binary_input %dm_nominal_input %dm_ordinal_input %dm_unary_input / level=nominal;
77    partition rolevar='_PartInd_'n (TRAIN='1' VALIDATE='0');
78    prune costcomplexity;
79    code file="&dm_file_scorecode." nocomppgm labelid=94430748;
80    ods output
81        CostComplexity = &dm_lib..pruning
82        VariableImportance = &dm_lib..varimportance TreePlotTable = &dm_lib..treeplot TreePerformance = &dm_lib..TreePerf
83        PredProbName = &dm_lib..PredProbName
84        PredIntoName = &dm_lib..PredIntoName
85    ;
86    ods exclude treeplottable OutputCasTables;
87  run;
88
```

27. Close the **Training Code** window and close the **Results**.
28. Right-click the **Data** node and select **Add child node** ➪ **Miscellaneous** ➪ **SAS Code**.

29. Click the **Open Code Editor** button.

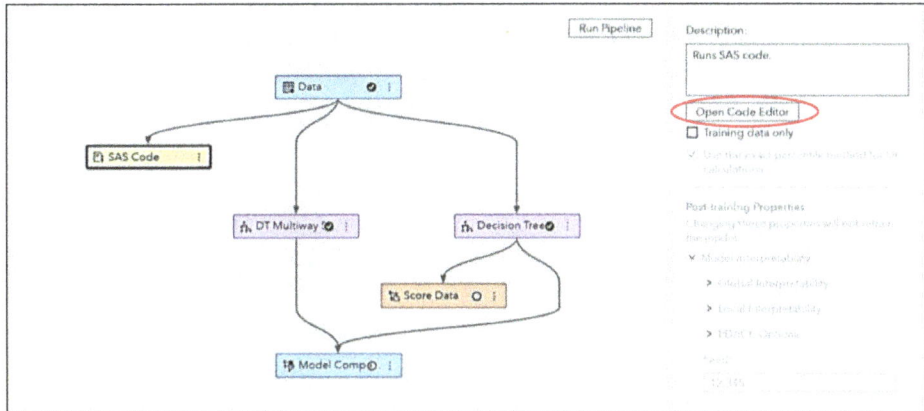

30. Paste (Ctrl+P) the Component Code in the **Training code** window.
31. Add an ODS GRAPHICS ON; statement, change ASSIGNMISSING to POPULAR, and add an RBAIMP option in the PROC TREESPLIT statement.

```
 1    *----------------------------------------------------------*;
 2     * Component Code;
 3    *----------------------------------------------------------*;
 4    ods graphics on;
 5 ⊖  proc treesplit data=&dm_datalib..'DM_2CSKLUS6H7NTJ428T07H1L9D1'n(&dm_data_caslib)
 6          maxdepth=10 numbin=10 minleafsize=20
 7          nsurrogates=0 maxbranch=2 assignmissing=POPULAR binmethod=QUANTILE
 8          pruningtable
 9          outmodel=&dm_datalib.._BFE9B3INOQN5V8X7TO9D2SE5W_model treeplot printtarget rbaimp;
10     grow IGR
11     ;
```

This creates a different tree in which the missing value is allocated to the largest branch and a variable importance table using random branch assignment is created.

32. Click **Close** and select **Save** to save the changes to the SAS Code node.
33. Right-click **SAS Code node** ⇨ **Move** ⇨ **Supervised Learning**.

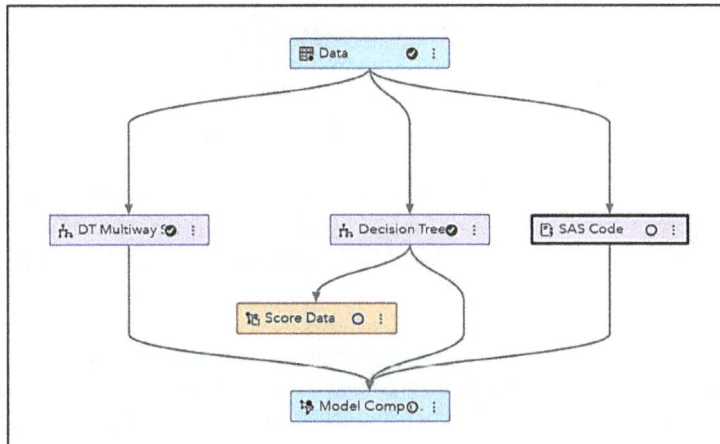

34. Click the **Run Pipeline** button.
35. Open the **Results** of the **SAS Code** node and scroll down.
36. Maximize the **Output** window and observe the variable importance tables.

Variable Importance				
Variable	Importance	Std Dev Importance	Relative Importance	Count
SAVBAL	242.59	0	1.0000	4
CD	108.79	0	0.4485	1
MM	76.6700	0	0.3160	2
DDABAL	65.0116	0	0.2680	3
DDA	26.3440	0	0.1086	2
CHECKS	5.9190	0	0.0244	2
ACCTAGE	5.7983	0	0.0239	3
IRA	5.4632	0	0.0225	1
LOC	1.2610	0	0.0052	1
INV	0.9127	0	0.0038	3
DEP	0.5723	0	0.0024	1
ILS	-0.3968	0	-0.002	1
ATM	-0.7040	0	-0.003	1
MOVED	-1.0759	0	-0.004	1
DEPAMT	-1.7123	0	-0.007	1
AGE	-1.8349	0	-0.008	1
MTG	-2.0271	0	-0.008	1
CDBAL	-2.2788	0	-0.009	1
INCOME	-2.6351	0	-0.011	1
BRANCH	-9.6800	0	-0.040	6

RBA Variable Importance				
	Training		Validation	
Variable	Importance	Relative Importance	Importance	Relative Importance
SAVBAL	0.3575	1.0000	0.3506	1.0000
DDABAL	0.2897	0.8103	0.2986	0.8517
MM	0.2883	0.8066	0.2910	0.8301
DDA	0.2719	0.7605	0.2804	0.7996
CD	0.2699	0.7550	0.2742	0.7819
ACCTAGE	0.2621	0.7331	0.2729	0.7785
BRANCH	0.2652	0.7417	0.2721	0.7760
CHECKS	0.2610	0.7300	0.2712	0.7736
LOC	0.2591	0.7248	0.2702	0.7706
INCOME	0.2601	0.7275	0.2695	0.7687
DEP	0.2622	0.7335	0.2692	0.7677
DEPAMT	0.2589	0.7242	0.2692	0.7677
INV	0.2590	0.7246	0.2690	0.7672
AGE	0.2586	0.7234	0.2686	0.7662
CDBAL	0.2585	0.7230	0.2683	0.7652
MOVED	0.2584	0.7227	0.2683	0.7652
IRA	0.2576	0.7205	0.2681	0.7647
ILS	0.2588	0.7240	0.2681	0.7647
MTG	0.2582	0.7223	0.2680	0.7642
ATM	0.2583	0.7225	0.2669	0.7613

The RBAIMP option created a variable importance table using random branch assignment (RBA) method. This table is created in addition to the normal variable importance table that is calculated using the residual sum of squares (RSS) error. Observe that the variable importance is moderately different in the two methods (for example, checking balance (**DDABAL**) and branch of bank (**BRANCH**) inputs are more important in RBA-based importance).

Note: Because of randomization, the RBA variable importance values might be slightly different from the ones shown above.

37. Close the Output window and close the Results of the SAS Code node.

End of Demonstration

Strengths and Weaknesses of Decision Trees

Decision trees are used to solve both classification and regression problems. However, one of the biggest drawbacks of decision trees is that they lead to overfitting of the data that we will discuss in the next chapter. Figure 4.18 summarizes some of the main advantages and disadvantages of a decision tree.

The strengths of decision trees are that they are easy to implement, and they are very intuitive. In fact, the results of the model are extremely easy to explain to non-technical personnel. Decision trees can be fit very quickly and can score new customers very easily.

Figure 4.18: Strengths and Weaknesses of Decision Trees

Unlike linear models such as linear regression and logistic regression, decision trees can handle nonlinear relationships between the target and the predictor variables without specifying the relationship in the model.

Missing values are handled because they are part of the prediction rules (for example, if income is missing, the customers can be put into their own input space rather than eliminated from the analysis).

Decision trees are also robust to outliers in the predictor variable values and can discover interactions. An interaction occurs when the relationship between the target and the predictor variable changes by the level of another predictor variable (for example, if the relationship between the target and income is different for males compared to females, decision trees would be able to discover it). It should be noted that interactions are nonlinear relationships.

On the other hand, like any other model, decision trees have several weaknesses. They are unstable models. That is, minor changes in the training data set can cause substantial changes in the structure of the tree.

Decision trees confront the curse of dimensionality by ignoring irrelevant predictor variables. However, decision trees have no built-in method for ignoring redundant predictors. Because decision trees can be fitted quickly and have a simple structure, this is usually not an issue for model creation. It can be an issue for model deployment though, in that decision trees might arbitrarily select from a set of correlated predictor variables. To avoid this problem, it is recommended that you reduce redundancy before fitting the decision tree.

A large decision tree can be grown until every node is as pure as possible. This decision tree is called the *maximal* tree, and it usually overfits the data.

Secondary Uses of Decision Trees

Decision trees can be used for several other analytical tasks beyond predictive modeling. Figure 4.19 summarizes some of the important secondary uses of decision trees.

Initial Data Analysis (IDA) and Exploratory Data Analysis (EDA)

Most experienced modelers use a variety of methods for exploring and examining the data before trying to fit a final model. Initial data analysis (IDA) might uncover problems with the inputs or target like the one (temporal infidelity) shown in the first demonstration in Chapter 2. IDA can be useful for getting a preliminary impression of the predictive power of the data. Both extremely deficient performance and extremely superior performance can indicate problems with the data.

Figure 4.19: Uses of Decision Trees Beyond Predictive Modeling

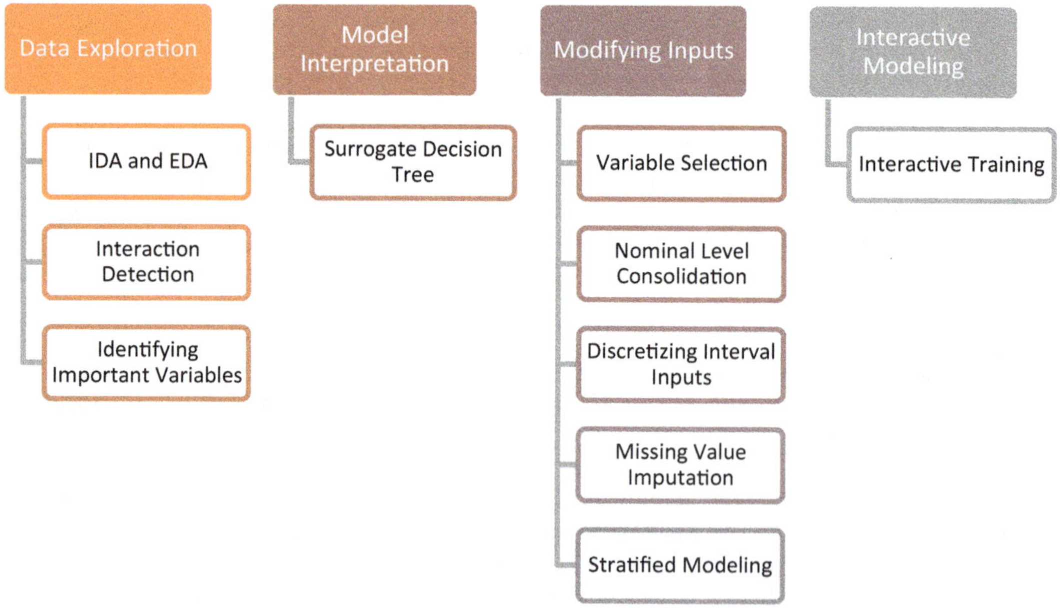

Decision trees are well suited for exploratory data analysis (EDA). With exploratory trees, you need to pay less attention to the issue of binary versus multi-way splits and to determining the right-sized tree. It is often useful to build several tree varieties and use different settings.

Decision trees are good for IDA and EDA because of these characteristics:

- interpretability
- no strict assumptions about the functional form of the model
- resistant to the curse of dimensionality
- robust to outliers in the input space
- no need for dummy variables for nominal inputs
- no need to impute missing values
- computationally fast (usually)

Trees are not robust to rare target events. If you are not careful with very unbalanced samples, the result is often no tree – only the root node.

Many predictive modelers are uncomfortable with trees as their final models. Often, trees are used as auxiliary tools for building more familiar models, such as logistic regression. Standard regression models are constrained to be linear and additive (on the link scale). They require more data preparation steps, such as missing value imputation and dummy-coding of nominal variables. Their statistical and computational efficiency can be severely affected by the presence of many irrelevant and redundant input variables.

Interaction Detection

The AID in CHAID stands for *automatic interaction detection*. Trees are often touted as useful for detecting interactions. After an interaction is detected, the appropriate interaction term (or terms) can be included in a regression model, such as

$\text{logit}(E(y)) = \beta_0 + \beta_1 x_1 + \beta_2 x_2 + \beta_3 x_3 + \beta_{13} x_1 x_3$.

Trees notice relationships from the interaction of inputs as shown in Figure 4.20. For example, the interval target might not correlate with input x_2 unless x_1 is less than 8. The tree noticed both inputs. x_2 has a split at 2.

AID is an exaggeration. Trees do not produce anything as straightforward as an ANOVA table that rigorously tests for an interaction effect. It is usually difficult to determine the strengths of interactions only by looking at a tree diagram. Trees might be better described as *automatic interaction accommodators*. Crossover (qualitative) interactions (effect reversals) are easier to detect by scanning splits on the same input in different regions of the tree. Magnitude (quantitative) interactions are rarely possible to detect.

Figure 4.20: Interactions of Inputs Noticed in a Decision Tree

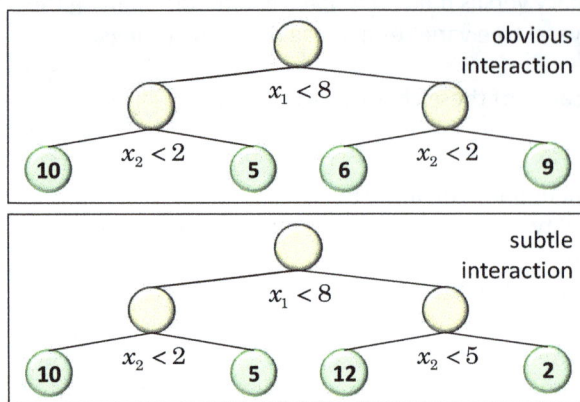

Identifying Important Variables

Intuitively, the variables used in a tree have various levels of importance. What makes a variable important is the strength of the influence and the number of cases influenced.

A formulation of variable importance that captures this intuition is as follows (Neville 1999):

1. Measure the importance of the model for predicting an individual. Specifically, let the importance for an individual equal the absolute value of the difference between the predicted value (or profit) of the individual with and without the model.
2. Divide the individual importance among the variables used to predict the individual, and then
3. Average the variable importance over all individuals.

The variable importance table was discussed earlier in this chapter.

Model Interpretation

The usual criticism of machine learning models (for example, neural networks, gradient boosting, and similar flexible models) is the difficulty in understanding the predictions.

This criticism stems from the complex parameterizations found in the model. Although it is true that little insight can be gained by analyzing the actual parameters of the model, much can be gained by analyzing the resulting prediction decisions.

Trees are sometimes used to help understand the results of other models. Entropy or another mechanism for generating a decision threshold defines regions in the input space corresponding

Figure 4.21: Decision Tree Surrogate Model

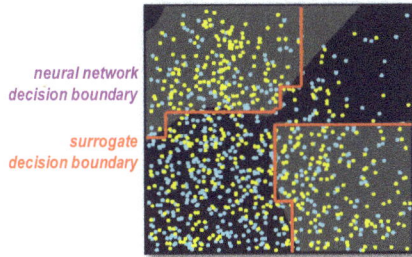

to the primary decision. You can approximate an inscrutable model's predictions with a decision tree surrogate model. The idea behind a surrogate model is building an easy-to-understand model that describes this region as shown in Figure 4.21. In this way, characteristics used by the neural network to make the primary decision can be understood, even if the neural network model itself cannot be.

Variable Selection

Among the most popular auxiliary uses of trees is modifying the set of potential input variables. The subset of variables that are selected by a decision tree can serve as promising inputs for other modeling methods, particularly nonlinear or nonadditive models such as neural networks. A variable that is important to a tree does not necessarily have a strong linear additive effect. Consequently, input selection using trees is conceptually more appropriate for other flexible models like neural networks.

You can always use the Decision Tree node with default settings to select inputs. However, this tends to select too few inputs for a subsequent model. Including surrogates rules and creating maximal tree result in more inputs being selected. When you use trees to select inputs for flexible models, it is better to err on the side of too many inputs rather than too few. The model's complexity optimization method can usually compensate for the extra inputs.

Changing the number of surrogates enables inclusion of surrogate splits in the variable selection process. Surrogate inputs are typically correlated with the selected split input. Although it is usually a bad practice to include redundant inputs in predictive models, many flexible models tolerate some degree of input redundancy. The advantage of including surrogates in the variable selection is to enable inclusion of inputs that do not appear in the tree explicitly but are still important predictors of the target.

Changing the Subtree method to Largest causes the tree algorithm to not prune the tree. As with adding surrogate splits to the variable selection process, it tends to add (possibly irrelevant) inputs to the selection list.

Figure 4.22: Input Selection by Tree

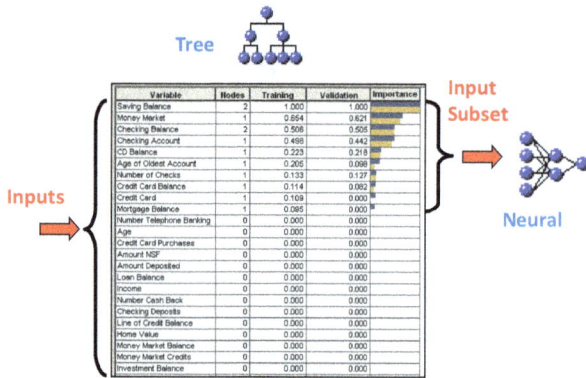

The variable importance table then includes a complete list of variables selected by the tree as illustrated in Figure 4.22. Only variables with nonzero values for Importance are the variables selected by the decision tree. This list includes surrogate variables that do not appear in the tree.

Nominal Levels Consolidation

Categorical inputs pose a major problem for parametric predictive models such as regressions and neural networks. Nominal inputs can be included in regression or neural networks models by using binary indicators (dummy variables) for each level. Because each categorical level must be coded by an indicator variable, a single input can account for more model parameters than all other inputs combined. This practice can cause a detrimental increase in dimensions when the nominal inputs have many levels (for example, ZIP codes).

Decision trees, on the other hand, thrive on categorical inputs. They can easily group the distinct levels of the categorical variable together and produce good predictions.

Figure 4.23: Collapsing Levels Using a Tree

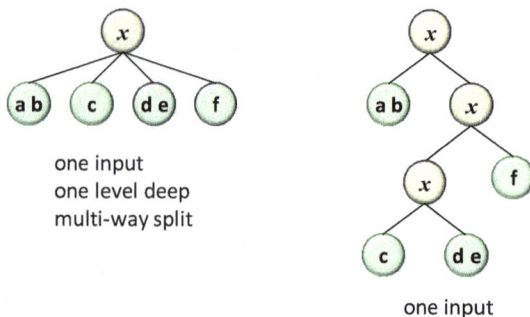

It is possible to use a decision tree to consolidate the levels of a categorical input. You simply build a decision tree using the categorical variable of interest as the sole modeling input. If a tree is fitted using only a single nominal input, the leaves represent subsets of the levels. The split search algorithm then groups input levels with similar primary outcome proportions. The IDs for each leaf replace the original levels of the input. Any type of tree can suffice, but a depth-one tree with a multi-way split is easier to interpret visually. See Figure 4.23.

> Collapsing levels based on subject-matter considerations is usually better than any data-driven approach. However, there are many situations where the knowledge about potentially important inputs is lacking.

Discretizing Interval Inputs

Trees are also used to bin levels of interval inputs to account for nonlinearity in linear models. The original input is then replaced by dummy variables for each bin. This method can unnecessarily inflate the dimension and should be used with caution (for example, with a strong linear trend, the tree would have many leaves and result in the needless addition of degrees of freedom) as illustrated in Figure 4.24. Discretizing interval inputs should usually be done in conjunction with diagnostic plots (for example, empirical logit plots).

The resulting model is a primitive spline function with discontinuous constant segments between the split values (knots). More sophisticated regression splines using higher order polynomials joined smoothly at the knots might be preferable.

Missing Value Imputation

To avoid the loss of data in regressions and neural networks due to complete-case analysis, missing values need to be imputed. Regression imputation (conditional mean imputation) is a

Figure 4.24: Binning Interval Inputs Using a Tree

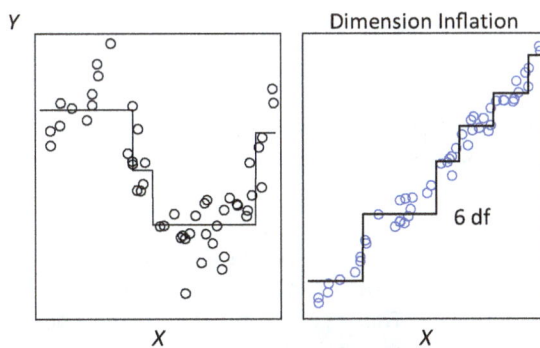

Figure 4.25: Imputing Missing Values Using a Tree

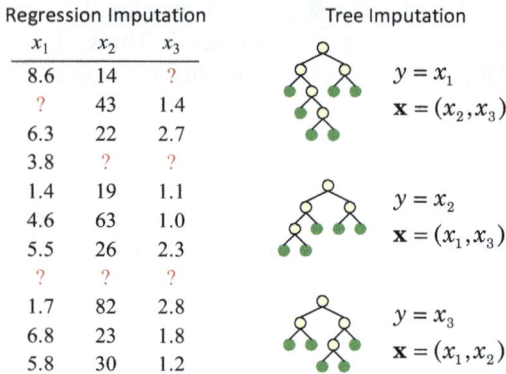

Regression Imputation			Tree Imputation
x_1	x_2	x_3	
8.6	14	?	
?	43	1.4	
6.3	22	2.7	
3.8	?	?	
1.4	19	1.1	
4.6	63	1.0	
5.5	26	2.3	
?	?	?	
1.7	82	2.8	
6.8	23	1.8	
5.8	30	1.2	

$$y = x_1$$
$$\mathbf{x} = (x_2, x_3)$$

$$y = x_2$$
$$\mathbf{x} = (x_1, x_3)$$

$$y = x_3$$
$$\mathbf{x} = (x_1, x_2)$$

well-regarded method. In regression imputation, the missing values of one input are predicted as a function of the other inputs. A (generalized) linear model is usually used where the target is the input to be imputed and the predictors are the other inputs. The model is fitted to the nonmissing cases of the input to be imputed. The fitted model is then used to predict the missing cases.

When there are several inputs with missing values, as illustrated in Figure 4.25, this method requires that several regression models be fitted and saved. One serious drawback is the complication that is caused by missing values on the other inputs. The other inputs used in the regression model to impute the missing values of one input need to be imputed first to avoid a complete-case analysis.

Trees are well-suited for regression imputation. They can be used in place of linear models to predict the nonmissing cases. They do not require that the other inputs be imputed. Furthermore, they make no assumptions on the functional form of the imputation model. When the input to be imputed is interval, a regression tree is used. Otherwise, a classification tree is used.

Stratified Modeling

The analyst preparing for a regression model faces another hidden pitfall when the data represent two populations. A different relationship between an input and the target might exist in the different populations.

The problem for the analyst is to recognize the need to perform two regressions instead of one. For this purpose, some analysts first create a small decision tree from the data and then run a separate regression in each leaf as shown in Figure 4.26. This is called *stratified regression*. Unfortunately, the tree usually will not split data the way the analyst hopes.

Figure 4.26: Stratified Regression Models

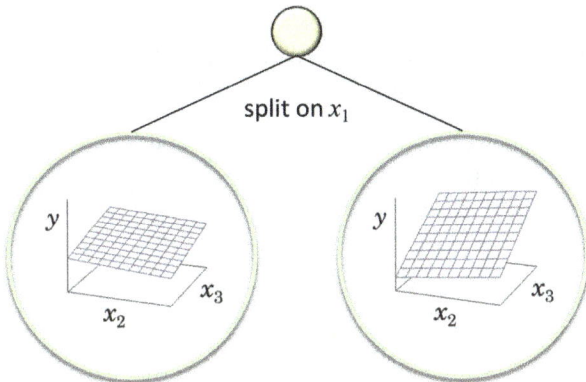

Interactive Training

Decision trees are one of those basic models that start with the data and automatically split that data into subsets to try to split it into pure subsets to be able to predict the target. So, it is naturally building out a set of rules for you. However, very often you need to be able to inject business rules domain knowledge or other types of information that the decision tree really does not know about. So, at times you want to intervene while training a decision tree. This can be accomplished by training decision trees interactively. Interactive decision trees in SAS Viya strike a balance between what the machine learning algorithm does for you and what humans want to inject into the overall decision process. The interactive decision tree walks you through a decision process by asking questions to lead you down the appropriate decision path. You can think of it as a user-friendly flow chart.

After you run a Decision Tree node, you can modify the splitting and pruning logic of nodes when you open the decision tree. The splitting logic that you can modify varies, depending on whether you want to modify a class input or interval input. Regardless of the type of input that you select, you can specify the following when splitting a node:

- Variable — Specifies the variable that you want to use to split the selected node. Variables are sorted by logworth. The logworth is used to determine how well a variable divides the data.
- Branches — Specifies the splitting criteria for each branch that stems from a node. You can also add branches and specify the splitting criteria for each of those branches.
- Missing values — Specifies how missing values are handled when splitting a node.

SAS Model Studio pipeline fits an "optimal" tree given the data in hand and the specified splitting and pruning settings as discussed in Chapters 2 and 3. Sometimes the user might want to modify this model based on business rules and prior experience to ease implementation and improve interpretability. Interactive training in SAS Visual Statistics enables the user to circumvent the automated, algorithm-driven tree growth and pruning partially or entirely.

Quiz

1. What tasks can you perform using a SAS Code node? (Select all that apply.)
   a. Incorporate new or existing SAS code into Model Studio pipelines.
   b. Write separate training code and DS1 scoring code.
   c. Make other SAS procedures available for use in your data mining analysis.
   d. Write SAS DATA steps to create customized scoring code, conditionally process data, or manipulate existing data sets.

Answers

1. a, b, c, and d.

Chapter 5: Tuning a Decision Tree

Model Complexity and Generalization

If you run a machine learning algorithm and it does not work as well as you were hoping, almost all of the time it is because you have either a high bias problem or a high variance problem. It is crucial to know whether you have bias or variance or a bit of both.

Fitting a model to data requires searching through the space of possible models. Constructing a model with good generalization requires choosing the right complexity. Selecting model complexity involves a trade-off between bias and variance as illustrated in Figure 5.1. An insufficiently complex model might not be flexible enough, which can lead to underfitting. *Underfitting* is systematically missing the signal (high bias). *Bias* is the difference between the average prediction of the model and the actual value that we are trying to predict. A model with high bias pays little attention to the training data and oversimplifies the model. It always leads to high errors on training and validation data.

A naïve modeler might assume that the most complex model should always outperform the others, but this is not the case. An overly complex model might be too flexible, which can lead to overfitting. *Overfitting* is accommodating nuances of the random noise in the sample (high variance). *Variance* is the variability of model prediction for a given data point or a value that tells us the spread of our data. A model with high variance pays a lot of attention to training data

Figure 5.1: Bias-Variance Trade-Off

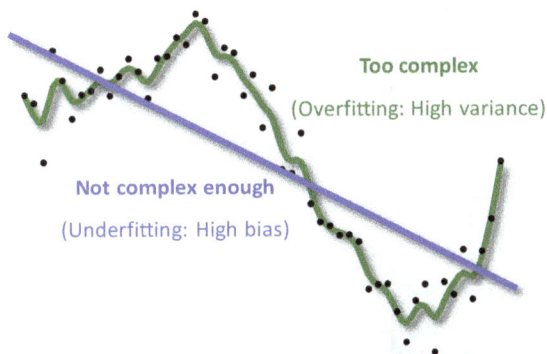

and does not generalize on the data that it has not seen before. As a result, such models perform very well on training data but have high error rates on validation data.

Underfitting happens when a model is unable to capture the underlying pattern of the data. These models usually have high bias and low variance. Overfitting happens when a model captures the noise along with the underlying pattern in data. These models have low bias and high variance. A model with the right amount of flexibility gives the best generalization.

A large decision tree can be grown until every node is as pure as possible. If at least two observations have the same values on the input variables but different target values, it is not possible to achieve perfect purity. The tree with the greatest possible purity on the training data is the *maximal classification* tree as represented in Figure 5.2.

The maximal tree is the result of overfitting. It adapts to both the systematic variation of the target (signal) and the random variation (noise). It usually does not generalize well on new (noisy) data as shown in Figure 5.3.

A small tree with only a few branches might underfit the data. It might fail to adapt sufficiently to the signal. This usually results in poor generalization.

Figure 5.2: Maximal Tree

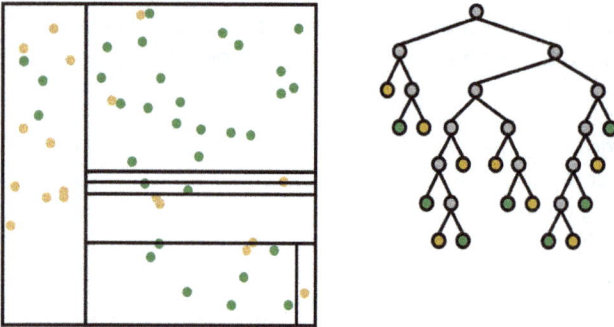

Figure 5.3: Overfitting

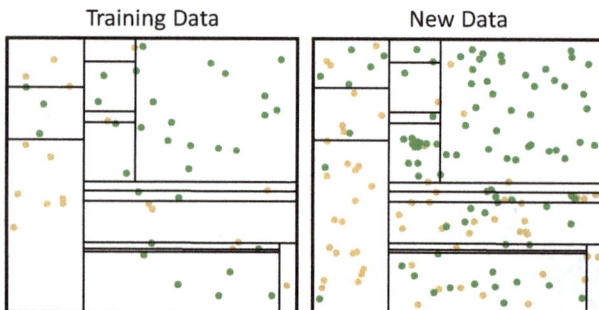

Figure 5.4: Underfitting

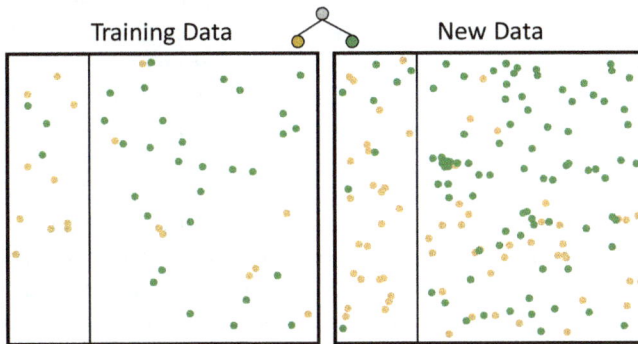

Tree complexity is a function of the number of leaves, the number of splits, and the depth of the tree. Determining complexity is crucial with flexible models like decision trees. A well-fit tree has low bias (adapts to the signal) and low variance (does not adapt to the noise). The determination of model complexity usually involves a tradeoff between bias and variance. An underfit tree that is not sufficiently complex has high bias and low variance, see Figure 5.4. In contrast, an overfit tree has low bias and high variance.

Pruning: Getting the Right-Sized Tree

You can create a classification or regression tree by first growing a tree as described in previous chapters. This usually results in a large tree that provides a good fit to the training data. The problem with this tree is its potential for overfitting the data: the tree can be tailored too specifically to the training data and not generalize well to new data. The solution is to find a smaller subtree that results in a low error rate on unseen instances. *Pruning* refers to the various methods for selecting tree complexity. There are two techniques for pruning – pre-pruning and post-pruning, shown in Figure 5.5. You can use either or both strategies to get a right-sized tree.

Figure 5.5: Pruning Strategies

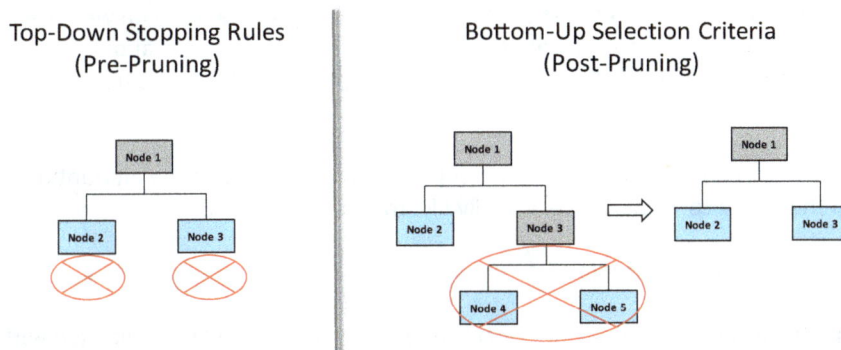

Pre-pruning prevents the generation of non-significant branches in a tree. It involves using some termination conditions to decide when it is desirable to terminate some of the branches prematurely as the tree is generated. When constructing the tree, some significant measures can be used to assess the goodness of a split. Pre-pruning is also called *forward pruning* or *top-down stopping*. Many practitioners contend that stopping rules cannot work. The inherent fallacy is the assumption that an appropriate threshold can be set without much understanding of the data.

The alternative is post-pruning. In post-pruning, a large decision tree is generated, and then non-significant branches are removed. Post-pruning a decision tree implies that you begin by creating the maximal tree and then adjust it with the aim of improving the classification accuracy on unseen instances. Post-pruning is also known as *backward pruning* and *bottom-up pruning*. All candidate sub-trees pruned from the large tree are available at the same time for comparison, and this gives retrospective pruning an advantage over a stopping rule that can consider only a single node at one time. The pruning process may evaluate entire subtrees instead of individual splits. An evaluation measure can be chosen that is more relevant to the end use of the tree than the splitting criterion. The proportion of cases misclassified is a common choice (Neville, 1999).

Stopping and/or pruning is necessary to avoid over-fitting the data. Top-down pruning is analogous to forward variable selection in regression. Bottom-up pruning is analogous to backward variable selection. After a tree is pruned, the ideal tree would exhibit low bias and low variance.

It is often prohibitively expensive to evaluate the error on all subtrees of the full tree. A more practical strategy is to focus on a sequence of nested trees that are obtained by successively pruning leaves from the tree. A simple example of pruning is shown above in which node 3's leaves (terminal nodes 4 and 5) are removed to create a nested subtree of the full tree. In the nested subtree, node 3 is now a leaf that contains all the observations that were previously in nodes 4 and 5. This process is repeated until only the root node remains. You can also achieve this by not allowing the tree to grow beyond node 3 (also known as stopping or pre-pruning).

Pre-Pruning Options

The size of a tree might be the most important single determinant of quality, more important than creating good individual splits. Trees that are too small do not describe the data well. Trees that are too large have leaves with too little data to make any reliable predictions about the contents of the leaf when the tree is applied to a new sample. Splits deep in a large tree can be based on too little data to be reliable.

Forward stopping rules can be based on limiting the depth of the tree, limiting the amount of fragmentation, and the statistical significance, visually illustrated in Figure 5.6.

Top-down pruning options in Model Studio:

- The liberal significance level of 0.2 (logworth=0.7) is the default. It can be changed with the **Significance level** option.

Figure 5.6: Top-Down Stopping Rules (Pre-Pruning)

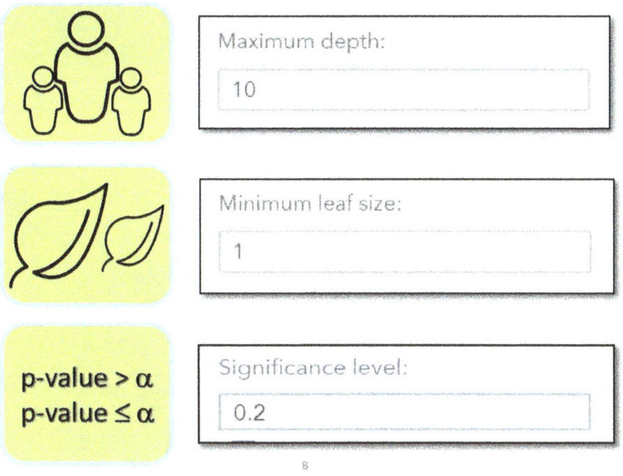

- The default maximum depth in the Decision Tree node is 10. The value can be changed with the **Maximum depth** option.
- The **Minimum leaf size** option specifies the minimum number of observations in the training data that each child of a split must contain in order for the split to be considered.
- To specify pre-pruning only, set **Subtree Method** to either **Cost complexity** or **Reduced-error** and **Selection method** to **Largest**.

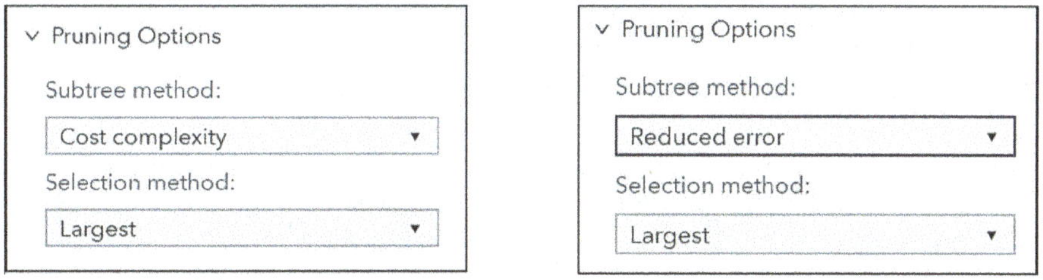

1.   This property combination will not perform any post-pruning.

An example is to not split a node if the number of cases drops below some threshold. If a chi-square or *F* test is used as the splitting criterion, then the *p*-value is a natural stopping rule. That is, stop growing if no splits are statistically significant. This is represented in Figure 5.7 where *m* is the number of splits. One problem with this method is that the effects of multiple selection invalidate the distribution theory of the tests. The *p*-values are typically too small.

The Bonferroni adjustments proposed by Kass (1980) have two uses:

- to equalize the split selection among inputs with different numbers of potential splits
- to correct the *p*-values for the effects of multiple selection

Figure 5.7: Top-Down Pruning Example

	$-\log_{10}(P)$	m	$-\log_{10}(mP)$	d	$-\log_{10}(2^d mP)$
	26.7	53	24.9	0	24.9
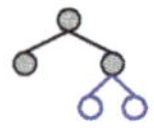	3.12	14	1.97	1	1.67
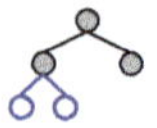	1.63	39	.039	1	−.26
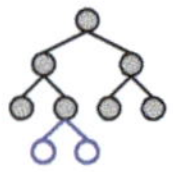	2.40	11	1.36	2	.76

The Bonferroni adjustments are only useful approximations that are applied because the exact statistical distributions of the tests are intractable. (With trees, they are not necessarily conservative approximations.)

Even if you could determine the correct *p*-value, there is no rational method for deciding what value is significant enough. In scientific applications of statistical inference, the custom is to use significance levels of 0.05 (logworth = 1.3) or 0.01 (logworth = 2).

Post-Pruning Options

In bottom-up (post) pruning, a large tree is grown and then branches are removed in a backward fashion using some model selection criterion. The bottom-up strategy of intentionally creating more nodes than can be used is also called *retrospective pruning*.

One of the simplest forms of pruning is reduced error pruning. Two other commonly used pruning algorithms for error minimization are C4.5's error-based pruning (Quinlan 1993) and CART's cost-complexity pruning (Breiman et al. 1984). Figure 5.8 summarizes their salient features along with relevant property options in the Decision Tree node in Model Studio.

This post-order traversal of the tree replaces a subtree by a single leaf node when the estimated error of the leaf replacing the subtree is lower than that of the subtree. This involves finding an honest estimate of error, which is defined as one that is not exceedingly optimistic for a tree that was built to minimize errors in the first place. The resubstitution error (error rate on the training set) usually does not provide a suitable estimate since a leaf node replacing a subtree will never have fewer errors on the training set than the subtree.

Cost-Complexity Pruning

Cost-complexity pruning is a widely used pruning method that was originally proposed by Breiman et al. (1984). You can request cost-complexity pruning for either a categorical or continuous target.

Figure 5.8: Bottom-Up Selection Criteria (Post-Pruning)

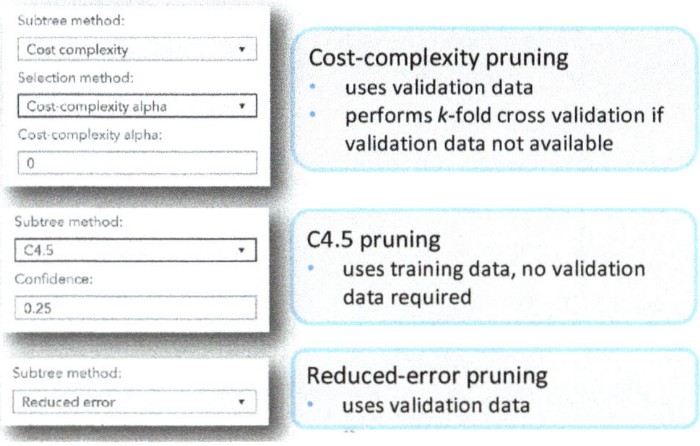

Cost-complexity pruning
- uses validation data
- performs *k*-fold cross validation if validation data not available

C4.5 pruning
- uses training data, no validation data required

Reduced-error pruning
- uses validation data

The cost-complexity pruning method helps prevent overfitting by making a trade-off between the complexity (size) of a tree and the error rate. Thus, large trees with a low error rate are penalized in favor of smaller trees. The cost complexity of a tree *T* is defined in Figure 5.9. It is also known as the CART pruning method.

For a categorical response variable, the misclassification rate is used for the error rate; for a continuous response variable, the residual sum of squares (RSS), also called the sum of square errors (SSE), is used for the error rate. Only the training data are used to evaluate cost complexity.

Breiman et al. (1984) show that for each value of α, there is a subtree of *T* that minimizes cost complexity. When $\alpha=0$, this is the full tree, T_0. As α increases, the corresponding subtree becomes progressively smaller, and the subtrees are in fact nested. Then, at some value of α, the root node has the minimal cost complexity for any α greater than or equal to that value. Optimal α is represented in Figure 5.10. Because there are a finite number of subtrees, each subtree corresponds to an interval of values of α; that is,

$[0, \alpha_1)$ = interval where T_0 (the full tree) has minimal cost complexity
$[\alpha_1, \alpha_2)$ = interval where T_1 has minimal cost complexity
$:::$ $:::$ $:::$
$[\alpha_m, \infty)$ = interval where T_m (the root node) has minimal cost complexity

Figure 5.9: Cost-Complexity of a Tree

$$CC(T) = R(T) + \alpha|T|$$

| cost-complexity of tree *T* | *T*'s error rate | cost of each leaf | number of leaves in *T* |

class target: MISC
interval target: RSS

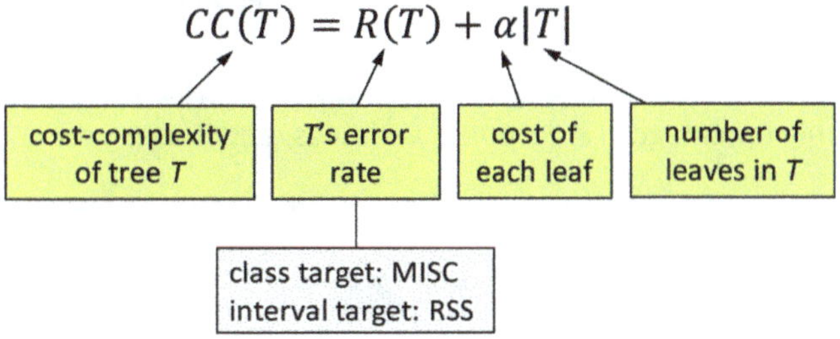

Figure 5.10: Cost-Complexity Parameter

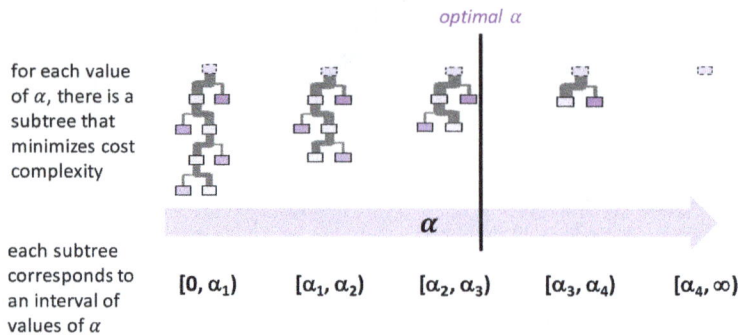

The decision tree in SAS Viya uses weakest-link pruning, as described by Breiman et al. (1984), to create the sequence of α_1, α_2, ..., α_m values and the corresponding sequence of nested subtrees, $T_1, T_2, ..., T_m$.

Finding the optimal subtree from this sequence is then a question of determining the optimal value of the complexity parameter α. This is performed either by using the validation partition or by using cross validation. In the first case, the subtree in the pruning sequence that has the lowest validation error rate is selected as the final tree. When there is no validation partition, k-fold cross validation can be applied to cost-complexity pruning to select a subtree that generalizes well and does not overfit the training data (Breiman et al. 1984; Zhang and Singer 2010).

C4.5 Pruning

Quinlan (1987) first introduced pessimistic pruning as a method of pruning classification trees. In this method, the estimate of the true error rate is increased by using a statistical correction to prevent overfitting. C4.5 pruning (Quinlan 1993) evolved from pessimistic pruning to use an even more pessimistic (that is, higher) estimate of the true error rate. An advantage of methods such as pessimistic and C4.5 pruning is that they enable you to use all the data for training instead of requiring a holdout sample.

In C4.5 pruning, the upper confidence limit of the true error rate based on the binomial distribution was used to estimate the error rate. (See Figure 5.11.) Decision trees in SAS Viya implement a C4.5 algorithm variant that uses the beta distribution in place of the binomial distribution to estimate the upper confidence limit.

> The C4.5 pruning method is available only for categorical response variables, and it uses only training data for tree pruning.

Figure 5.11: Pessimistic Pruning in C4.5

Pessimistic error estimates of leaves in a subtree

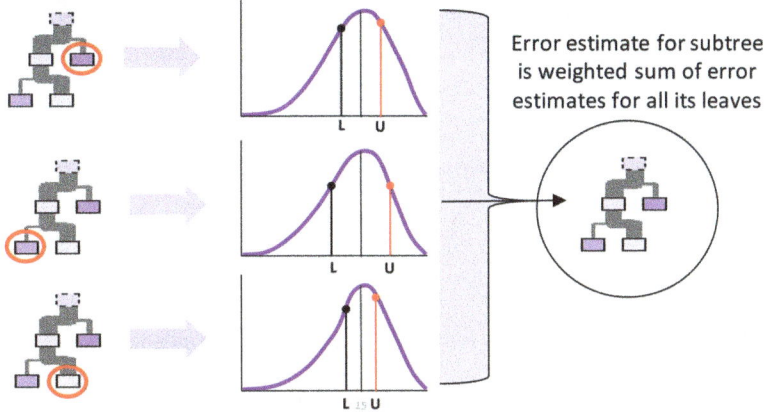

Error estimate for subtree is weighted sum of error estimates for all its leaves

Figure 5.12: Simplified Illustration of Steps Involved in C4.5 Pruning

1. For each node in tree that has only leaves as children, create a candidate subtree by pruning those leaves.

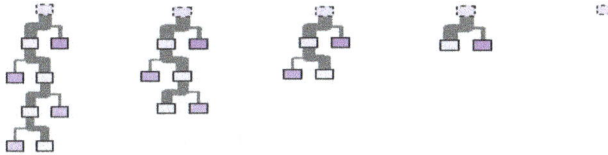

2. For each candidate subtree, calculate its 'pessimistic' prediction error.

$$E_0 \qquad E_1 \qquad E_2 \qquad E_3 \qquad E_4$$

3. Select the candidate subtree that has the largest decrease (or smallest increase) in prediction error.

$$\Delta_1 = E_1\text{-}E_0 \qquad \Delta_2 = E_2\text{-}E_1 \qquad \Delta_3 = E_3\text{-}E_2 \qquad \Delta_4 = E_4\text{-}E_3$$

The C4.5 pruning method follows these steps:

1. Grow a tree from the training data table, and call this full, unpruned tree T_0.
2. Set $i = 0$, and do the following until T_i is only the root node:
   a. For each leaf (terminal node) in the tree T_i, solve the following equation for p_l (which is the adjusted prediction error rate for leaf l):

$$\alpha = 1 - \frac{\Gamma(N_l + 1)}{\Gamma(F_l + 1)\,\Gamma(N_l - F_l)} \int_0^{p_l} v^{F_l}(1-v)^{N_l - F_l + 1}\, dv$$

Here the confidence level α is the value of the confidence level, F_l is the number of failures (misclassified observations) at leaf l, N_l is the number of observations at leaf l, and the function $\Gamma(x)$ is defined as

$$\Gamma(x) = \int_0^\infty v^{x-1} e^{-v} dv$$

b.   Given these values of p_l, use the following formula for the prediction error E_i of tree T_i

$$E_i = \sum_{l \in T_i} N_l p_l$$

c.   For each node in tree T_i that has only leaves as children, create a candidate subtree by pruning those leaves.

d.   For each candidate subtree, use the equations from steps 2 and 3 to calculate its prediction error. Then select the candidate subtree that has the largest decrease (or smallest increase) in prediction error, E_i. Let this be the next subtree in the sequence, T_{i+1}.

e.   Set $i = i+1$.

3. Calculate the change in error between each pair of consecutive subtrees, $\Delta_i = E_i - E_{i-1}$ for each $i = 1, ..., m$.

4. Find the smallest integer j such that $\Delta_j > 0$.

5. Select the subtree T_{j-1} as the final subtree.

In this way, in each node, an upper confidence limit of the number of misclassified data is estimated assuming a Beta distribution around the observed number misclassified. The confidence limit then serves as an estimate of the error rate on future data. The pruned tree minimizes the sum over leaves of upper confidences.

Reduced-Error Pruning

Quinlan's reduced-error pruning (1987) performs pruning and subtree selection based on minimizing the error rate in the validation partition at each pruning step and then in the overall subtree sequence. The error rate is based on the misclassification rate for a categorical response variable and on the ASE for a continuous response.

For any subtree, T, in a tree grown from 1 to n leaves, define its complexity or size (number of leaves) as L and define $R(L)$ as the validation set misclassification cost.

In Model Studio, the pruning process starts with the maximal tree T_{max} with L leaves. The maximal tree is denoted as T_L. Reduced-error pruning creates a sequence of subtrees from the largest tree, T_0, to the root node, T_m. Construct a series of smaller and smaller trees $T_L, T_{L-1}, T_{L-2}, ..., T_1$, so that the following is true:

Figure 5.13: Steps 1 and 2 in Reduced-Error Pruning

1. Grow a maximal tree.

2. Prune to create optimal sequence of subtrees. *

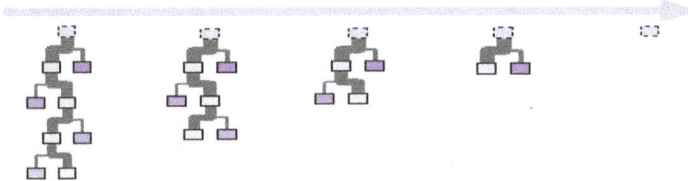

* The trees in the series of subtrees are not necessarily nested.

For every value of H_i, where $1 \leq H_i \leq L$, consider the class T_{Hi} of all subtrees of size H_i. Select the subtree in the series that minimizes $R(T_{Hi})$. These steps are pictorially represented in Figure 5.13.

The subtree that has the smallest validation error is then selected as the final subtree as shown by a dotted line in Figure 5.14. Pruning could be stopped as soon as the error begins to increase in the validation data as originally described by Quinlan; continuing to prune to create a subtree sequence back to the root node enables you to select a smaller tree that still has an acceptable error rate.

Top-down (pre) pruning is usually faster but is considered less effective than bottom-up (post) pruning. Breiman and Friedman, in their criticism of the FACT tree algorithm (Loh and

Figure 5.14: Step 3 in Reduced-Error Pruning

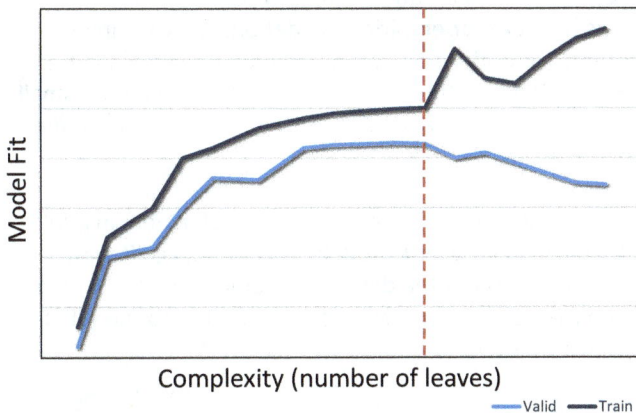

Vanichsetakul 1988), discussed their experiments with stopping rules as part of the development of the CART methodology:

"Each stopping rule was tested on hundreds of simulated data sets with different structures. Each new stopping rule failed on some data set. It was not until a very large tree was built and then pruned, using cross validation to govern the degree of pruning, that we observed something that worked consistently."

Pruning Requirements

Pruning reduces the size of decision trees by removing parts of the tree that do not provide power to predict or classify instances. Bottom-up (post) pruning has two requirements:

- A method for honestly measuring performance
  The purpose of tree-based models is generalization, which is the performance of the predictions on new data. Evaluating the model on the same data the model was fit on usually leads to an optimistically biased assessment. Honest assessment, which is highly related to the bias-variance tradeoff, involves calculating error metrics from scoring the model on data that were not used in any way during the training process.
- A relevant model selection criterion
  To compare many models, an appropriate fit statistic (or statistics) must be selected. These statistics are typically measured on data that were not used in any way during the training process. For a series of models, which could be generated by an automatic selection routine, it is conceivable to plot a fit measure against some index of complexity. For decision tree models, this index is equivalent to the number of leaves in the model.

Honest Assessment

The simplest strategy for correcting the optimism bias is data splitting, where a portion of the data is used to fit the model and the rest is held out for empirical validation. You can simply get the data split into training and validation sets (or optionally have a test data partition also). The validation data is used for model comparison. Data splitting is inefficient when the data is small. Removing data from the training set can degrade the fit. Furthermore, evaluating performance on a single validation set can give imprecise results.

A more efficient remedy, but more computationally expensive, is k-fold cross validation. In k-fold cross validation, performance measures are averaged over k models. Each model is fit with $(k-1)/k$ of the data and assessed on the remaining $1/k$ of the data. The average over the k holdout data sets is then used to honestly estimate the performance for the model fitted to the full data set. (Cross validation is discussed later in this chapter.)

Model Selection Criteria

Model performance characteristically follows a fairly straight forward trend. As the complexity increases (that is, as more leaves are added in a tree) the fit on the training data gets better. After a point, the fit might plateau, but on the training data, the fit gets better as model complexity increases. Some of this increase is attributable to the model capturing relevant trends in the data. Detecting these trends is the goal of modeling. Some of the increase, however, is due to the model identifying vagaries of the training data set. This behavior has been called overfitting. Because these vagaries are not likely to be repeated in the validation data or in future observations, it is reasonable to want to eliminate those models. Hence, the model fit on the validation data, for models of varying complexity, is also plotted. The typical behavior of the validation fit line is an increase (as more complex models detect more usable patterns) followed by a plateau, which might finally result in a decline in performance, as shown in Figure 5.14. The decline in performance is due to overfitting. The plateau just indicates more complicated models that have no fit-based arguments for their use. A reasonable rule would be to select the model associated with the complexity that has the highest validation fit statistic.

Plotting the training and validation results together permits a further assessment of the model's generalizing power. Typically, the performance deteriorates from the training data to the validation data. This phenomenon, sometimes known as *shrinkage*, is an additional statistic that some modelers use to get a measure of the generalizing power of the selected model. For example, the rule used to select a model from the many different models plotted might be: "Choose the simplest model that has the highest validation fit measure, with no more than 10% shrinkage from the training to the validation results."

For classification problems, the most appropriate measures of generalization depend on the number of correct and incorrect classifications and their consequences.

Figure 5.15: Subtree Selection

Leaves	Misc	ASE	Subtree
5	0.10 (0.12)	0.17 (0.15)	
4	0.10 (0.12)	0.18 (0.14)	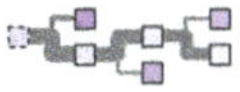
3	0.11 (0.09)	0.19 (0.16)	
2	0.12 (0.09)	0.20 (0.16)	
1	0.41 (0.36)	0.48 (0.46)	

*The number in the braces is for validation data.

For many purposes, including analyses with interval targets, average square error was found to work very well as a general method for selecting a subtree on the validation data. Figure 5.15 shows subtree selection for a binary target. A two-leaf or a four-leaf subtree might be selected by the decision tree algorithm because the subtree with two leaves has the minimum misclassification rate, whereas the subtree with four leaves has the minimum average square error on the validation data.

Tree models are selected based on validation data if it is present. Common selection criterion is Average Square Error or Misclassification for interval or categorical targets, respectively.

User Specification of Subtree

You might want to select a different tree from the one selected by default when you use cost-complexity or reduced-error pruning to create the sequence of subtrees. For example, you might have a subtree that has a slightly larger error but is smaller, and thus simpler, than the subtree that has the minimum error according to reduced-error pruning. You can override the selected subtree and instead select the subtree that has *n* leaves and was created by cost-complexity or reduced-error pruning, where *n* is specified in the **Number of leaves** property when you set the **Selection method** to **N**. In addition, if you are using cost-complexity pruning, you can override the selected subtree by using the **Cost-complexity alpha** property. These property options are shown in Figure 5.16.

Alternatively, you might want to select the largest tree that is created by specifying **Largest** in the **Selection method** property. You can still see the statistics for the sequence of subtrees that are created according to the specified pruning error measure, even though the largest (unpruned) tree is selected as the final subtree.

Confusion Matrix

The confusion matrix is a crosstabulation of the actual and predicted classes that displays four counts: true positives (n_{TP}), true negatives (n_{TN}), false positives (n_{FP}), and false negatives (n_{FN}) as shown in the left side of the Figure 5.17. A true positive (or TP) is a case known to be a primary outcome and also predicted as a primary outcome. A true negative (or TN) is a known secondary

Figure 5.16: Model Studio Options for Overriding Selected Subtree

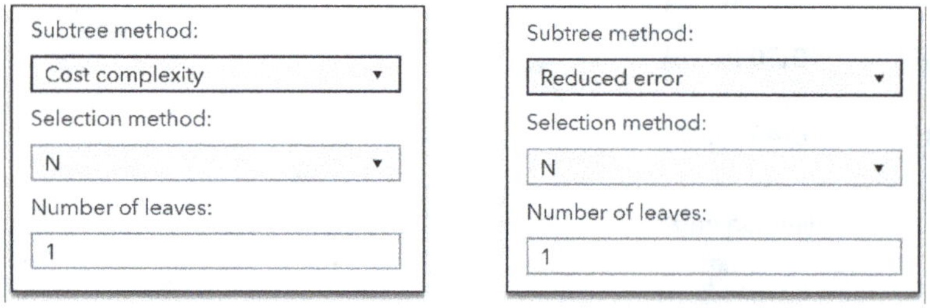

Subtree method:

| Cost complexity ▾ |

Selection method:

| N ▾ |

Number of leaves:

| 1 |

Subtree method:

| Reduced error ▾ |

Selection method:

| N ▾ |

Number of leaves:

| 1 |

Figure 5.17: Event Classification in Confusion Matrix

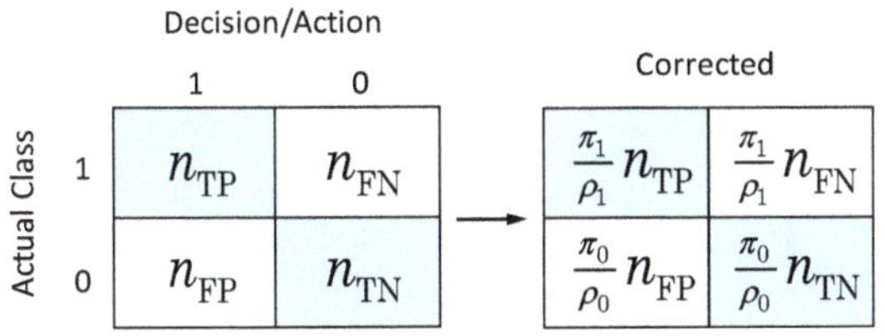

$$\text{Accuracy} = \tfrac{1}{n}\left(\tfrac{\pi_0}{\rho_0} n_{TN} + \tfrac{\pi_1}{\rho_1} n_{TP}\right)$$

Model Studio automatically adjusts assessment measures, assessment graphs, and prediction estimates for this bias.

case predicted as a secondary case. A false positive (or FP) is a case that is predicted as a primary outcome but is actually a known secondary outcome. And a false negative (or FN) case is a case that is predicted as a secondary outcome but known to be a primary outcome.

The counts need to be adjusted if the class proportion in the training sample (ρ_j) is not the same as in the population (π_j). This is shown on the right side of Figure 5.17. It is common practice to try to balance the classes, particularly with rare outcomes.

A confusion matrix presupposes an allocation (decision) rule. A primitive rule allocates cases to the class with the greatest posterior probability. For binary targets, this corresponds to a 50% cutoff on the posterior probability.

Model Studio automatically adjusts assessment measures, assessment graphs, and prediction estimates for this bias. After running the pipeline, you can examine the score code. The score code contains a section titled **Adjust Posterior Probabilities**. This code block modifies the posterior probability by multiplying it by the ratio of the actual probability to the event-based sampling values specified above.

Tree Accuracy

The accuracy of a tree can be calculated as a weighted average of the accuracy in each leaf. The weights are the node sizes as shown in Figure 5.18.

Accuracy is obtained by multiplying the proportion of observations falling into each leaf by the proportion of those correctly classified in the leaf.

For example, to calculate training accuracy for the tree shown in Figure 5.19, 42% of the observations fall into the left branch and all are classified as a 1. Of those, 85% are classified correctly. Consequently, the accuracy in the left branch would be (0.42)(0.85). Similarly, 58% of

Figure 5.18: Accuracy in a Decision Tree

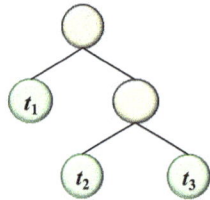

$$\text{Accuracy} = \tfrac{1}{n}\big(n_1(t_1)\,acc(t_1) + n_2(t_2)\,acc(t_2) + n_3(t_3)\,acc(t_3)\big)$$

Figure 5.19: Maximize Accuracy

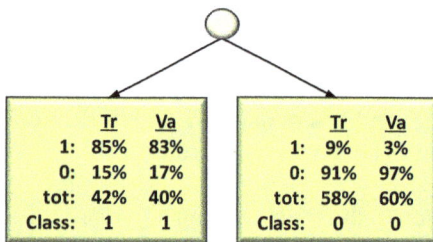

	Tr	Va
1:	85%	83%
0:	15%	17%
tot:	42%	40%
Class:	1	1

	Tr	Va
1:	9%	3%
0:	91%	97%
tot:	58%	60%
Class:	0	0

Training Accuracy = (0.42)(0.85) + (0.58)(0.91) = 0.88

Validation Accuracy = (0.40)(0.83) + (0.60)(0.97) = 0.91

the observations fall into the right branch and all are classified as a 0. Of those, 91% are classified correctly. The accuracy in the right branch would be (0.58)(0.91). The training accuracy of the tree would be (0.42)(0.85) + (0.58)(0.91) = 0.88. The validation accuracy is calculated in a similar manner.

- Classifying according to the maximum posterior probability is not always reasonable, particularly with rare target events (for example, all the cases might be allocated to the same class).
- Simply adding up the correct classifications does not account for the eventuality that some correct classifications can be more valuable than others or that some misclassifications can be costlier.

Both issues are intertwined. If suitable profits and losses can be specified, then statistical decision theory can be used to determine the optimal decision rule that maximizes profit or minimizes loss.

> Currently, a profit or loss matrix is not directly available in SAS Viya. However, you can programmatically calculate this using a SAS Code node in Model Studio.

Demo 5.1: Pruning Decision Trees Using Validation Data

This demonstration builds further on the Insurance_ClassTree project using **insurance_part** data. You will explore and compare CART's cost-complexity pruning and reduced error pruning using holdout data.

1. Ensure that the **Insurance_ClassTree** project is open in Model Studio and you are on the Pipelines tab.
2. Select the **DT Multiway Split** node and observe some of the criterion for recursively splitting parent nodes into child nodes as the tree was grown and some of the pruning criterion.

   Splitting Options:
   o   significance level of chi-square test was set up at 0.01,
   o   growth process continued until the tree reached a maximum depth of 10, and
   o   smallest number of training observations that a leaf can have was at 20.

   These were essentially the pre-pruning stopping rules. The post-pruning rules are mentioned below.

   Pruning Options:
   o   cost complexity is the subtree method, and
   o   selection method is the largest tree.
3. Open the **Results** of the DT Multiway Split node and focus on the Pruning Error Plot.

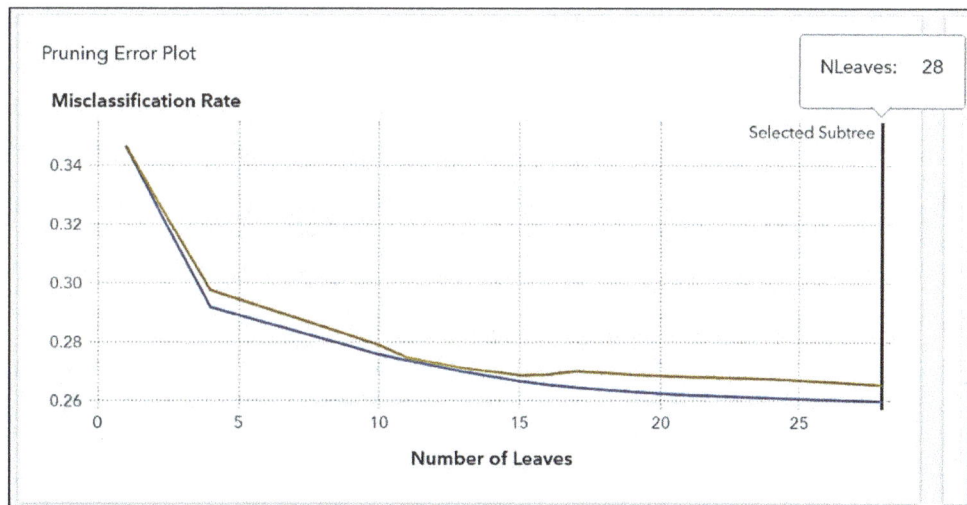

This plot displays the misclassification rates for the training and validation data as the tree is pruned.

Why is the tree with 28 leaves selected (out of 49 leaves) as the final tree? Is it because it has the lowest misclassification rate of 0.2654 on the validation data?

4. Maximize the **Output** window and scroll down to see the **Cost Complexity Pruning** table.

The Cost Complexity Pruning table contains the pruning results of cost complexity pruning that includes the cost-complexity parameter (α) values, information about the number of leaves and the error for cost-complexity pruning when validation data are used. This table is generated when you perform cost complexity pruning, which is the default.

		Misclassification Rate	
Cost Complexity Pruning			
Alpha	Number of Leaves	Training	Validation
0	28	0.2598	0.2654
0.00030	24	0.2610	0.2674
0.00032	21	0.2619	0.2681
0.00052	19	0.2630	0.2688
0.00074	17	0.2644	0.2700
0.00096	16	0.2654	0.2688
0.00118	15	0.2666	0.2686
0.00162	13	0.2698	0.2711
0.00196	11	0.2737	0.2747
0.00199	10	0.2757	0.2790
0.00267	4	0.2917	0.2976
0.01823	1	0.3464	0.3463

The parameter (α) indexes a sequence of progressively smaller subtrees that are nested within a large tree.

The parameter value $\alpha=0$ corresponds to the fully grown tree, and positive values control the trade-off between complexity (number of leaves) and fit to the training data, as measured by average misclassification rate. Recall that you opted for Largest as the selection method and the default cost-complexity parameter value is 0. Therefore, the tree is selected at $\alpha=0$ as the tuning parameter for cost-complexity pruning. Incidentally, this selected tree (with 28 leaves) also has the smallest validation misclassification rate. You can specify a different value of the cost-complexity alpha.

5. Close the **Results** of the DT Multiway Split node.
6. Under Splitting Options, change the **Significance level** from 0.01 to **0.20**, **Maximum depth** from 10 to **20**, and **Minimum leaf size** from 20 to **10**.

Under Pruning Options, change the **Selection method** from Largest to **Automatic**.

Significance level:

0.2

☑ Bonferroni

Maximum number of branches:

4

2   6   10

Maximum depth:

20

Minimum leaf size:

10

⌄ Pruning Options

Subtree method:

Cost complexity ▾

Selection method:

Automatic

The growth process continues until the tree reaches a maximum depth of 20. The purpose of a liberal *p*-value of 0.2 for the chi-square and *F* tests is to grow a tree that is sufficient for pruning but not as computationally burdensome as the maximal tree. Smaller node size further helps in creating a larger tree. The result is often a large tree that overfits the data and is likely to perform poorly in predicting future data. A recommended strategy for avoiding this problem is to prune the tree to a smaller subtree that minimizes prediction error.

7. Run the **DT Multiway Split** node and open the **Results**.

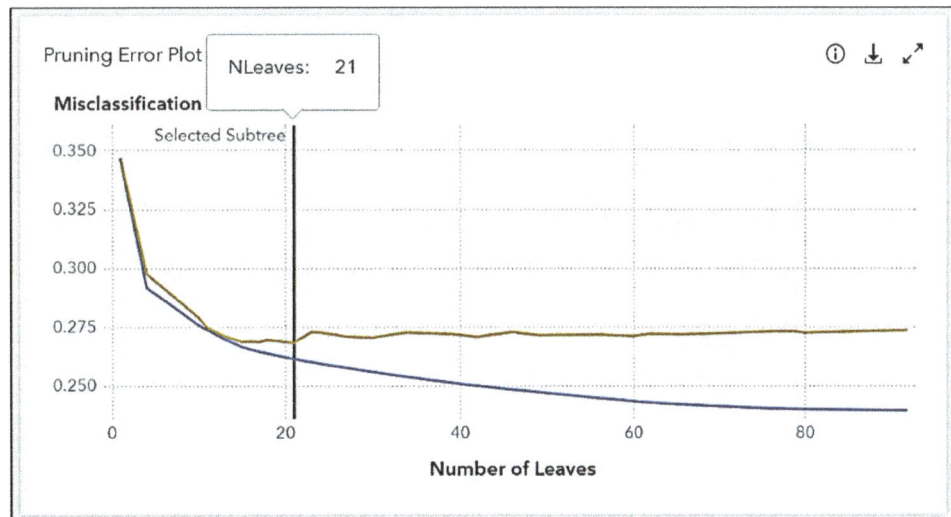

Pruning Error Plot   NLeaves: 21

Misclassification

Selected Subtree

0.350

0.325

0.300

0.275

0.250

0   20   40   60   80

Number of Leaves

A smaller, but not maximal, tree is selected. In this case, the pre-pruning resulted in a 148-leaf tree. Its size is limited by depth (20) and significance level (0.2). Subtrees of this large tree are then examined using the cost-complexity criterion. In this case, the 21-leaf subtree is selected because it has the smallest misclassification rate.

Cost Complexity Pruning			
Alpha	Number of Leaves	Misclassification Rate	
		Training	Validation
0	92	0.2395	0.2735
0.00004	84	0.2398	0.2728
0.00006	80	0.2400	0.2724
0.00007	79	0.2401	0.2729
0.00010	76	0.2404	0.2729
0.00015	69	0.2414	0.2721
0.00017	66	0.2419	0.2717
0.00022	62	0.2428	0.2719
0.00030	60	0.2434	0.2711
0.00033	56	0.2447	0.2716
0.00036	49	0.2472	0.2714
0.00037	46	0.2483	0.2728
0.00041	42	0.2500	0.2707
0.00044	39	0.2513	0.2719
0.00049	34	0.2537	0.2726
0.00054	30	0.2559	0.2704
0.00057	27	0.2576	0.2709
0.00059	24	0.2593	0.2726
0.00066	23	0.2600	0.2728
0.00070	21	0.2614	0.2683
0.00076	18	0.2637	0.2695
0.00081	17	0.2645	0.2686
0.00089	16	0.2654	0.2688
0.00118	15	0.2666	0.2686
0.00162	13	0.2698	0.2711
0.00196	11	0.2737	0.2747
0.00199	10	0.2757	0.2790
0.00267	4	0.2917	0.2976
0.01823	1	0.3464	0.3463

8. Close the **Results** of the DT Multiway Split node.
9. Select the **Decision Tree** node now.

10. Open the **Results** of the Decision Tree node.

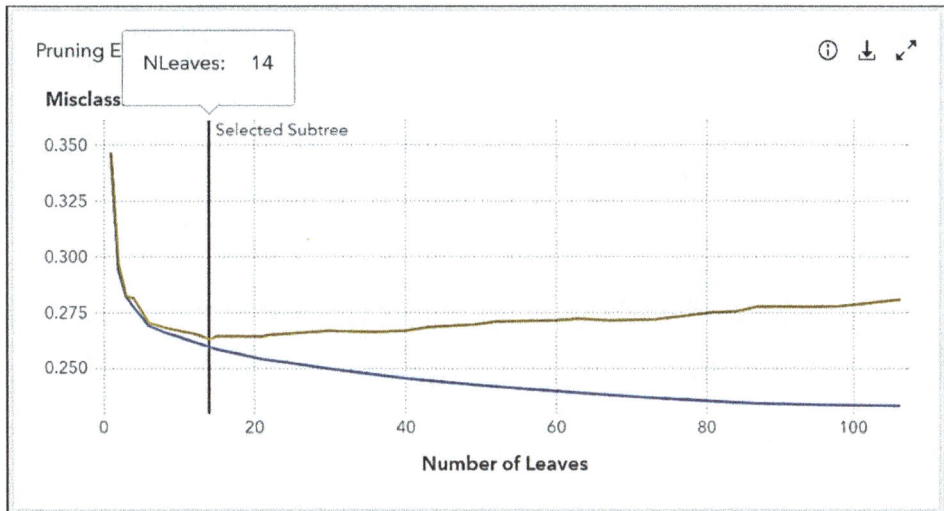

Recall that the subtrees of a 172-leaf large tree were examined using the cost-complexity criterion. Large trees with a low error rate were penalized in favor of smaller trees resulting in a 14-leaf final tree with ASE and MISC of 0.1815 and 0.2633, respectively.

11. Keep the Decision Tree node selected, and under Pruning Options, change the **Subtree method** to **Reduced error**.

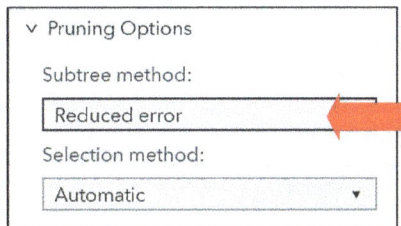

12. Click the **Run Pipeline** button.

13. Open the **Results** of the **Decision Tree** node.

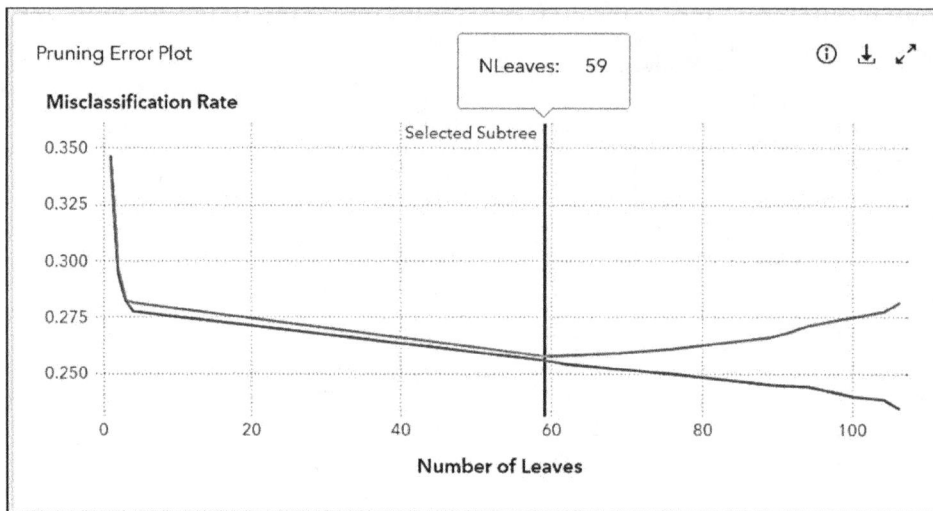

This is a different tree than what you got earlier using cost-complexity, which is default. In this case, the 59-leaf subtree is selected from 172-leaf trees. Subtrees are now examined using the reduced error criterion. The ASE and MISC have marginally improved from 0.1815 to 0.1769 and 0.2633 to 0.2578, respectively. This improvement comes with a cost of almost increasing the number of leaves by more than 75%. Is it worth it?

Remember that the cost-complexity pruning method helps prevent overfitting by making a trade-off between the complexity (size) of a tree and the error rate, whereas reduced error pruning takes in to account only the error rate.

14. Close the **Results** of the **Decision Tree** node.
15. Open the **Results** of the **Model Comparison** node.

Champi...	Name	Algorith...	KS (You...	Misclas...	Misclas...	Root Av...	Averag...
⊡	Decision Tree	Decision Tree	0.4382	0.2578	0.2578	0.4206	0.1769
	SAS Code	SAS Code	0.4259	0.2667	0.2667	0.4262	0.1817
	DT Multiway Splits	Decision Tree	0.4008	0.2683	0.2683	0.4255	0.1811

16. Close the **Results** of the **Model Comparison** node.
    The default method for determining the right-sized tree in the Decision Tree node is a blend of pre-pruning and post-pruning settings. The two pruning approaches are not mutually exclusive.

End of Demonstration

Cross Validation

Data partitioning is a simple but costly approach to model validation. When the holdout set is small, performance measures can be unreliable. You can choose not to split and build a CHAID-type tree using the *p*-value associated with the chi-square or *F* statistics as a forward stopping (stunting) rule.

Alternatively, you can split more efficiently using *k*-fold cross validation (BFOS 1984; Ripley 1996; Hand 1997). In five-fold cross validation, for example, the data are split into five equal sets. The model is refit on each four-fifths of the data using the remaining one-fifth for assessment. The five assessments are averaged. In this way, all the data are used for both training and assessment. A five-fold cross validation along with relevant properties in Model Studio is illustrated Figure 5.20. In Model Studio, the TREESPLIT procedure computes cross validation estimates of a fit statistic for each subtree in the subtree sequence. It then uses those estimates to select the best subtree as the final model.

The final estimate of the fit statistic is the weighted average of the fit statistic computed for all *k* trees. Typically, the weights are the number of observations in the validation data.

BFOS (1984) proposed 10-fold cross validation for decision trees. This is a feature of CART methodology. The complexity parameter, α, is used to map the optimal cross validated subtree to the final subtree fit to the entire data set.

The **Cross validation folds** property specifies the number of cross validation folds to use for cost-complexity pruning when there is no validation data. Values range from 2 to 20. The default value is 10.

The **1–SE rule** property specifies whether to perform the one standard error rule when performing cross validated cost complexity pruning. By default, this is deselected.

Figure 5.21 is the plot of unpartitioned **insurance** data on cross validation cost-complexity. This plot is displayed only when cross validation cost-complexity pruning is performed. INS is the

Figure 5.20: Five-Fold Cross Validation

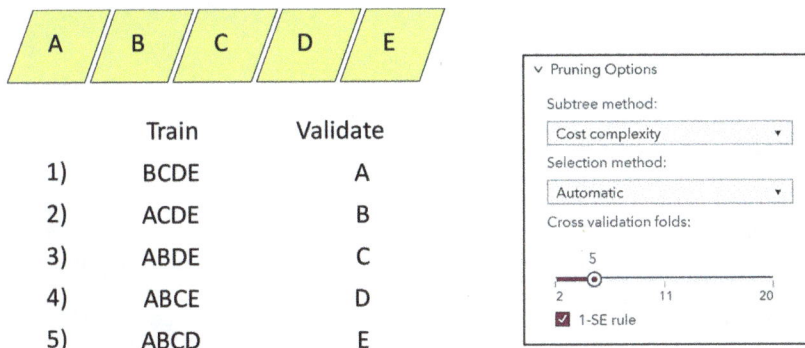

Figure 5.21: Cross Validation Cost-Complexity Pruning with the 1-SE Rule

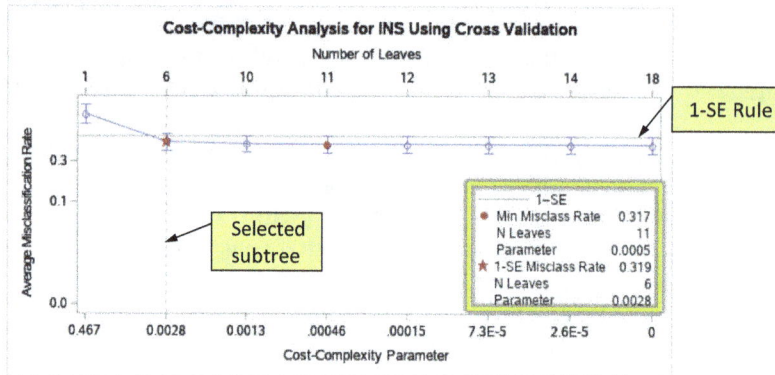

target variable, which is categorical and therefore the plot displays the misclassification rate. For an interval target, this plot displays ASE as a function of the number of leaves in a subtree. Error bars are also given, indicating the average error rate plus or minus one standard error. A vertical reference line is drawn to indicate which subtree was selected as the final model. A horizontal reference line is drawn to represent the "1-SE Rule" even if it is not used for subtree selection.

The example shows that the minimum average misclassification rate, which is obtained by 10-fold cross validation, is 0.317. On the plot, this value is indicated by a filled-in circle. Information about the 1-SE misclassification rate is also included.

Breiman's 1-SE rule chooses the parameter that corresponds to the smallest subtree for which the misclassification rate is less than one standard error above the minimum misclassification rate (Breiman et al. 1984). The parameter value that corresponds to the 1-SE rule is indicated by a star. The value 0.0028 corresponds to a tree that has only six leaves. The dotted vertical line indicates the chosen tree.

The following code is used to produce the plot shown in Figure 5.21:

```
libname mycas cas;
libname local '/home/student/casuser/VBBF';

data mycas.insurance;
  set local.insurance;
run;

ods graphics on;

proc treesplit data=mycas.insurance;
class ins inv loc mm mtg moved nsf res sdb sav atm branch cc cd dda dirdep
    hmown ils ira inarea;
model ins = age acctage atmamt cashbk ccbal ccpurc atm branch cc cd cdbal
    crscore checks dda ddabal dep depamt dirdep hmown hmval ils ilsbal ira
```

```
     irabal inarea income inv invbal loc locbal lores mm mmbal mmcred mtg
     mtgbal moved nsf nsfamt pos posamt phone res sdb sav savbal teller;
grow igr;
prune costcomplexity(leaves=SE);
run;
```

Notice that the entire data set (without partition) is used for creating the decision tree. Therefore, cross validation is automatically performed. Observe that the LEAVES=SE suboption was specified.

Demo 5.2: Performing Cross Validation in a Regression Tree

In this demonstration, you return to the regression tree created earlier on the **housing** data set and perform a 10-fold cross validation cost-complexity pruning. Recall that the regression tree that you last created was created interactively on the entire data (data was not partitioned) and no pruning was performed.

1. Click the **View all projects** icon to return to the **Housing_RegTree** project.

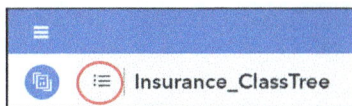

2. Double-click the **Housing_RegTree** tile (or single-click on the name of the project) to open the project.

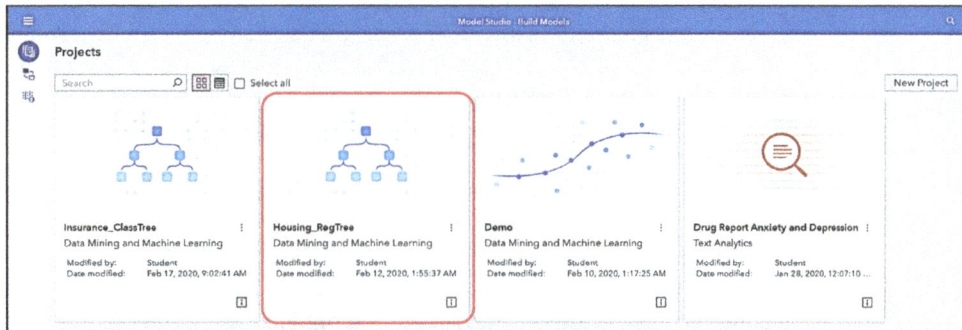

3. Click the **Pipelines** tab to access the pipeline created earlier.
4. Select **Interactive-Model Pipeline**. Select the **Model Comparison** node and open the **Results**.

Model Comparison				
Champion	Name	Algorithm Na...	Average Squared Error	Root Average Square...
★	Decision Tree	Decision Tree	11.6002	3.4059

Recall that the ASE and RASE (root average squared error) of this interactively created decision tree were 11.6002 and 3.4059, respectively.

5. Close the **Results** of the **Model Comparison** node.
6. Right-click the **Decision Tree** node and select **Enable Properties** if they were not previously enabled. Select **Yes** for the pop-up box if that appears.
7. To optimize the complexity of this regression tree, calibrate several pre-pruning and post-pruning options.

   Under Splitting Options, change the **Maximum depth** from 5 to **10**, **Minimum leaf size** from 5 to **20**, and **Number of interval bins** from 50 to **100**.

   Under Pruning Options, observe that the **Selection method** is set to **Automatic**. This ensures that the cost-complexity pruning is performed consuming 10-fold cross validation. Recall that you did not partition the housing data set.

The tuner partitions all the data into 10 subsets (folds). For each fold, a new model is trained on all folds except the selected (holdout) fold (that is, it is trained on nine folds) and then validated using the selected (holdout) fold. The objective function value is averaged over the set of fold validation executions to obtain a single error estimate value.

Note: For efficiency, the cross-validation process might be terminated before all 10 folds are evaluated. Cross validation is terminated under the following conditions: the validation score is two times worse than the current best score after the first fold, the validation score is 1.75 times worse than the current best score after two folds, or the validation score is 1.5 times worse than the current best score after three or more folds.

8. Run the **Decision Tree** node and open the **Results**.

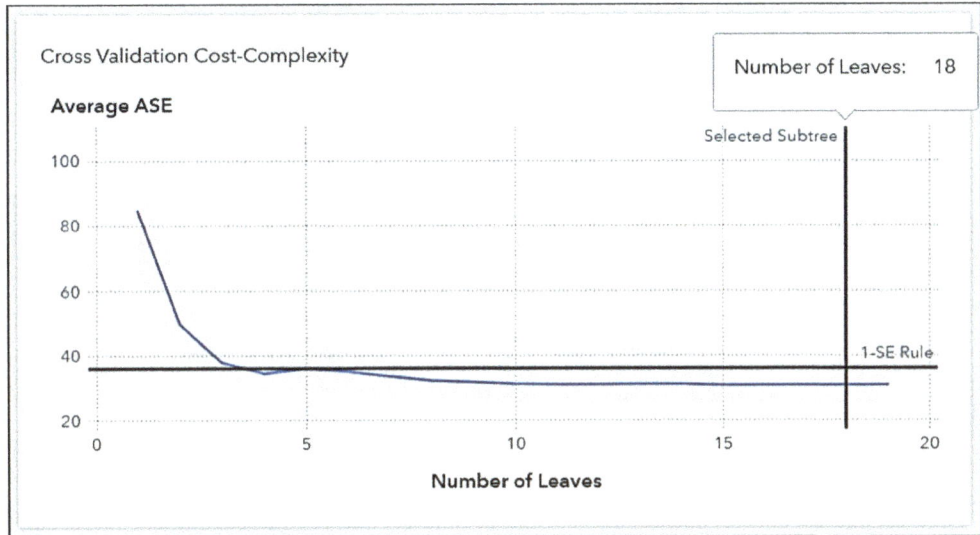

Cross Validation Cost-Complexity

Number of Leaves: 18

An 18-leaf tree is selected as the optimum complexity tree based on 10-fold cross-validation cost-complexity pruning because the final estimate of the ASE, which is the weighted average of the ASE computed for all 10 trees, is the lowest across all trees. A horizontal reference line representing the "1-SE Rule" is not used for subtree selection; however, it reveals that if used, the final tree would be much simpler with less than five leaves.

9. Switch to the **Assessment** tab and Maximize the **Fit Statistics** table.

Target N...	Data Role	Sum of ...	Average...	Divisor f...	Root Av...
MEDV	TRAIN	506	19.7106	506	4.4397

The ASE and RASE of 19.7106 and 4.4397 reveal that the previous tree created interactively was overfitted since it had an ASE of 11.6002 and RASE of 3.4059 (on the training data, the data was not partitioned).

10. Close the **Results** of the Decision Tree node.
11. Under Pruning Options, check the **1-SE rule** box.

Cross validation folds:

10

2   11   20

☑ 1-SE rule

12. Run the **Decision Tree** node and open the **Results**.

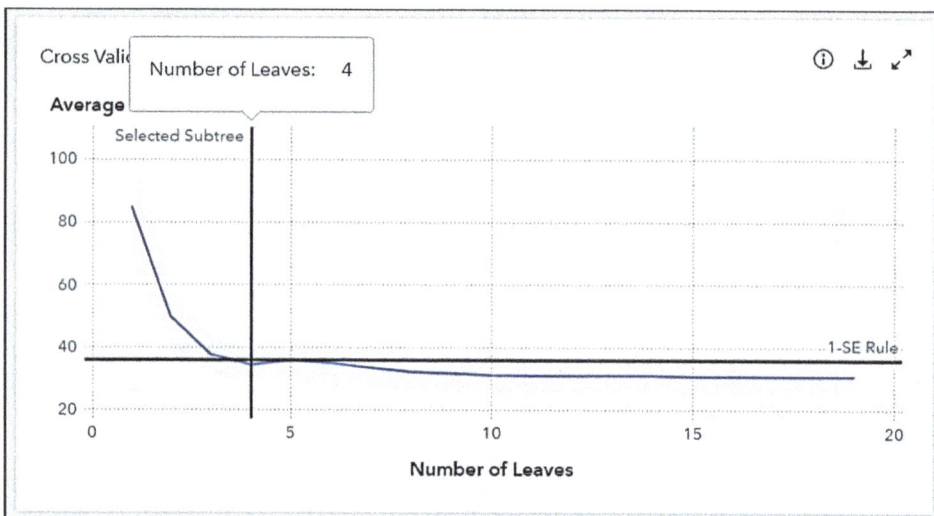

A much simpler tree with four leaves is selected. The 1-SE rule chooses the cost-complexity parameter that corresponds to the smallest subtree for which the ASE is less than one standard error above the minimum ASE.

		10-Fold Cross Validation Assessment of Pruning Parameter				
			Average Square Error			
N Leaves	Pruning Parameter		Min	Avg	Standard Error	Max
19	0	.	23.3576	30.9149	5.1311	40.3987
18	0.0493	.	23.3576	30.8547	5.1185	40.3987
17	0.0839	.	22.9481	30.8983	5.2327	40.4208
16	0.1376	.	23.4264	30.8957	5.1400	40.9619
15	0.1903	.	22.4186	30.8990	5.2131	40.8145
14	0.2318	.	22.2284	31.1982	5.1243	40.8145
13	0.2341	.	22.2284	31.1982	5.1243	40.8145
12	0.2424	.	22.2284	31.1124	5.1197	40.8145
11	0.2967	.	22.2284	31.1588	5.0065	41.1222
10	0.3854	.	22.3548	31.2789	5.1220	42.4228
9	0.4602	.	22.3548	31.8900	5.3450	41.9771
8	0.6847	.	21.4865	32.3653	5.5827	41.9771
7	1.2467	.	21.4865	33.6858	6.0375	43.1667
6	1.8213	.	21.1900	35.1789	6.2714	43.1667
5	2.0088	.	23.7980	36.0358	6.1361	45.6728
4	3.2692	*	24.0684	34.5947	5.7794	43.0774
3	8.2599	.	29.3850	37.8845	5.4394	45.4456
2	21.9504	.	38.7768	49.9018	9.2215	69.9045
1	3734.4	.	56.8851	84.6823	17.6752	125.2
	* Selected pruning parameter					

13. Switch to the **Assessment** tab and maximize the **Fit Statistics** table.

Target Name	Data Role	Sum of Freq...	Average Sq...	Divisor for A...	Root Averag...
MEDV	TRAIN	506	28.8852	506	5.3745

14. The price for this simpler tree is paid in terms of increased ASE and RASE.
15. Close the **Results** of the Decision Tree node.

End of Demonstration

Autotuning

To create a good decision tree model or other tree-based models, many choices must be made when deciding on their parameters. The usual approach is to apply trial-and-error methods to find the optimal combination of properties for the problem at hand. Often, a data scientist chooses algorithms based on practical experience and personal preferences. This is reasonable, because usually there is no unique and relevant solution to create a good tree-based machine learning model. Many algorithms have been developed to automate manual and tedious steps of the machine learning pipeline. Still, it requires a lot of time and effort to build a machine learning model with trustworthy results.

A substantial portion of this manual work relates to finding the optimal set of hyperparameters for a chosen modeling algorithm. Hyperparameters are the parameters that define the model applied to a data set for automated information extraction.

Autotuning is a two-step process as illustrated in Figure 5.22. First, you can enable the Perform Autotuning property available in the Decision Tree node (and in the Forest and Gradient Boosting nodes) in Model Studio and determine which hyperparameters to tune. Then, select the search method. Autotuning seeks to minimize or maximize a chosen objective function (typically a measure of model error) by using search methods. Several search options are available in SAS Viya.

The autotuning capability in SAS Visual Data Mining and Machine Learning takes advantage of the SAS local search optimization (LSO) framework to provide a flexible and effective hybrid search strategy. The default search strategy begins with a Latin hypercube sample (LHS), which provides a more uniform sample of the hyperparameter space than a grid or random search provides. The best samples from the LHS are then used to seed a genetic algorithm (GA), which crosses and mutates the best samples in an iterative process. This generates a new population of model configurations for each iteration. Importantly, SAS Viya can evaluate the LHS samples in parallel, and the GA population at each iteration can also be evaluated in parallel. Alternate search methods include a single LHS, a purely random sample, a grid search, and a Bayesian search method.

Figure 5.22: Autotuning a Decision Tree Model

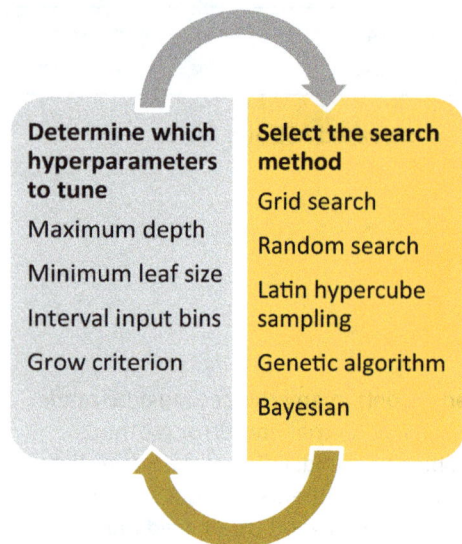

Determine which hyperparameters to tune	Select the search method
Maximum depth	Grid search
Minimum leaf size	Random search
Interval input bins	Latin hypercube sampling
Grow criterion	Genetic algorithm
	Bayesian

Demo 5.3: Comparing Various Tree Settings and Performance

In this demonstration, you return to the **Insurance_ClassTree** project created on the **insurance_part** data. Recall that the regression tree that you last created was created interactively on the entire data (data was not partitioned) and no pruning was performed.

1. Select the **View all projects** icon to return to the Insurance_ClassTree project.

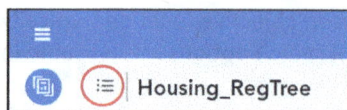

2. Click the **Insurance_ClassTree** tile to open the project.
3. Click the **Pipelines** tab and press **+** to add a new pipeline.
4. In the New Pipeline window, name the pipeline **Autotuned DT**. Ensure that **Blank Template** is selected under **Template.**
5. Click **OK**.

6. Add five Decision Tree nodes and rename them as **CHAID-like**, **CART-like**, **CHAID-like+Validation**, **C4.5**, and **Autotuning DT**. Your pipeline should look like the one below.

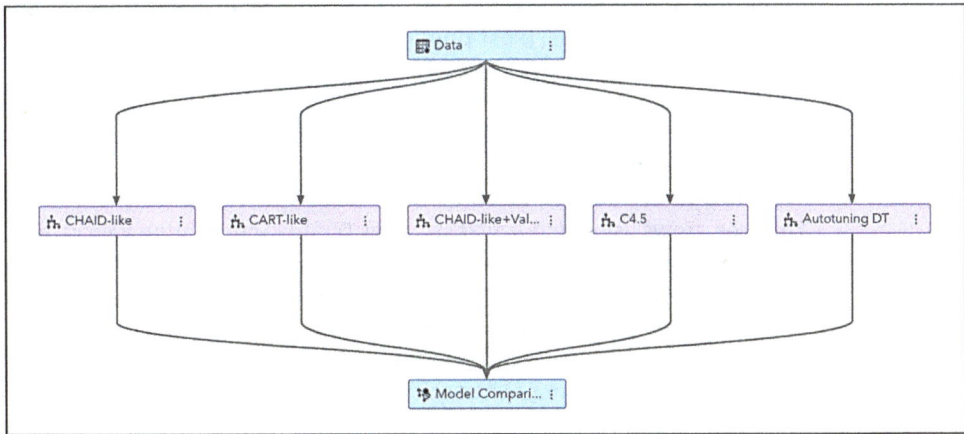

7. Configure the first four decision tree nodes as below:
   a. **CHAID-like (top-down pruning only)**
      Splitting Options: Class target criterion: **CHAID**; Significance Level: **0.0001**; Bonferroni: **Yes**; Maximum depth: **6**; Minimum leaf size: **150**; Maximum number of branches: **6**.
      Pruning Options: Selection method: **Largest**.
      This tree is configured to approximate CHAID. Because no validation is used to tune the model, a small significance level is chosen for larger data sets. (The tree is not pruned because of Method: Largest.) Multi-way splits and Bonferroni adjustments are CHAID-like. Because multi-way split trees can be large and bushy, splitting is constrained to leaf sizes of at least 150.
   b. **CART-like**
      Splitting Options: Class target criterion: **Gini**; Minimum leaf size: **60**; Surrogate rules: **2**.
      Pruning Options: Default.
      This tree is configured to approximate CART. The Gini index is used as the splitting criterion. Because validation is used to select the final model, the default maximum depth of 10 enables a larger sequence of subtrees that will be pruned using the cost-complexity criterion. CART trees tend not to grow as bushy (wide) as CHAID trees that allow for multi-way splits. Parent and child size is reduced from the above CHAID-like settings. The addition of surrogates is also a CART-like feature.
   c. **CHAID-like+Validation**
      Splitting Options: Class target criterion: **CHAID**; Significance Level: **0.01**; Bonferroni: **Yes**; Maximum number of branches: **4**.
      Pruning Options: Subtree method: **Reduced error**.

An advantage of decision tree modeling in SAS Viya is the additional flexibility that is available through a generalized tree node. For example, this tree is a blend of CHAID-like (CHAID criterion, small significance level, multi-way splits, Bonferroni adjustment) and CART-like (post-pruning of larger trees using validation data) settings. (Post-pruning is accomplished by using the reduced error misclassification rate.) The use of the default maximum depth of 10 constrains the tree from growing too large.

d.   **C4.5**

Splitting Options: Default

Pruning Options: Subtree method: **C4.5**.

This tree grows using *depth-first* strategy. Default information gain ratio is the splitting criterion to use for determining best splits on inputs. The C4.5 algorithm considers all the possible tests that can split the data and selects a test that gives the best information gain (that is, highest gain ratio). The C4.5 algorithm allows pruning of the resulting decision trees.

The settings above were applied to illustrate some (but not all) characteristics of available tree types. These settings do not result in exact replication of algorithms. Some settings were also chosen to balance the requirements of satisfactory performance and small tree size (interpretability).

8. Select the last Decision Tree node that was renamed as **Autotuning DT**.
9. Slide the **Perform Autotuning** button to the right (the on position).

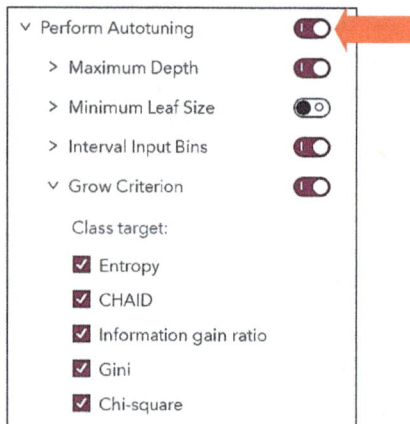

There are reasonable defaults set for these hyperparameters so that you can take no action and use these defaults. However, these defaults are not based on the data. Selecting better hyperparameters using the actual data to improve your model performance can be done by autotuning.

Autotuning will be done for the maximum depth parameter, the number of interval input bins, and the grow criterion. You can specify whether to autotune a hyperparameter (for example, minimum leaf size is not autotuned by default, but you can autotune it).

In general, performing autotuning can increase run time.

10. Click the **Run Pipeline** button.
11. Open the **Results** of the **Model Comparison** node.

Model Comparison				⤓ ⤢
Champion	Name	Algorithm Name	KS (Youden)	Misclassification Rate
⊡	Autotuning DT	Decision Tree	0.4271	0.2590
	CART-like	Decision Tree	0.4203	0.2693
	C4.5	Decision Tree	0.4116	0.2705
	CHAID-like+Validation	Decision Tree	0.3841	0.2698
	CHAID-like	Decision Tree	0.3730	0.2812

Autotuning significantly improved the accuracy of the resulting model with no additional effort from the modeler. An autotuned decision tree outperformed all other trees. However, that might not be true always. It has the highest KS (Youden)-statistic, that was the default objective function to optimize for tuning parameters for a nominal target.

Note: Because of parallel processing in the autotuning process, your results of Autotuning DT might differ from the ones shown above.

12. Switch to the **Assessment** tab and maximize the **Lift Reports**. Change the **View Chart** to **Cumulative Lift** (drop-down) and **Display** to **One Partition** (radio button). Select **Validate** from the drop-down menu. Adjust the view to focus on some top percentages.

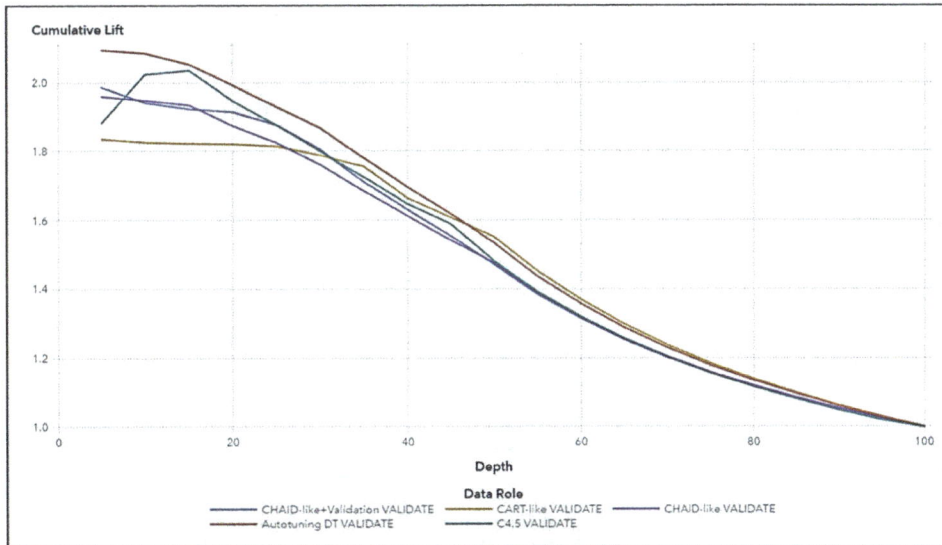

The autotuned tree comes out to be the best. The CHAID-like tree and CHAID-like tree with validation data are also not inferior to the autotuned one, whereas the CART-like fare worse in top 25% and the C4.5 tree is having highly inconsistent behavior in the top 20% of the cases.

13. Close the **Lift Reports** chart.
14. Maximize the **ROC Reports**. Change the **View Chart** to **ROC** (drop-down) and **Display** to **One Partition** (radio button). Select **Validate** from the drop-down menu.

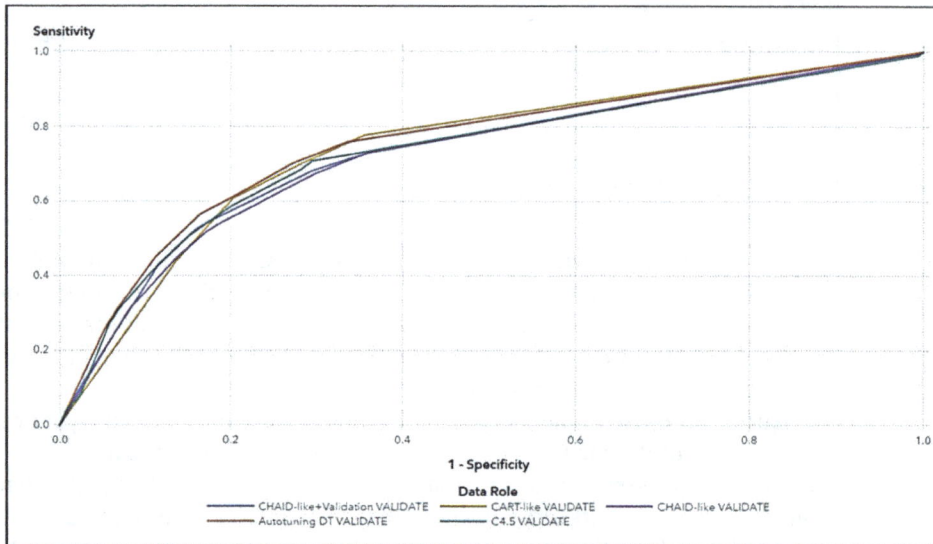

The autotuned tree is outperforming all others. The CART-like and C4.5 trees have better performance than the CHAID-like tree and the CHAID-like tree with validation data.
15. Close the **ROC Reports** chart.
16. Maximize the **Fit Statistics** table. Below are some of the performance statistics on validation data.

Name	Misclassification Rate	Average Squared Error	Gini Coefficient	Area Under ROC
Autotuning DT	0.2590	0.1800	0.5047	0.7523
CART-like	0.2693	0.1839	0.4850	0.7425
C4.5	0.2705	0.1891	0.4609	0.7304
CHAID-like+Validation	0.2698	0.1886	0.4514	0.7257
CHAID-like	0.2812	0.1905	0.4450	0.7225

Although the modeling objective is more focused on classification, other statistics might be of interest to the modeler. If decision making is important, misclassification rate might be most relevant. If a model is to be used primarily for ranking cases, the C-statistic (the area under the receiver operator curve) might be more relevant. If good probability estimates are required, the average square error of estimates should be considered.

The autotuned tree gives the smallest misclassification rate and average square error, as well as the highest ROC index and Gini coefficient.

	CHAID-like	CART-like	CHAID-like + Validation	C4.5	Autotuning DT
Split Criterion	Chaid	Gini	Chaid	IGR	CHISQUARE
Pruning Method	Cost Complexity	Cost Complexity	Reduced Error	C45	Cost Complexity
Max Branches per Node	6	2	4	2	2
Max Tree Depth	6	10	10	10	10
Tree Depth Before Pruning	6	10	8		10
Tree Depth After Pruning	4	6	6	10	6
Number of Leaves Before Pruning	22	83	52		374
Number of Leaves After Pruning	16	9	30	84	16

The C4.5 tree, with 84 leaves, is large compared to the others. The C4.5 algorithm grows the decision tree for all combinations of each attribute, the level of the decision tree generated is high, which generates a substantial number of decision rules. It would have grown even farther if we had not restricted the maximum depth at 10. C4.5 trees are susceptible to outliers.

The CART-like tree has grown truly large (83 leaves), and there are nine leaves after pruning. Many times, CART might have an unstable decision tree. Insignificant modification of the training sample such as eliminating several observations causes changes in decision tree: increase or decrease of tree complexity, changes in splitting variables and values.

The final pruned trees in CHAID-like (with and without validation) and CART-like trees are much simpler, having a depth of only four to six levels as against the C4.5 tree with a depth of 10 levels. The CART-like tree could be preferred even with slightly reduced performance if its size enables simpler interpretations.

Try fitting a tree with the entropy criterion that is used in machine learning (for example, C4.5/5.0) tree algorithms. How does it perform?

17. Minimize the **Fit Statistics** table and close the **Results** of the Model Comparison node.
18. Open the **Results** of the **Autotuning DT** node.
19. Maximize the **Output** window.

Best Configuration	
Evaluation	27
Maximum Tree Levels	11
Maximum Bins	20
Criterion	CHISQUARE
Kolmogorov-Smirnov	0.4271415015

SAS Visual Data Mining and Machine Learning automated the selection of hyperparameter values using an intelligent optimization-based methodology. Chi-square criterion comes out to be a preferred splitting criterion in cost-complexity pruning.

20. Close the **Results** of the **Autotuning DT** node.

End of Demonstration

Quiz

1. Which of the following are pre-pruning decision tree options that limit tree growth?
   a.   significance level
   b.   maximum depth
   c.   minimum leaf size
   d.   all of the above

2. How can you avoid overfitting in decision tress?
   a.   using CHAID for stopping the tree growth and no pruning after that (a purely pre-pruning approach)
   b.   pruning the maximal tree and not using tree growth stopping before that (a purely post-pruning approach)
   c.   using CHAID for stopping the tree growth and then pruning that tree (a hybrid approach of pre-pruning and post-pruning)
   d.   all the above

3. Which of the following statements is false?
   a.   Pruning a tree using average square error gives you a result that is equivalent to using an overall Gini index.
   b.   Ten-fold cross validation is a feature of the CART methodology.
   c.   Subtree selection using validation data can be eliminated by choosing Method=Largest or Method=N in the Method property.
   d.   None of the above

4. If you do not have any clue about the decision tree hyperparameters, autotuning could be a good starting point.
   a.   True
   b.   False

Answers

| 1. d | 2. d | 3. c | 4. True |

Chapter 6: Ensemble of Trees: Bagging, Boosting, and Forest

"Small changes in the training data can cause large changes in the topology of the tree. However, the overall performance of the tree remains stable."

– Breiman et al. (1984)

Instability in a Decision Tree

Decision trees are among the simplest, yet most powerful, forms of models, both conceptually and in their interpretation, with the entire data set of observations being recursively partitioned into subsets along different branches so that similar observations get grouped together at the terminal leaves of the decision tree. However, the methods of constructing decision trees from data suffer from a major problem of instability. The symptoms of instability include variations in the predictive accuracy of the model and in the model topology. Instability can be revealed not only by using disjoint sets of data, but even by replacing a small portion of training cases, like in the cross-validation procedure. If an algorithm is unstable, the cross-validation results become estimators with high variance, which means that an algorithm fails to make a clear distinction between persistent and random patterns in the data. The instability problem of decision tree classification algorithms is that minor changes in input training samples can cause dramatically large changes in output classification rules.

Decision trees are unstable models. That is, minor changes in the training data can cause large changes in the topology of the tree. However, the overall performance of the tree remains stable (Breiman et al. 1984). For example, Figure 6.1 shows a scatter plot of two inputs and a binary target with the two outcomes green and gold. The plot also contains lines showing the results of recursive partitioning. This tree has an accuracy of 81%. Changing the outcome of one case from green to gold results in a completely different tree, as shown by the change in the results of recursive partitioning. However, the accuracy is almost identical: 80%.

The instability results from the substantial number of univariate splits that are considered and the fragmentation of the data. At each split, there are typically several splits on the same and

Figure 6.1: Instability Illustration

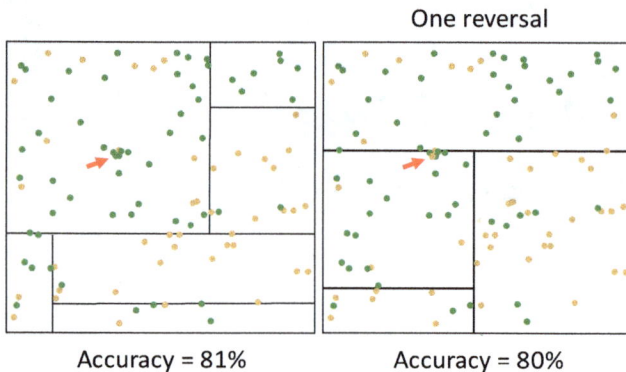

Accuracy = 81%   Accuracy = 80%

Figure 6.2: Competitor Splits

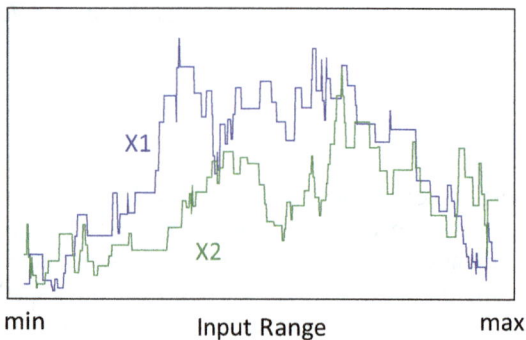

different inputs that give similar performance (competitor splits) as represented in Figure 6.2. This plot shows the logworth for all values of two input variables, X1 and X2, which are candidates for a single split. The input range is on the X axis and logworth is on the Y axis. A slight change in the data can easily result in a different split being chosen. This produces different subsets in the child nodes. The changes in the data are larger in the child nodes. The changes continue to cascade down the tree.

Perturb and Combine (P&C)

Methods were devised to take advantage of the instability of trees to create models that are more powerful. *Perturb and combine (P&C)* methods generate multiple models by manipulating the distribution of the data or altering the construction method, such as changing the tree settings, and then averaging the results (Breiman 1998). Any unstable modeling method can be used, but trees are most often chosen because of their speed and flexibility.

Perturb and combine methods are used to create ensemble models in two steps.

Step 1: Perturb

The perturb step creates different models by manipulating the distribution of the data or altering the construction method.

The three pairs of plots shown in Figure 6.3 represent three different trees created during the perturb step. These trees are based on data shown earlier, which has a binary target (with green and gold outcomes) and two inputs. For each tree, the top plot is a scatterplot of the data with lines showing the splits. The bottom plot shows the decision predictions from the tree. Notice that the data points are identical in all three scatterplots. From this, we can infer that the perturb method in this example involves altering the construction method, such as changing the splitting criterion, instead of manipulating the data distribution.

Some common perturbation methods include:

- resample
- subsample
- add noise
- adaptively reweight
- randomly choose from the competitor splits.

Step 2: Combine

The combine step then uses the predictions of the models built in the perturb step to create a single prediction. Different models then can be combined to create a unison model, a concept typically called *ensembling*. An ensemble model is the combination of multiple models. The combinations can be formed with the following methods:

- voting on the classifications
- weighted voting where some models have more weight
- averaging (weighted or unweighted) the predicted values

Figure 6.3: Perturb Step

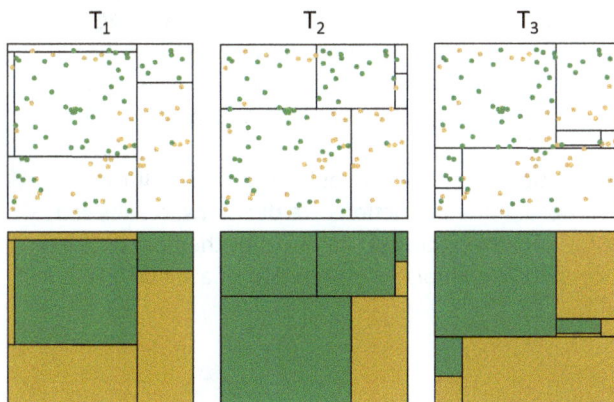

Figure 6.4: Combine Step

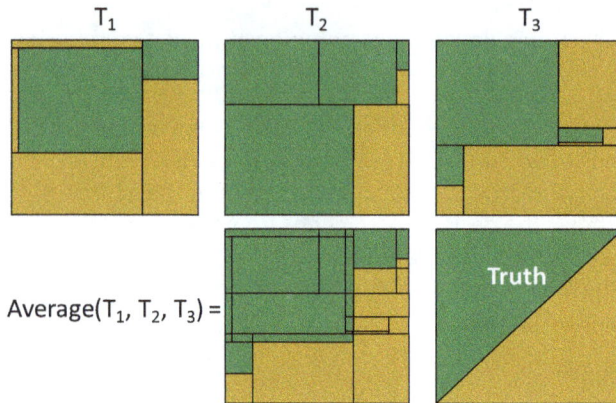

Perturb and combine methods can be used with any unstable algorithm but they are mostly used with trees.

Figure 6.4 contains a graph of the ensemble model for the considered example. Graphical representations show that ensembles of trees have decision boundaries of much finer resolution than would be possible with a single tree. The decision boundaries of a single tree can be viewed as step functions. The benefit of ensembling many trees together is that, by adding more steps, the steps themselves are smoothed out.

Ensemble methods are a very active area of research in the fields of machine learning and statistics. Many other P&C methods were devised.

The attractiveness of P&C methods is their improved performance over single models. Bauer and Kohavi (1999) demonstrated the superiority of P&C methods with extensive experimentation. One reason why simple P&C methods give improved performance is variance reduction. If the base models have low bias and high variance, then averaging decreases the variance. In contrast, combining stable models can negatively affect performance. The reasons why adaptive P&C methods work go beyond simple variance reduction. They are the topic of much research (for example, see Breiman 1998). Graphical explanations show that ensembles of trees have decision boundaries of much finer resolution than would be possible with a single tree (Rao and Potts 1997).

A new case is scored by running it down the multiple trees and averaging the results. If the predictions are estimates, the results are averaged. If the predictions are decisions, results can be combined by voting or averaging the posterior probabilities as the basis for the final decision. Multiple models need to be stored and processed. The simple interpretation of a single tree is lost.

"Bagging goes a long way towards making a silk purse out of a sow's ear, especially if the sow's ear is twitchy. ...What one loses, with the trees, is a simple and interpretable structure. What one gains is increased accuracy."

– Breiman (1996)

Ensemble Models

The ensemble approach relies on combining many relatively weak simple models to obtain a stronger ensemble prediction. Ensembling creates a new model by combining the predictions from multiple models to improve accuracy, reduce bias and variance, and provide robust models in the presence of new data. For prediction estimates and rankings, this is usually done by averaging. When the predictions are decisions, this is done by voting.

Ensembling is in fact an aggregation of more than one model where the final prediction is a combination of the predictions from the component models. An ensemble model creates a single consensus prediction as illustrated in Figure 6.5 by a bold dark blue line along with many component models represented by thin gray lines given the dots as observed points.

The commonly observed advantage of ensemble models is that the combined model is better than the individual models that compose it. It is important to note that the ensemble model can be more accurate than the individual models only if the individual models disagree with one another.

Bagging and boosting are common ensemble methods.

Bagging

Bagging consists of creating several samples to train models (for example, decision trees) in parallel and then combining their predicted probabilities as represented in Figure 6.6. The idea is to average many noisy but unbiased models to reduce the variance of estimated prediction function.

Figure 6.5: Ensemble Model

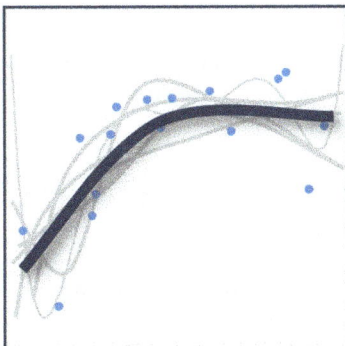

Figure 6.6: Parallel Ensemble Approach

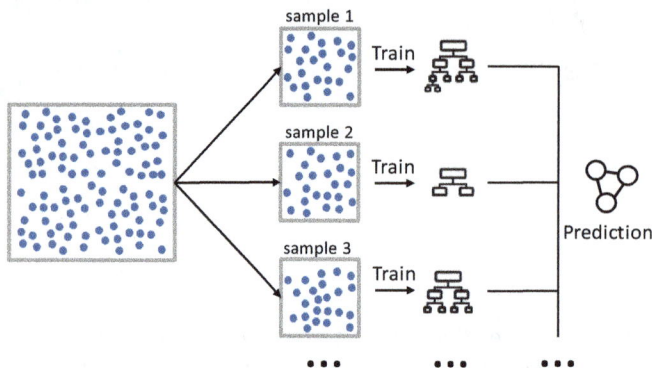

Bagging, as the name alludes, takes repeated unweighted samples with replacement of the data to build models and then combine them. Think of your observations like grains of wild rice in a bag. Your objective is to identify the black grains because they have a resale price 10x greater when sold separately.

1. Take a scoop of rice from the bag.
2. Use your scoop of rice to build a model based on the grain's characteristics, excluding that of color.
3. Write down your model classification logic and fit statistics.
4. Pour the scoop of rice back into the bag.
5. Shake the bag for good measure and repeat.

How big the scoop is relative to the bag, and how many scoops you take, will vary by industry and situation, but you can use 25–30% of your data and take 7–10 samples. This results in a likelihood that every observation will be included one to two times in the model.

(The above rice example is adapted from a SAS blog "How ensemble models help make better predictions" by Jared Dean, https://blogs.sas.com/content/subconsciousmusings/2014/05/01/ how-ensemble-models-help-make-better-predictions/*)*

> Bagging stands for "bootstrap aggregating."

Bootstrapping is a statistical resampling technique that involves random sampling of a data set with replacement. It is often used as a means of quantifying the uncertainty associated with a machine learning model.

Bagging (*bootstrap aggregation*) is the original P&C method (Breiman 1996).

1. Draw *K* bootstrap samples.
   A *bootstrap sample* is a random sample of size *n* drawn from the empirical distribution of a sample of size *n*. That is, the training data are resampled with replacement. Some

Figure 6.7: Bagging (Bootstrap Aggregation) Example

case	k=1 freq	k=2 freq	k=3 freq	k=4 ... freq
1	1	0	3	1
2	0	1	1	1
3	2	0	0	2
4	0	2	2	0
5	2	2	0	1
6	1	1	0	1

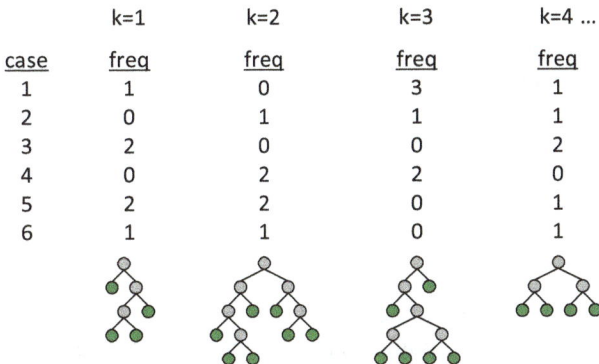

of the cases are left out of the sample, and some cases are represented more than once.

2. Build a tree on each bootstrap sample.
   Pruning can be counterproductive (Bauer and Kohavi 1999). Large trees with low bias and high variance are ideal.
3. Combine the predictions by voting or averaging.
   For classification problems, take the mean of the posterior probabilities or take the plurality vote of the predicted class. Bauer and Kohavi (1999) found that averaging the posterior probabilities gave slightly better performance than voting. Take a mean of the predicted values for regression.
   Breiman (1996) used 50 bootstrap replicates for classification and 25 for regression and for averaging the posterior probabilities. Bauer and Kohavi (1999) used 25 replicates for both voting and averaging.

A simple example is illustrated in Figure 6.7. From a training data set that has six cases, the first step is to draw four bootstrap samples. Each sample has six cases. For bagging, a bootstrap sample is drawn with replacement. This means that some of the cases might be left out of the sample, and some cases might be represented more than once. For example, in the first bootstrap sample, cases 3 and 5 appear twice, cases 1 and 6 appear once, and cases 2 and 4 do not appear at all.

Next, a tree on each bootstrap sample is built. In the bagging method, trees are grown independently of each other so that they can be built in parallel in a distributed computing environment like SAS Viya. This makes bagging a fast method compared to other P&C methods. Lastly, the predictions from each tree are combined by voting or averaging.

Boosting

One of the biggest challenges in creating decision trees is selecting the optimal size of the tree, particularly when data is scarce or target events are rare. The idea of *boosting* was proposed as a sequential iteration of several decision trees, without the need to prune them. This algorithm

Figure 6.8: Sequential Ensemble Approach

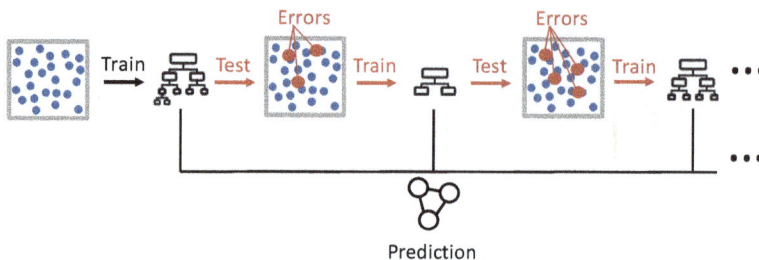

was formulated under the premise that trees do not need to be very strong predictors because misclassified observations are used to train models over iterations, and the models learn to catch misclassified observations. A classification tree or a regression tree or for that matter any other predictive model is called a "weak" learner when it performs poorly – its accuracy is above chance, but just barely. Many instances of such weak learners are then being pooled (via boosting, bagging, and so on) together to create a "strong" ensemble model.

Boosting is like bagging except that the observations in the samples are now weighted. To follow the previous rice problem, after step 3 you would take the grains of rice that you had incorrectly classified (for example, black grains that you said were non-black or non-black grains that you thought were black) and place them aside. You would then take a scoop of rice from the bag and leave some room to add the grains that you had incorrectly classified. By including previously misclassified grains at a higher rate, the algorithm has more opportunities to identify the characteristics for correct classifications. This is the same idea behind giving more time to review flashcards of facts that you did not know than those you did.

For what it is worth, we tend to use bagging models for prediction problems and boosting for classification problems. By taking multiple samples of the data and modeling over iterations, you allow factors that are otherwise weak to be explored. This provides a more stable and generalizable solution. When model accuracy is the most important consideration, ensemble models will be your best bet.

> Boosting stands for an approximation of the sequential bootstrap aggregation.

> "Boosting is a machine learning ensemble meta-algorithm for primarily reducing bias, and also variance."
>
> *– Breiman (1996)*

> "The term boosting refers to a family of algorithms that are able to convert weak learners to strong learners."
>
> *– Zhou (2012)*

Figure 6.9: Boosting (Arcing) Example

	k=1		k=2		k=3		k=4 ...
case	freq	m	freq	m	freq	m	freq
1	1	1	1.5	1	.5	2	.97
2	1	0	.75	0	.25	0	.06
3	1	1	1.5	2	4.25	3	4.69
4	1	0	.75	1	.5	1	.11
5	1	0	.75	0	.25	0	.06
6	1	0	.75	0	.25	1	.11

Arcing (adaptive resampling and combining) methods are examples of boosting. They sequentially perturb the training data based on the results of the previous models. Cases that are incorrectly classified are given more weight in subsequent models. Arc-x4 (Breiman 1998) is a simplified version of the AdaBoost (adaptive boosting, also known as Arc-fs) algorithm of Freund and Schapire (1996). Both algorithms give similar performance (Breiman 1998, Bauer and Kohavi 1999).

At the kth step, a model (decision tree) is fit using weights for each case. For the ith case, the Arc-x4 weights (that is, the selection probabilities) are

$$p(i) = \frac{1 + m(i)^4}{\sum \left(1 + m(i)^4 \right)},$$

where $0 \leq m(i) \leq k$ is the number of times that the ith case is misclassified in the preceding steps. Unlike bagging, pruning the individual trees improves performance (Bauer and Kohavi 1999).

Each tree that boosting creates is dependent on the tree from the previous iteration. Across iterations, as shown in Figure 6.9, the algorithm keeps a cumulative count of misclassifications. Each case is then weighted based on whether its misclassification count increases in the current iteration. If the misclassification count increases, the weight increases. If the misclassification count remains the same, the weight decreases. The weights influence the likelihood that the case is selected in the sample for the next iteration.

The weights are incorporated either by using a weighted analysis or by resampling the data such that the probability that the ith case is selected is $p(i)$. For convenience, the weights can be normalized to frequencies by multiplying by the sample size, n. Bauer and Kohavi (1999) found that resampling performed better than reweighting for arc-x4 but did not change the performance of AdaBoost. AdaBoost uses a different (more complicated) formula for $p(i)$. Both formulas put greater weight on cases that are frequently misclassified.

The process is repeated *K* times, and the *K* models are combined by voting or averaging the posterior probabilities. AdaBoost uses weighted voting where models with fewer misclassifications, particularly of the hard-to-classify cases, are given more weight. Breiman (1998) used *K*=50. Bauer and Kohavi (1999) used *K*=25.

Arcing improves performance to a greater degree than bagging, but the improvement is less consistent (Breiman 1998, Bauer and Kohavi 1999).

Comparison of Tree-Based Models

It would be interesting to compare the accuracy of three tree-based models: a single decision tree, a tree-based ensemble model based on bagging, and a tree-based ensemble model based on boosting. The three plots shown in Figure 6.10 are based on a data set with two inputs.

A decision tree model is created through recursive partitioning, so all tree-based models use step functions to split the input space. The plot for the single decision tree shows that there are few steps, and those steps are large. Therefore, few steps are used to classify the target, which can decrease the accuracy.

The plots for the two ensemble models have more steps and those steps are smaller. Because more steps are used to classify the target, accuracy can be increased compared to a single tree.

Notice that boosting smooths the prediction surface more than bagging because boosting emphasizes the misclassified cases during the training.

Bagging and boosting decrease the variance of your single estimate as they combine several estimates from different models. Therefore, the result might be a more robust model. It is

Figure 6.10: Single, Bagged, and Boosted Tree

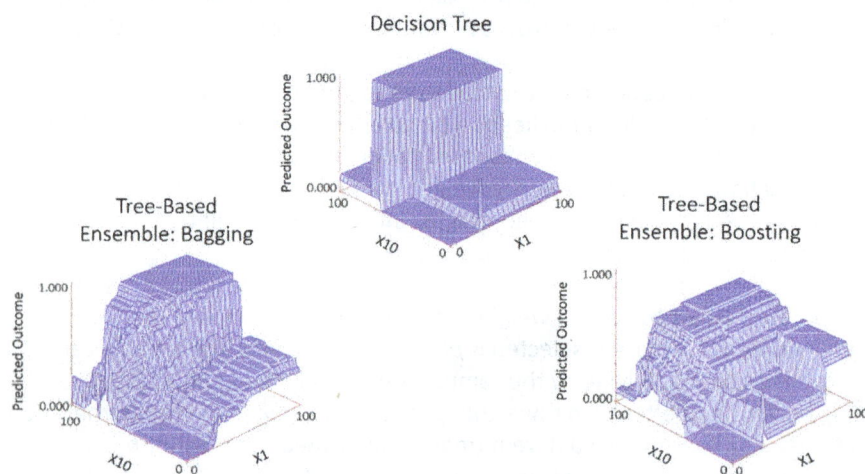

Table 6.1: Bagging versus Boosting

	Bagging	Boosting
Sampling	Randomly chosen unweighted cases	Higher weight for misclassified cases
Goal	Minimize variance	Increase accuracy
Method used	Random subspace	Gradient descent
Combining models	Weighted average	Weighted majority weight
Example Algorithms	Forest	Gradient Boosting

generally recommended that you use bagging when the goal is to reduce the variance of a decision tree classifier. Boosting is used to create a collection of predictors. Table 6.1 summarizes bagging and boosting features against salient characteristics.

If the problem is that the single model gives a very low performance, bagging will seldom get a *low bias*. However, boosting could generate a combined model with smaller errors as it optimizes the advantages and reduce drawbacks of the single model. On the other hand, if the difficulty of the single model is *overfitting,* then bagging is the best option. Boosting might not help to avoid overfitting; in fact, this technique is faced with this problem itself.

> Bagging and boosting are nondeterministic algorithms, and therefore your results might differ from those presented in the upcoming demonstrations in chapters 6 through 9.

Forest Models

Leo Breiman (2011) contributed to the work on how classification and regression trees and ensemble of trees fit to bootstrap samples (bagging). This paid off in part for bridging the gap between statistics and computer science in machine learning.

A forest model is essentially a bagging algorithm that creates a predictive model by ensemble of classification or regression trees. A forest model is an ensemble learning method for classification and regression tasks that operates by constructing a multitude of decision trees at training time and predicting the class that is most of the classes (classification) or average prediction (regression) of the individual trees.

The need for random forest models surfaced after discovering that the bagging algorithm results in correlated trees. Let us say a data set has a very strong predictor, along with other moderately strong predictors. In bagging, a tree grown every time would consider the very strong predictor at its root node, thereby resulting in trees like each other. Unfortunately, averaging several highly correlated trees does not lead to a significant reduction in variance.

The main difference between random forest and bagging is that random forest induces more variation in trees by considering only a subset of predictors at a split. This results in trees with different predictors at top split, thereby resulting in uncorrelated trees and more reliable average output.

Combining Trees in a Forest

Tuning your forest is an important step in your modeling process in order to obtain the most accurate, useful, and generalizable model. The Forest node in Model Studio or in general the FOREST procedure in SAS Viya provides the ability to tune your random forest through several options. SAS has set the defaults for these options to be most generally effective, but further adjustment can usually lead to better model accuracy. So that leads us to the main point of this tip: what are the most effective ways to tune a random forest? Two of these most important features are characterized in Figure 6.11.

First, a forest is an ensemble of many decision trees, so it stands to reason that the number of trees will have a significant effect on the resulting model accuracy. The number of trees has a significant positive effect on the model accuracy. That is, as you add more trees, your ensemble model becomes more accurate and generalizable. The **Number of trees** property specifies the number of trees in the forest. The default value is 100.

Another important consideration is how to calculate the predicted probability of the target levels for a nominal target. The predicted level is the level that has the highest predicted probability. This option affects the scoring and fit statistics of the forest model. The **Class target voting**

Figure 6.11: Important Characteristics for Combining Trees in Forest

Two distinguishing characteristics of decision trees ensembled in a forest:

- *How many trees to train to make the forest?*

- *How to combine the predictions from different trees?*

Number of trees:
40

Class target voting method:
Probability
Majority
Probability

method property specifies the method to use for calculating the predicted probability for a class target. The possible values include probability (default) and majority.

The **Probability** option specifies that the predicted probability of each target level is equal to the probability of that level averaged over each tree in the forest.

The **Majority** option specifies that the predicted probability of each target level is equal to the number of trees in the forest that predicted that level as the target, divided by the total number of trees in the forest.

Perturbing Trees in a Forest

Recall that bagging takes bootstrap samples of the rows of training data. All columns are considered for splitting at every step. The forest algorithm samples the rows *and* the columns at each step. The difference in trees in a forest model is illustrated in Figure 6.12. The forest algorithm perturbs the training data more than the bagging algorithm. This increased variation among the trees in the ensemble often leads to improved predictive accuracy.

The decision trees in a random forest are overtrained by letting them grow to a large depth and small leaf size. The theory behind this approach is that averaging the predicted probabilities of many overtrained trees is more robust than using a single fine-tuned decision tree.

The data used to train each tree is a random sample of the complete data set (with replacement). The input variables that are considered for splitting each node are a random subset of all variables (as opposed to all variables being candidates for defining the splitting rule), reducing bias toward the most influential factors and allowing secondary factors to play a role in the model.

The variation among the trees can be controlled by properties like the ones described in Figure 6.13.

Figure 6.12: Difference in Trees

Trees in the forest differ from each other in two ways.

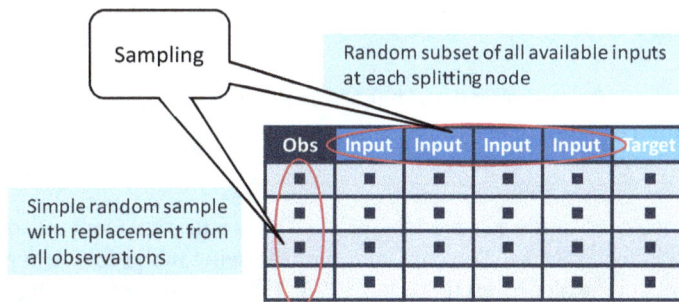

Figure 6.13: Important Characteristics for Perturbing Trees in a Forest

Perturbation in trees can be
regulated by the following
characteristics:

- *What proportion of the training
  observations should be selected to
  train a tree?*

- *How many variables should be
  considered for each split?*

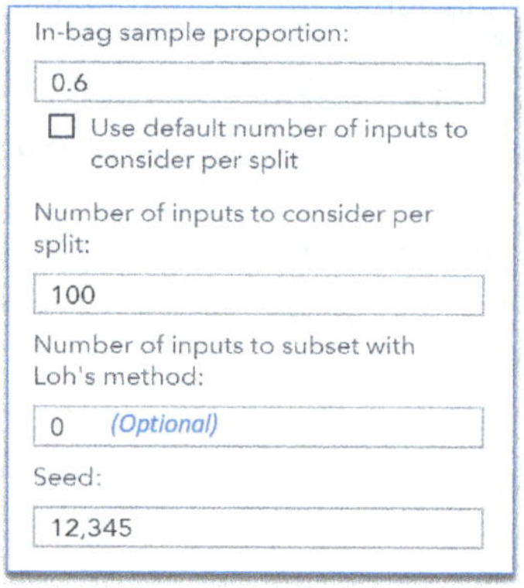

In-bag sample proportion:

0.6

☐ Use default number of inputs to
consider per split

Number of inputs to consider per
split:

100

Number of inputs to subset with
Loh's method:

0   *(Optional)*

Seed:

12,345

The **In-bag sample proportion** property specifies the fraction of the random bootstrap sample
of the training data to be used for growing each tree in the forest, where the number is a value
between 0 and 1. The default value is 0.6.

The **Use default number of inputs to consider per split** property specifies whether to use the
default number of inputs to consider per split, that is, the square root of the number of inputs.
By default, this option is selected.

The **Number of inputs to consider per split** property specifies the number of input variables that
are randomly sampled to use per split. This option is available only if you deselect **Use default
number of inputs to consider per split**. The default value is 100.

The **Number of inputs to subset with Loh's method** property specifies the number of input
variables to further sample using the Loh's method. When set to 0, no further sampling of inputs
is performed. The default value is 0.

The **Seed** property specifies the seed for generating random numbers. This option is used to
select training observations for each tree and to select candidate variables in each node to split
on. The default value is 12345.

Reducing Correlation of the Predictions of the Trees

A decision tree in a forest trains on new training data that are derived from the original training
data. Using different data to train different trees reduces the correlation of the predictions of the
trees, which in turn should improve the predictions of the forest.

As illustrated in Figure 6.14, the Forest node samples the original data with replacement to
create the training data for an individual tree. The convention of sampling with replacement
originated with Leo Breiman's bagging algorithm (Breiman 1996, 2001). The word bagging stems

Figure 6.14: In-Bag and Out-Of-Bag Samples

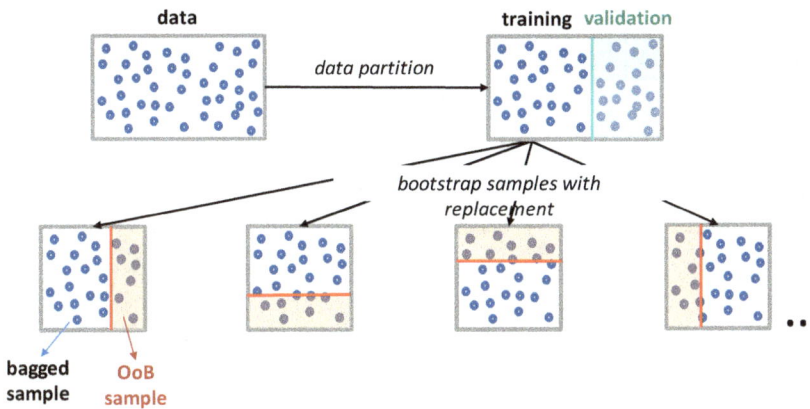

from "bootstrap aggregating," where "bootstrap" refers to a process that uses sampling with replacement. Breiman refers to the observations that are excluded from the sample as out-of-bag (OOB) observations. Therefore, observations in the training sample are called the bagged observations, and the training data for a specific decision tree are called the bagged data.

Estimating the goodness of fit of the model by using the training data is usually too optimistic; the fit of the model to new data is usually worse than the fit to the training data. Estimating the goodness of fit by using the out-of-bag data is usually too pessimistic at first. With enough trees, the out-of-bag estimates are an unbiased estimate of the generalization fit. Some model assessments such as the iteration plots are computed using the out-of-bag sample as well as all the training and validation data.

Loh's Method

The Forest node trains a decision tree by splitting the bagged data, then splitting each of the resulting segments, and so on recursively until some constraint is met.

Splitting involves the following subtasks:

1. selecting candidate inputs
2. computing the association of each input with the target
3. searching for the best split that uses the most highly associated inputs

Figure 6.15 represents this process. For example, if you have j input variables in the training data, the Forest node randomly selects m candidate input variables independently in every node, where m is the value of the **Number of inputs to consider per split** property. If you specify L as the value of the **Number of inputs to subset with Loh's method** property and $L < m$, then the node chooses the best L input variables from the m variables according to the criterion described by Loh in a series of papers (Loh and Shih 1997; Loh 2002, 2009). This method selects

Figure 6.15: Loh's Method

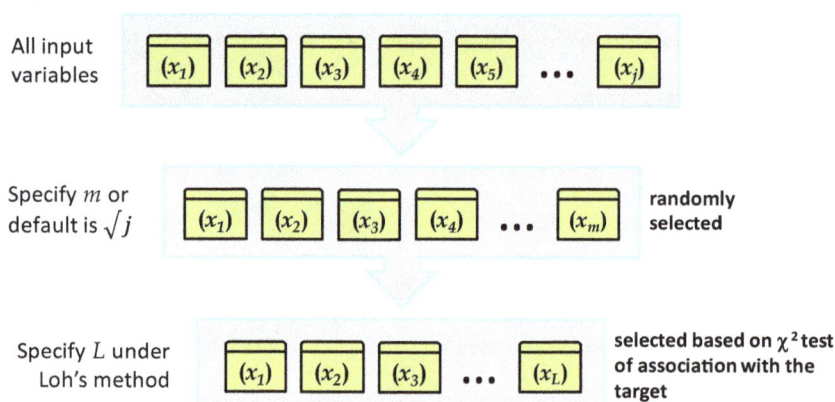

the number of variables that have the smallest p-value of a chi-square test of association in a contingency table. These variables are selected from the *m* randomly selected variables that are chosen for a single decision tree in the forest.

A split search is performed on all *L* or *m* variables, and the best rule is kept to split the node. The reason for searching fewer input variables for a splitting rule instead of searching all inputs and choosing the best split is to improve the prediction on new data. An input that offers more splitting possibilities provides the search routine more chances to find a spurious split. Loh and Shih (1997) demonstrate the bias toward spurious splits that result. They also demonstrate that preselecting the input variable and then searching only on that one input reduces the bias. You can choose to preselect a few input variables by using the **Number of inputs to subset with Loh's method** option.

The split search seeks to maximize the reduction in the gain for a nominal target and the reduction in variance of an interval target.

Splitting Options You Already Know

By now you are already familiar with the splitting options for individual trees. They have been discussed in Chapter 3. Here is a quick discussion of such options in the context of forest models.

Splitting Criteria

Figure 6.16 represents splitting criteria options in the Forest node in Model Studio. Visibly, many of the tree splitting options in the Forest node are similar to that of the Decision Tree node. You already know the following tree split options.

Class target criterion and **Interval target criterion** specify the splitting criterion to use for determining the best splits on inputs given a class/interval target. Exactly the same options are

Figure 6.16: Class and Interval Target Criteria

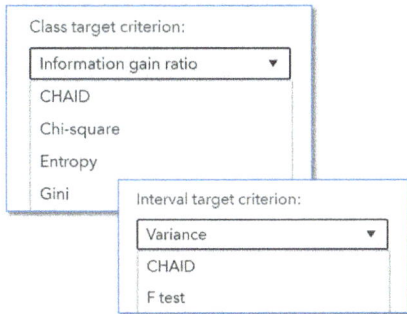

available that you have seen in decision tree, that is, **CHAID**, **Chi-square**, **Entropy**, **Gini**, and **Information gain ratio** for a class target and **CHAID**, **F test**, and **Variance** for an interval target.

Branches and Depth

Figure 6.17 represents the number of branches and depth options in the Forest node in Model Studio.

Maximum number of branches specifies the maximum number of branches to consider for each node split in the tree. Possible values range from 2 to 5. The default value is 2.

Maximum depth specifies the maximum depth for each generated tree within the forest. Possible values range from 1 to 50. The default value is 20.

Leaf Size

Another important tuning parameter in forests is the minimum leaf size, which specifies the smallest number of observations a node can have in a decision tree. If the splitting rule results in a child node with fewer observations than this number, the node is not split. In a Decision Tree node, you control this by specifying the minimum leaf size. Here in forests, this can be controlled

Figure 6.17: Number of Branches and Depth Options

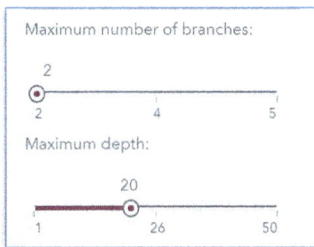

Figure 6.18: Leaf Size Options

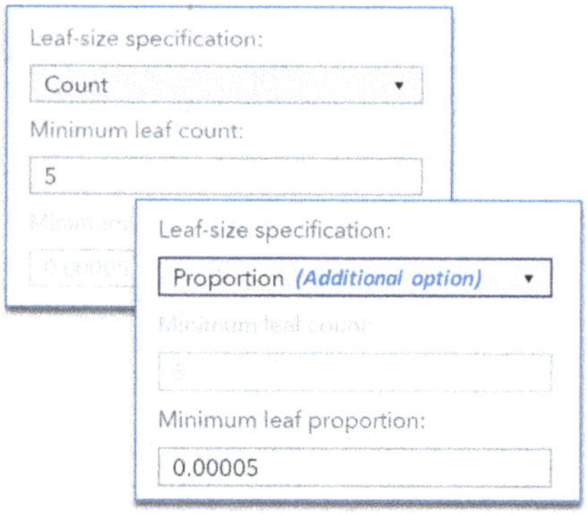

either by setting an absolute value or as a percentage of the total set of observations in the node being split. This is represented in Figure 6.18.

Leaf-size specification specifies the method for determining the minimum leaf size. Possible values are **Count** and **Proportion**. The default value is **Count.**

Minimum leaf count specifies the smallest number of training observations that a new branch can have. That number is expressed as the count of the available observations. This option is available when **Leaf-size specification** is set to **Count**. Possible values range from 1 to 64. The default value is 5.

Minimum leaf proportion specifies the smallest number of training observations that a new branch can have. That number is expressed as a fraction of the training observations with a nonmissing target value. This option is available when **Leaf-size specification** is set to **Proportion**. The default value is 0.00005.

Missing Values

Missing values are treated in a similar fashion in each of the trees in a forest model as in an individual decision tree. However, here missing observations cannot be assigned in a separate or largest or most correlated branch. Figure 6.19 represents the available options.

Figure 6.19: Missing Values Options

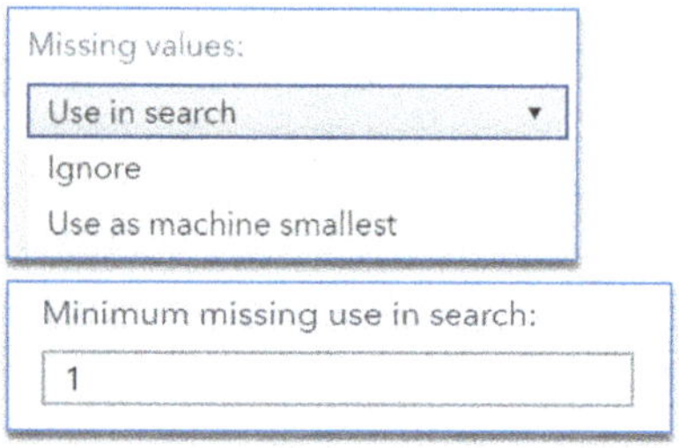

Missing values specifies the action to take when there are values missing. There are limited options for handling missing values in forest as compared to decision tree. You can specify **Ignore, Use as machine smallest,** or **Use in search**. The default value is **Use in search**.

Minimum missing use in search specifies a threshold for using missing values in the split search. The default value is 1.

Interval Binning

The same binning options for interval inputs are available in the Forest node as what you have seen in Decision Tree node. Figure 6.20 represents these options.

Number of interval bins specifies the number of bins that are used for interval inputs. Bin size is (maximum value – minimum value)/(Number of interval bins). The default value is 50.

The interval bin method specifies the method used to bin the interval input variables. Select **Bucket** to divide input variables into evenly spaced intervals based on the difference between maximum and minimum values. Select **Quantile** to divide input variables into approximately equal sized groups. The default value is **Quantile**.

Measuring Variable Importance

The importance of a variable is the contribution that it makes to the success of the model. Variable importance is also useful for selecting variables for a subsequent model. The comparative importance between the selected variables does not matter. Researchers often seek speed and simplicity from the first model and seek accuracy from the subsequent model. Despite this tendency, a forest is often more useful than a simpler regression as a first model when you want interactions because variables contribute to the forest model through interactions.

Several authors have demonstrated that using a forest to first select variables and then using only those variables in a subsequent forest produces a final forest that predicts the target better than only training a forest without the variable selection.

As described in Figure 6.21, the variable importance for the forest is the average of the variable importance across all trees in forest model. The forest model in SAS Viya implements three methods for computing variable importance, which were described in Chapter 4. By default,

Figure 6.20: Binning Options

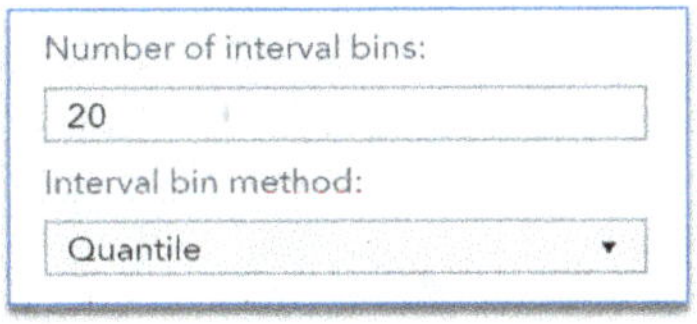

Figure 6.21: Variable Importance in Forest Model

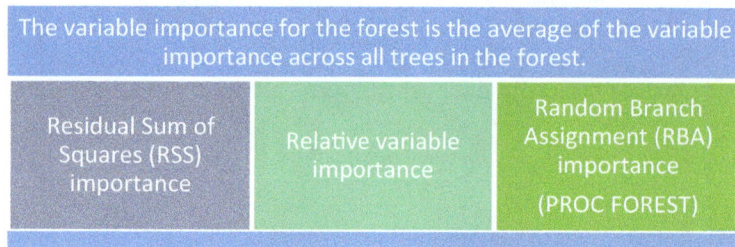

The variable importance for the forest is the average of the variable importance across all trees in the forest.		
Residual Sum of Squares (RSS) importance	Relative variable importance	Random Branch Assignment (RBA) importance (PROC FOREST)

the variable importance is calculated by using the change in the residual sum of square errors. You can use the FOREST procedure to calculate the variable importance by random branch assignment (RBA) by specifying the RBAIMP option.

For a single tree in a forest, the RSS-based metric measures variable importance based on the change in RSS when a split is found at a node. The RSS variable importance for the forest is the average of the RSS variable importance across all trees in the forest.

The random branch assignment (RBA) method computes the importance of an input variable by comparing how well the data fits the predictions before and after they are modified. For an interval target, the RBA importance of squared error loss is computed. For a nominal target, the misclassification rate as the loss function is computed.

Demo 6.1: Building a Default Forest Model

In this demonstration, you create a forest model using Forest node in Model Studio. You continue working with the **insurance** data set and observe some of the default options in the Forest node.

1. Ensure that the **Insurance_ClassTree** project is open in Model Studio.
2. Click the **Pipelines** tab and press (**+**) to add a new pipeline.
3. In the New Pipeline window, name the pipeline **Forest Models**. Ensure that **Blank Template** is selected under **Template**.
4. Click **OK**.
5. Right-click the **Data** node and select **Add child node** ⇨ **Supervised Learning** ⇨ **Forest**.
6. Click the **Run Pipeline** button to produce a forest model with all the default settings.
7. Open the **Results** of the Forest node. Maximize the **Error Plot** window.
   A forest model with 100 trees is created.

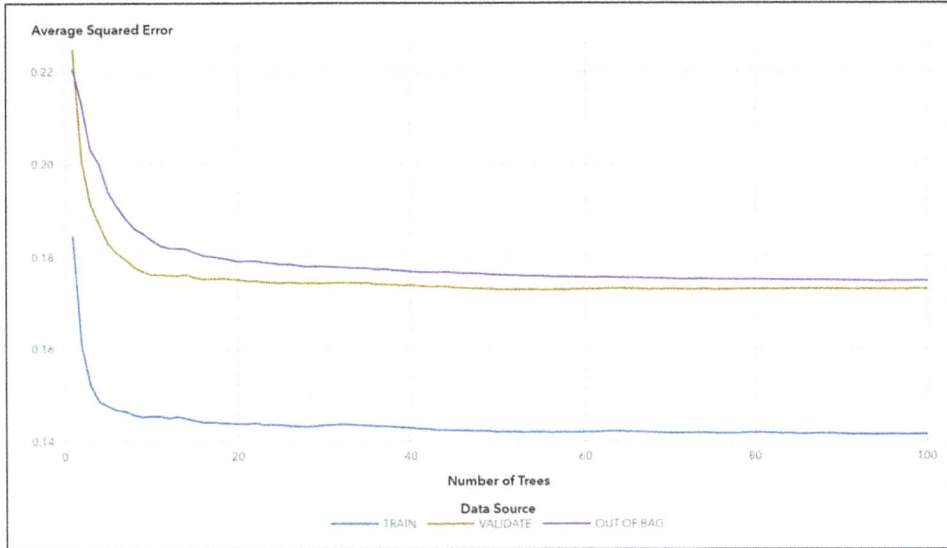

The error plot displays the error statistics for the forest model, giving the average squared error as a function of the number of trees. The average squared error is given for each of the data roles, as well as for the out-of-bag data. The training error typically decreases as the number of trees increases, but the error for the VALIDATE partition gives you an indication of how well your model generalizes.

8. From the drop-down menu in the upper left corner, change the error plot from Average Squared Error to **Misclassification Rate** to examine the misclassification rate as a function of the number of trees because you have a binary target.

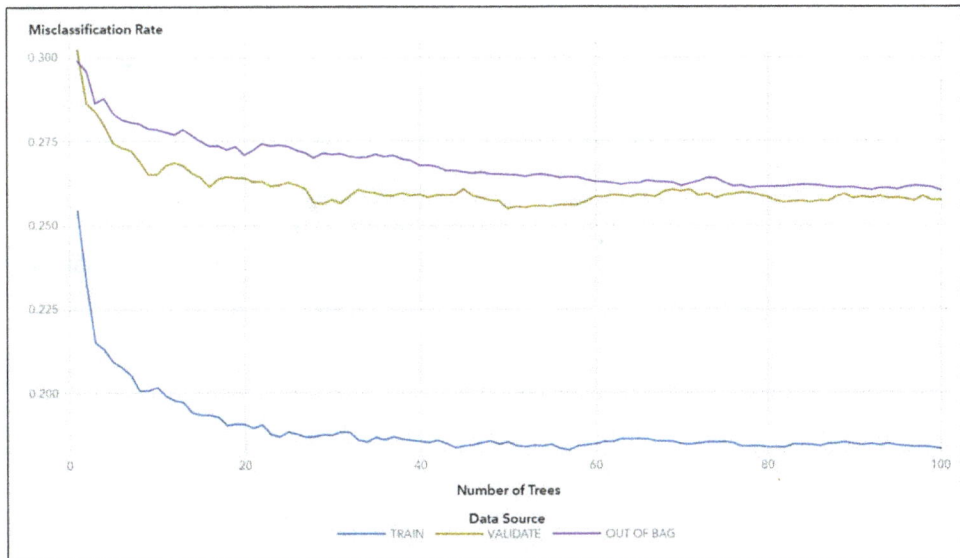

The plot contains the misclassification rate for the ensemble of number of trees as a single predictive algorithm. This plot confirms what you expect – apart from the individual decision trees being overfit by design (significant difference between train and out-of-bag), as you add more trees, your ensemble model becomes more accurate and generalizable.

Note that there is an obvious point of diminishing returns (flattening of the curve) past around 50 trees, and by about 60 trees, you are not gaining much accuracy for the extra cost (time and memory) of training. These rough thresholds will be different for each problem, and it is your decision how finely you want to modify the model by adjusting other hyperparameters. The point is to generalize well, so fine tuning beyond some minuscule gain (in MISC, or other measure) is not worthwhile because the trees are independent (unlike gradient boosting), and a forest with 60 trees would be expected to perform equally well as one with 100 trees. You would want to verify this by running a separate forest with the number of trees set to 60.

Forests generally do not overfit. The out-of-bag (or validation) performance of forest does not decrease (due to overfitting) as the number of trees increases. Hence, after a certain number of trees the performance tends to stay around a certain value. This is quite evident from the chart above.

At any number of trees on the X axis above, you can score the training data based on that many trees in your forest model. The misclassification rate is calculated by aggregating over all observations classified correct/incorrect using the majority voting rule for each observation at the terminal node that each observation belongs to. For each tree, this majority voting rule implies that the posterior probability threshold is 0.50.

Note: If you want to see this plot computed on some different threshold, you can change this value to any specific value, 0.01 for example, using the Decision Tree Action Set interface:

a) After you get the model, call **forestScore**. When passing in the **modelTable**, you can use WHERE clause to select which subset of trees will be used to score the data, such as **_treeid_** > 10. Loop each subset by changing the WHERE clause and calling **forestScore** with **encodedName=True** and a **casout** for each subset.

b) Using these casout tables, recompute the misclassification rate using a computed **var** by simply passing in your cutoff value.

If the data is big, you can use only one **casout** and recompute the misclassification rate immediately and keep using the same **casout** for other subsets.

9. Minimize the **Error Plot** and return to the Results of the Forest node.
10. Maximize the **Variable Importance** table.

Variable Label	Role	Variable Name	Training Importa...	Importance Standard Dev...	Relative Importance
Saving Balance	INPUT	SAVBAL	289.0960	42.6676	1
Checking Balance	INPUT	DDABAL	107.6821	11.6783	0.3725
Branch of Bank	INPUT	BRANCH	99.5404	2.3544	0.3443
CD Balance	INPUT	CDBAL	84.0894	27.7103	0.2909
Age of Oldest Account	INPUT	ACCTAGE	58.6203	2.0897	0.2028
Checking Account	INPUT	DDA	54.4333	38.0470	0.1883
Certificate of Deposit	INPUT	CD	51.6566	38.6648	0.1787
Home Value	INPUT	HMVAL	47.3457	1.6454	0.1638
Credit Score	INPUT	CRSCORE	46.3539	1.6431	0.1603
Age	INPUT	AGE	45.9614	1.6050	0.1590
Amount Deposited	INPUT	DEPAMT	43.6054	6.0121	0.1508
ATM Withdrawal Amount	INPUT	ATMAMT	43.0923	4.6575	0.1491
Income	INPUT	INCOME	42.7691	1.5889	0.1479

The table displays variable importance based on residual sum of square errors explained earlier. The rows are sorted by the importance measure. Note that unlike decision trees, the Importance Standard Deviation column has nonzero values. The variance of the importance is taken over partially independent trees in the forest. A conclusion from fitting the forest model to these data is that SAVBAL is the most important predictor of annuity insurance; however, it is not consistently important across several trees in the forest (having the highest importance standard deviation).

11. Minimize the **Variable Importance** table.
12. Scroll down in the Results window and maximize the **Properties** table.
    Observe the default values of Number of trees (100), Maximum depth (20), Minimum leaf size (5), In-bag training fraction (0.60), and Number of inputs to consider per split (default, which is the square root of the number of inputs to consider per split—that is, $\sqrt{47} \cong 7$ inputs in this case). This supersedes varsToTry=100, as shown below.

Property Name	Property Value
train	true
maxTrees	100
voteMethod	PROBABILITY
criterionMethod	IGR
iCriterionMethod	VARIANCE
maxBranch	2
maxDepth	20
leafSpec	COUNT
leafSize	5
leafProp	0.0001
missingValue	USEINSEARCH
minUseInSearch	1
intervalBins	50
intBinMethod	QUANTILE
trainFraction	0.6000
defaultVarsPerTree	true
varsToTry	100
loh	0

In the next demonstration, you modify these properties and try improving the forest model.

13. Minimize the **Properties** table and close the **Results**.
14. Open the **Results** of the Model Comparison node. Click the **Assessment** tab and maximize the **Fit Statistics** table.

Statistics Label	Train: Forest	Validate: Forest
Area Under ROC	0.8686	0.7918
Average Squared Error	0.1416	0.1730
Divisor for ASE	13,550	5,807
Formatted Partition	1	0
Gamma	0.7726	0.6211
Gini Coefficient	0.7372	0.5836
KS (Youden)	0.5758	0.4576
KS Cutoff	0.4000	0.3000
Misclassification at Cutoff	0.1833	0.2574
Misclassification Rate	0.1833	0.2574
Misclassification Rate (Event)	0.1833	0.2574
Multi-Class Log Loss	0.4491	0.5212

The default forest model has a ROC Index of 0.7918, an ASE of 0.1730, and MISC of 0.2574.

15. Close the **Results** of the Model Comparison node.

End of Demonstration

Tuning a Forest Model

Key differentiators in tree-based models are the splitting rules. The forest algorithm involves several hyperparameters controlling the structure of each individual tree (for example, minimum leaf size or maximum depth) and the size of the forest (for example, the number of trees) as well as its level of randomness (for example, number of inputs to consider per split or in-bag sample proportion). They are listed in Figure 6.22. The values of these tuning parameters must be optimized carefully because the optimal values are dependent on the data set at hand. Optimality here refers to a certain performance measure (for example, misclassification rate, average squared error, loss function) that must be chosen beforehand and evaluated on an out-of-bag sample to avoid overfitting.

Apart from the tuning parameters mentioned above, binning of interval variables is also one of the important tuners. Here are some recommendations:

- Using quantile bins instead of bins of equal width often results in a better fit, especially when some inputs have outliers.
- Using more bins sometimes improves and rarely degrades accuracy. However, training takes longer with more bins, and often much longer with many more bins.

Figure 6.22: How to Tune a Forest?

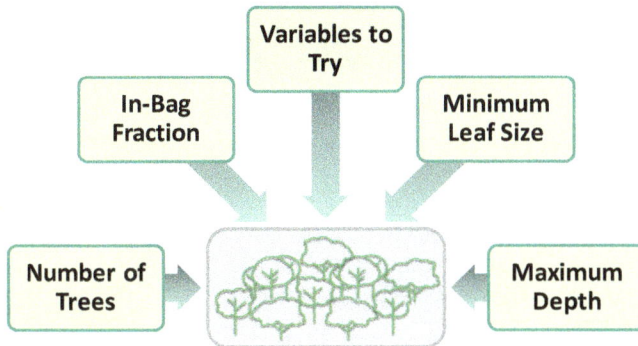

It is well known that in most cases, a forest works reasonably well with the default values of the hyperparameters specified in SAS Viya. Nonetheless, tuning the hyperparameters can improve the performance of your forest model.

"Hyperparameters and Tuning Strategies for Random Forest," by Probst et al. (2019) is a good read for a detailed discussion on forest parameter tuning.

Tuning Parameters Manually

Number of Trees

The number of trees in a forest is a parameter that is not tunable in the classical sense but should be set sufficiently high (Díaz-Uriarte and De Andres 2006; Oshiro et al. 2012; Probst and Boulesteix 2017).

The number of trees needed to obtain optimal performance depends on the characteristics of the data set at hand. Using many data sets, Oshiro et al. (2012) and Probst and Boulesteix (2017) empirically showed that the biggest performance gain can often be achieved when growing the first 100 trees. The convergence behavior can be investigated by inspecting the out-of-bag curves showing the performance for a growing number of trees in a forest.

The forest attains a lower out-of-bag or validation error only by variance reduction. Therefore, increasing the number of trees in the ensemble will not have any effect on the bias of your model. A higher number of trees will reduce only the variance of your model. You can achieve a higher variance reduction by reducing the correlation between trees in the ensemble. Note that the convergence rate of forest not only depends on the data set's characteristics but also on other hyperparameters that are discussed ahead.

Parameter tuning of number of trees in forest is summarized in Figure 6.23.

Figure 6.23: Tuning Number of Trees

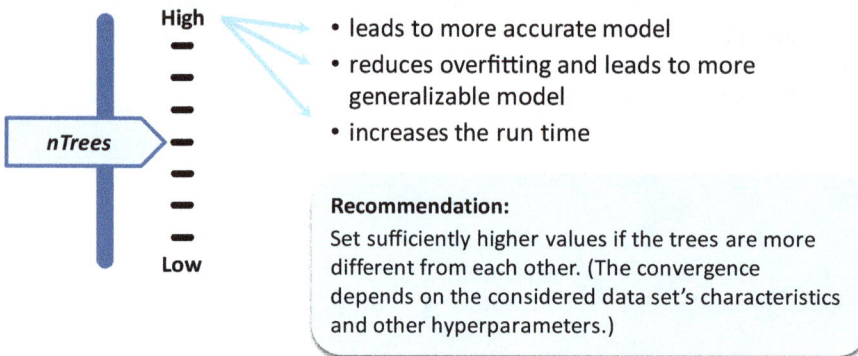

High
- leads to more accurate model
- reduces overfitting and leads to more generalizable model
- increases the run time

nTrees

Recommendation:
Set sufficiently higher values if the trees are more different from each other. (The convergence depends on the considered data set's characteristics and other hyperparameters.)

Low

In-Bag Sample Proportion

Decreasing the in-bag sample proportion leads to more diverse trees and thereby lower correlation between the trees, which has a positive effect on the prediction accuracy when aggregating the trees. However, the accuracy of the single trees decreases, because fewer observations are used for training. Therefore, the choice of the in-bag sample proportion is a trade-off between stability and accuracy of the trees. The optimal value is problem dependent and can be estimated with the out-of-bag predictions. It is generally observed that setting it to lower values give better results and reduces the run time (Martínez-Muñoz and Suárez 2010).

Parameter tuning of in-bag sample proportion for each of the individual trees in the forest is summarized in Figure 6.24.

Number of Inputs to Consider Per Split

The number of inputs to consider per split is one of the most important tuning parameters in the forest. Lower values lead to more different, less correlated trees, producing a more stable model

Figure 6.24: Tuning In-Bag Sample Proportion

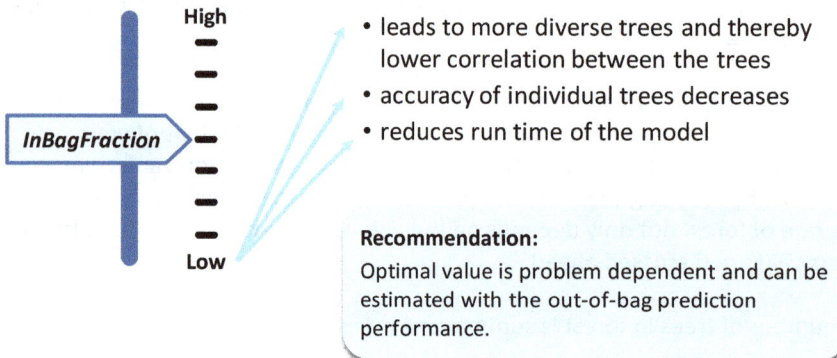

High
- leads to more diverse trees and thereby lower correlation between the trees
- accuracy of individual trees decreases
- reduces run time of the model

InBagFraction

Recommendation:
Optimal value is problem dependent and can be estimated with the out-of-bag prediction performance.

Low

after aggregating those diverse trees. A lower number of inputs to consider per split also tend to better exploit inputs with moderate effect on the target than would be masked by inputs with a strong effect if those had been candidates for splitting. However, lower values of several inputs to consider per split also lead to individual trees that perform worse because they are built based on suboptimal variables (that were selected out of a small set of randomly drawn candidates): non-important variables might be chosen. You must deal with a trade-off between stability and accuracy of the single trees (Probst et al. 2019).

The optimal number of inputs to consider per split is highly influenced by the real number of relevant input variables. If there are many relevant predictors, the number of inputs to consider per split should be set small, because then not only the strongest influential variables are chosen in the splits but also less influential variables, which can provide small but relevant performance gains. On the other hand, if there are only a few relevant variables out of many, the number of inputs to consider per split should be set high so that the algorithm can find the relevant variables (Goldstein et al. 2011).

Computation time decreases approximately linearly with a lower number of inputs to consider per split values (Wright and Ziegler 2017) because most of the forest's computing time is devoted to the selection of the split variables.

Parameter tuning of number of inputs to consider per split in a tree in the forest is summarized in Figure 6.25.

Minimum Leaf Size (Count or Proportion)

The minimum leaf size tuning parameter specifies the minimum number of observations in a terminal node. Setting it lower leads to trees with a larger depth, which means that more splits are performed until the terminal nodes. If you have more noise variables in your data, you need to set a higher value of minimum leaf size (Segal 2004).

Figure 6.25: Tuning the Number of Inputs to Consider per Split

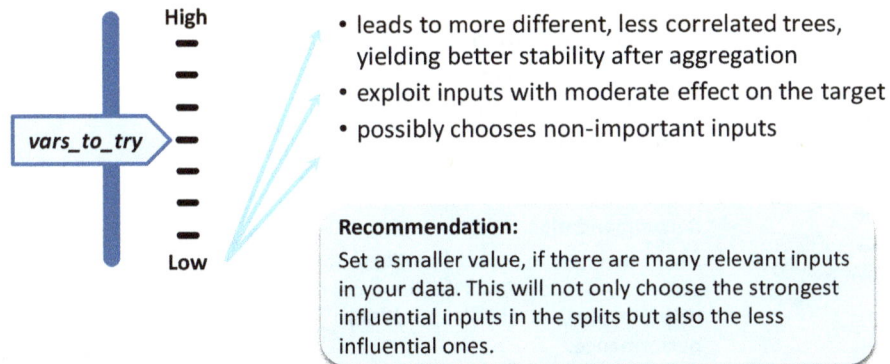

High

vars_to_try

Low

- leads to more different, less correlated trees, yielding better stability after aggregation
- exploit inputs with moderate effect on the target
- possibly chooses non-important inputs

Recommendation:
Set a smaller value, if there are many relevant inputs in your data. This will not only choose the strongest influential inputs in the splits but also the less influential ones.

The effect of the minimum leaf size is quite interesting. If you let the trees split down to leaf nodes with a single observation, that usually results in the most accurate random forest models. However, smaller leaf sizes mean larger (deeper) trees, so accuracy again comes at the cost of computation time and memory. A smaller leaf size usually makes the model more prone to capturing noise in training data.

The computation time decreases approximately exponentially with increasing leaf size (Probst et al. 2019). In a large sample data set, it might be helpful to set this parameter to a value higher than the default, as it decreases the run time substantially, often without substantial loss of prediction performance (Segal 2004).

Note that hyperparameters other than the minimum leaf size can be considered to control the size of the trees (for example, the strongly related hyperparameter is the maximal depth of the trees, which is the maximum number of splits until the terminal node).

Parameter tuning of minimum leaf size in a tree in the forest is summarized in Figure 6.26.

Maximum Depth

The deeper the tree, the more splits it has, and it captures more information about the data. However, it can cause overfitting in the individual trees. If you select smaller values of minimum leaf size, larger values of maximum depth are recommended so that the tree can grow deeper and vice versa. With lower tree depth, the tree might even fail to recognize useful signals from the data. Deeper trees usually take more time to run.

Parameter tuning of maximum depth for each of the individual trees in the forest is summarized in Figure 6.27.

Figure 6.26: Tuning Minimum Leaf Size (Count or Proportion)

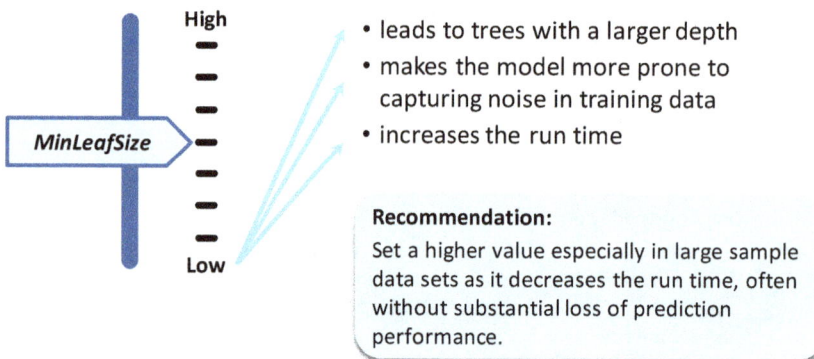

High

MinLeafSize

Low

- leads to trees with a larger depth
- makes the model more prone to capturing noise in training data
- increases the run time

Recommendation:
Set a higher value especially in large sample data sets as it decreases the run time, often without substantial loss of prediction performance.

Figure 6.27: Tuning Maximum Depth

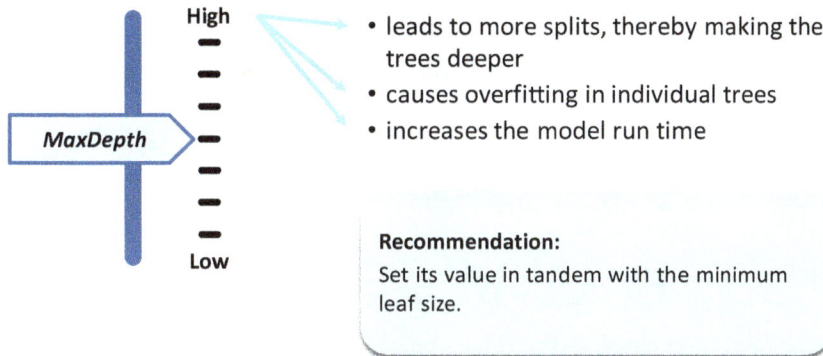

High
- leads to more splits, thereby making the trees deeper
- causes overfitting in individual trees
- increases the model run time

MaxDepth

Low

Recommendation:
Set its value in tandem with the minimum leaf size.

Autotuning Hyperparameters

Autotuning specifies whether to perform autotuning of any forest parameters. You can determine which hyperparameters to be tuned. Figure 6.28 lists forest hyperparameters that can be autotuned.

If **Perform Autotuning** is selected, the following options are available:

Maximum Depth specifies whether to autotune the maximum depth parameter.

Minimum Leaf Size specifies whether to autotune the minimum leaf size parameter.

Number of Interval Bins specifies whether to autotune the number of interval bins parameter.

Number of Trees specifies whether to autotune the tree number parameter.

In-bag Sample Proportion specifies whether to autotune the in-bag sample proportion parameter.

Figure 6.28: Autotunable Parameters in Forest

- Maximum Depth
- Minimum Leaf Size
- Number of Interval Bins
- Number of Trees
- In-bag Sample Proportion
- Number of Inputs per Split

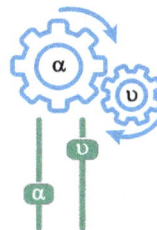

Number of Inputs per Split specifies whether to autotune the number of inputs per split parameter.

If any of the above options are selected, the following option is available for each one of them:

Initial value specifies the initial value for autotuning that hyperparameter. The default value is set for that hyperparameter. Use the **From** and **To** options to specify the range. The default **From** value and the default **To** value is set for that hyperparameter.

> Performing autotuning can substantially increase run time.

You can perform cross validation in the **General Options** of **Perform Autotuning** in the Forest node. However, if your data is already partitioned, then that partition is used and all the other options are ignored: Validation method, Validation data proportion, and Cross validation number of folds. If not, then in the Autotuning options, you can select **K-fold cross validation** and then specify the number of partition folds in the cross-validation process.

The FOREST Procedure

```
PROC FOREST < options > ;
CROSSVALIDATION < options > ;
GROW criterion ;
ID variables ;
INPUT variables < / LEVEL=NOMINAL | INTERVAL > ;
OUTPUT OUT=CAS-libref.data-table < option > ;
PARTITION partition-option ;
TARGET variable < / LEVEL=NOMINAL | INTERVAL > ;
```

The PROC FOREST statement invokes the procedure.

The CROSSVALIDATION statement performs k-fold cross validation to find the average estimated validation error. You cannot specify this statement if you specify either the AUTOTUNE statement or the PARTITION statement.

The GROW statement specifies the criterion by which to split a parent node into child nodes. As it grows the tree, PROC FOREST calculates the specified criterion for each predictor variable and then splits on the predictor variable that optimizes the specified criterion.

The ID statement lists one or more variables that are to be copied from the input data table to the output data tables that are specified in the OUT= option in the OUTPUT statement and the RSTORE= option in the SAVESTATE statement.

The INPUT statement names input variables that share a common option. You can specify the INPUT statement multiple times.

The OUTPUT statement creates an output data table that contains the results of PROC FOREST.

The PARTITION statement specifies how observations in the input data set are logically partitioned into disjoint subsets for model training, validation, and testing.

The TARGET statement names the variable whose values PROC FOREST tries to predict.

Demo 6.2: Tuning a Forest Model

In this demonstration, you try to improve the previously created default forest model by tuning certain parameters. You continue working with the **insurance** data set and experiment with some of the options in the forest node.

1. Ensure that the **Insurance_ClassTree** project is open in Model Studio, and you are on the **Forest Models** pipeline.
2. Expand the **Tree-splitting Options** properties and observe that **Use default number of inputs to consider per split** is checked by default.

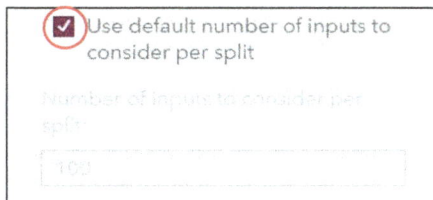

Number of inputs to consider per split is one of the key aspects of forests, as this generally affects the reduction in bias toward the most influential variables. Consequently, the question is, how many variables should be considered at each node? This option is available only if you deselect **Use default number of inputs to consider per split**.

Instead of manually changing this option over numerous training runs, we will use a programmatic approach to study the effect of modifying this option.

3. Right-click the **Data** node and select **Add child node** ⇨ **Miscellaneous** ⇨ **SAS Code**. Rename the SAS code to **#Inputs Evaluation**. Your pipeline should look like the one below.

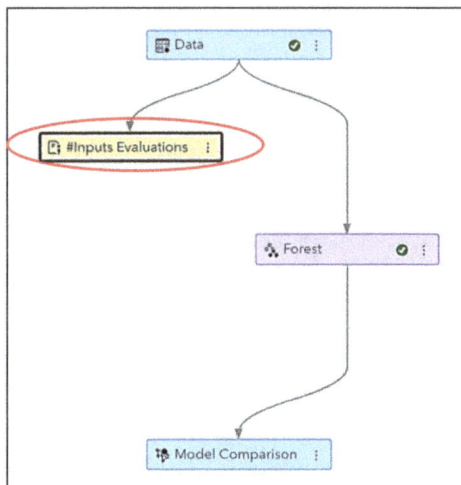

4. While the #Inputs Evaluations node selected, click the **Open Code Editor** button on your left.

5. Ensure that your cursor is on the **Training code** window, click the **Load source code file** icon (the one on the extreme left), and navigate to **D:\Workshop\Winsas\VBBF**.

6. Select the **Forest_Var_Tries_Evaluation.sas** file that contains the following code:

```
%macro forestStudy (nVarsList=10,maxTrees=200);
%let nTries = %sysfunc(countw(&nVarsList.));
/* Loop over all specified number of variables to try */
%do i = 1 %to &nTries.;
%let thisTry = %sysfunc(scan(&nVarsList.,&i));
%put &nTries.;

/* Run Forest for this number of variables */
proc forest data=&dm_data ntrees=&maxTrees. vars_to_try=&thisTry.;
  partition rolevar='_PartInd_'n (TRAIN='1' VALIDATE='0');
  target 'INS'n / level=nominal;
  input %dm_interval_input / level=interval;
  input %dm_binary_input %dm_nominal_input %dm_ordinal_input %dm_
unary_input / level=nominal;
  id 'IDNUM'n;
  ods output FitStatistics=&dm_lib..fitstats_vars&thisTry.;
run;

/* Add the value of varsToTry for these fit stats */
data &dm_lib..fitstats_vars&thisTry.;
length varsToTry $ 8;
set &dm_lib..fitstats_vars&thisTry.;
varsToTry = "&thisTry.";
run;

/* Append to the single cumulative fit statistics table */
proc append base=&dm_lib..fitStats1
data=&dm_lib..fitstats_vars&thisTry. force;
run;

%end;
%mend forestStudy;
%forestStudy(nVarsList= 3 7 10 15 25 47, maxTrees=100);

data &dm_data_outfit;
set &dm_lib..fitstats1;
run;
%dmcas_report(reportType=table,dataSet=&dm_lib..fitStats1);
%dmcas_report(reportType=Seriesplot,dataSet=&dm_lib..fitStats1,
x=Trees,y=miscOob,group=varsToTry,description=Out of Bag
Misclassification Rate);
```

The macro is used to take a list of the number of inputs to try (nVarsList) and the number of trees to train in the forest. The macro loops over all the numbers in the list, calling PROC FOREST multiple times (for different numbers of inputs to consider) and gathering the fit statistics from the output. The fit statistics of all the PROC FOREST models are saved into a data set that are used later to visualize the performance of the models.

7. Click **Open** to load the program.
8. Click **Close** button to close the editor and then click **Save**.
9. Run the **#Inputs Evaluation** node and open the **Results**.
10. Focus on the **Out of Bag Misclassification Rate** plot.

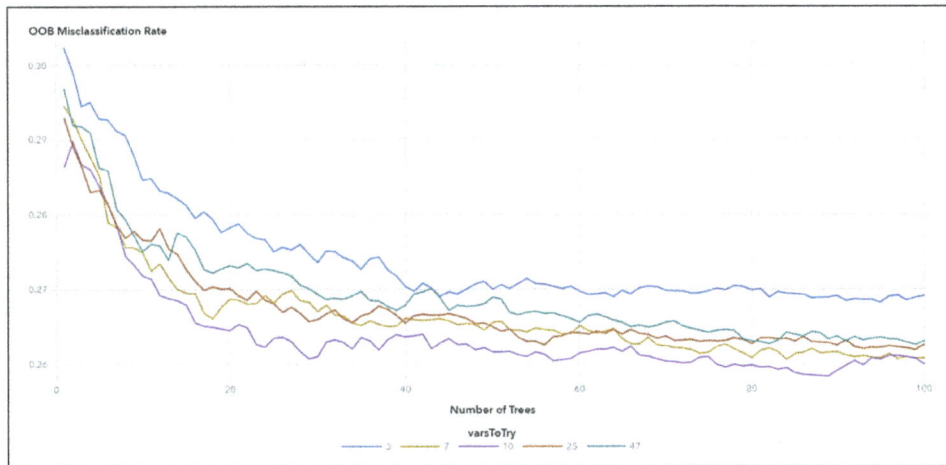

The effect of changing the number of inputs to use is not as directly intuitive as the effect of the number of trees. The plot suggests that for a smaller number of trees (which, of course, is not recommended for a forest) you might want to consider more inputs at each split – this makes sense, since with a small number of trees you need to make sure the most influential variables are considered. With a larger number of trees, the plot indicates that an intermediate number of inputs will provide the best model; anything too low carries the risk of missing out on the most influential inputs too often, and anything too high allows the most influential inputs to dominate too often, missing out on important secondary factors that allow the model to generalize better. Although it's fair to say that an intermediate number of variables to try is likely the most effective, the actual value/range will be data dependent, as the number of features and their relative influence will vary. Note that Breiman suggests that the square root of the number of variables be used as a good default; the data set used for this tip contained 47 inputs, which would lead you to start with seven variables to consider.

11. Close the **Results**.
12. Rename the Forest node to **Forest Default**.

13. Right-click the **Data** node and select **Add child node** ⇨ **Supervised Learning** ⇨ **Forest**. Rename this Forest node to **Forest Tuned**. Your pipeline should look like the one below.

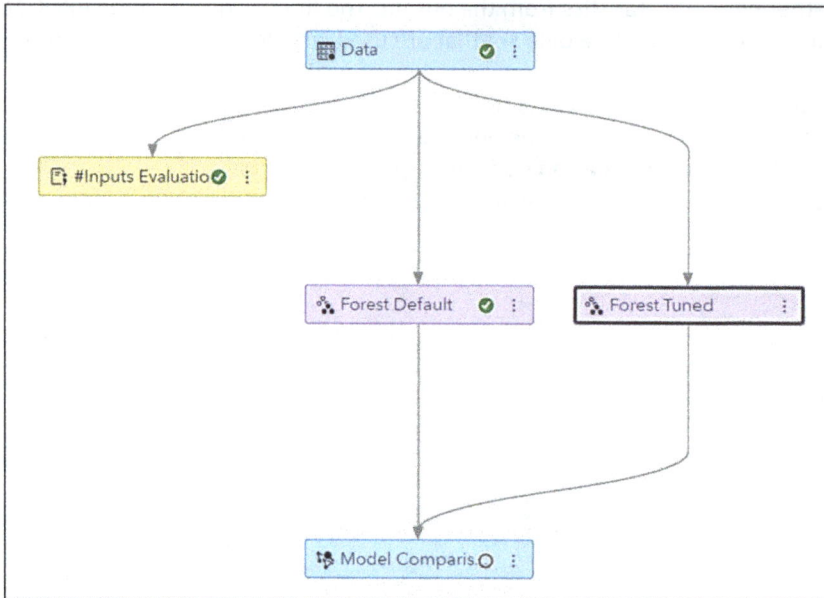

14. Change the **Number of trees** from 100 to **500**.

Number of trees:
```
500
```

The convergence in your forest model is not only data dependent but might also depend on other hyperparameters. Lower in-bag sample proportion, higher minimum leaf size, and a smaller number of inputs to consider per split values lead to less correlated trees. These trees are more different from each other and are expected to provide different predictions. Therefore, you need more trees to get accurate predictions for each observation.

Let's try to calibrate our model on similar lines.

15. Expand the **Tree-splitting Options** and change the **Class target criterion** from Information gain ratio to **Entropy, Maximum depth** from 20 to **30, Minimum leaf size** from 5 to **20**, and **In-bag sample proportion** from 0.60 to **0.40**.

16. Deselect **Use default number of inputs to consider per split** and enter **5** for **Number of inputs to consider per split**.

A small number of inputs to consider per split will reduce the variance of the ensemble but will also increase the bias of an individual tree in the ensemble. Let us see how this would turn out.

17. Hit **Run Pipeline** button.

18. Open the **Results** of the Forest Tuned node and maximize the **Error Plot**. Change **View Chart** to **Misclassification Rate**.

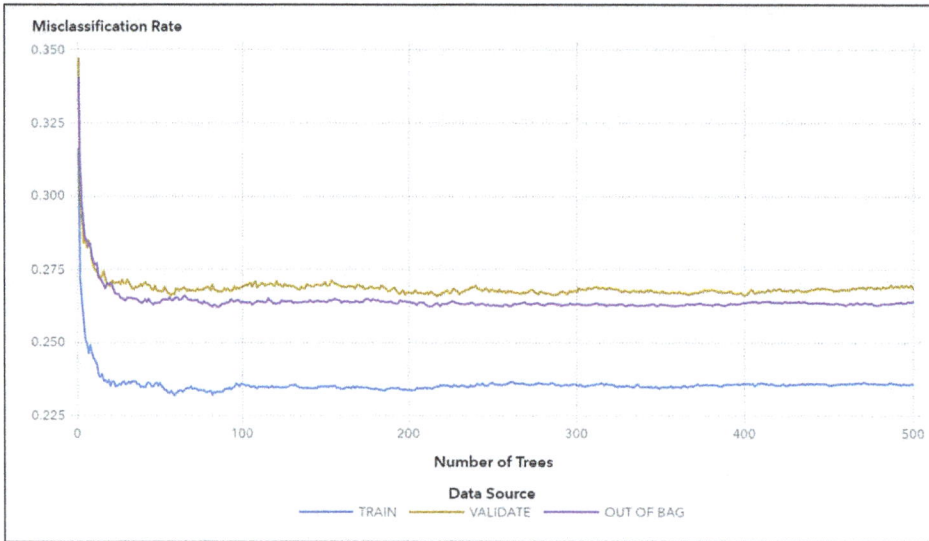

Creating 500 trees was a conservative choice. A lower number of trees would be acceptable from an efficiency point of view.

19. Change the **Number of trees** from 500 to **200**.
20. Click the **Run Pipeline** button again to run the pipeline with fewer tress in the forest.
21. Open the **Results** of the Model Comparison node.

Champion	Name	Algorithm Name	KS (Youden)	Misclassification Rate
⊡	Forest Tuned	Forest	0.4649	0.2673
	Forest Default	Forest	0.4576	0.2574

Both the models have comparable performance. However, the Forest Tuned model has a slightly better KS (Youden) statistic than the Forest Default model. It would be interesting to see the other performance measures as well.

You can further modify certain properties and search for more performance gain. Instead, we will use autotuning functionality in SAS Viya.

22. Close the **Results** of the Model Comparison node.

23. Right-click the **Data** node and select **Add child node** ⇨ **Supervised Learning** ⇨ **Forest**. Rename this Forest node to **Forest Autotuned.** Your pipeline should look like the one below.

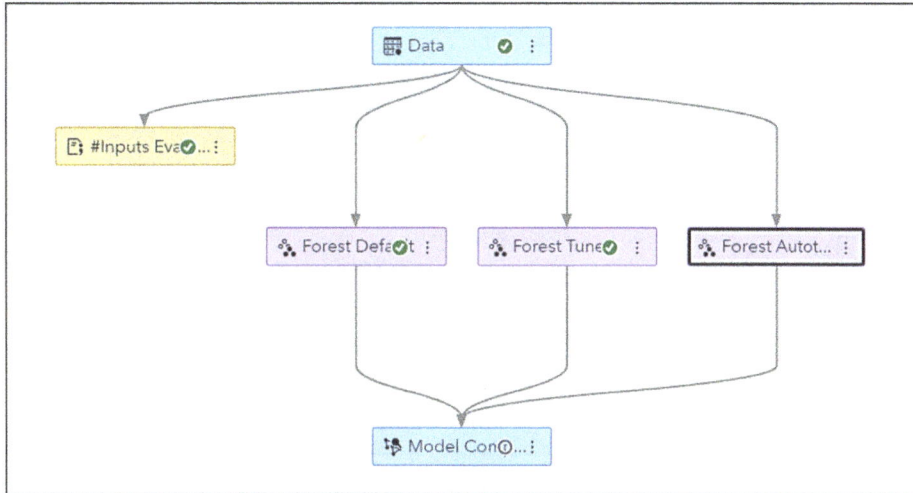

24. Expand the **Perform Autotuning** properties and set it to **On** using the slider button.

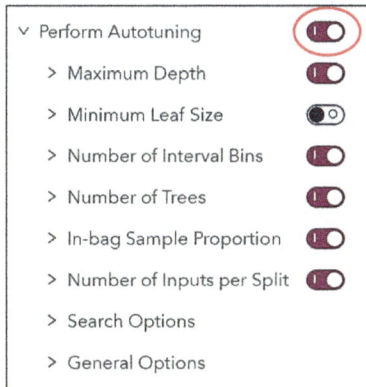

Note that maximum depth, minimum leaf size, number of interval bins, number of trees, in-bag sample proportion, and number of inputs per split are autotunable. By default, all of them are set to **On** except the minimum leaf size. You can expand each one of them and change their autotuning properties. However, no change is required for this run.

25. Click the **Run Pipeline** button.

26. Open the **Results** of the Forest Autotuned node and maximize the **Error Plot** with **Misclassification Rate** chosen in the **View chart**.

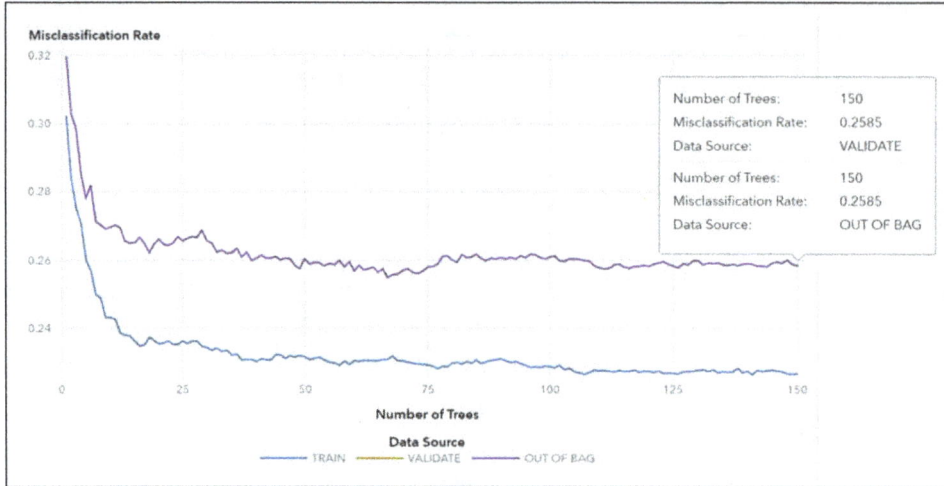

Autotuning gives a forest with a greater number of trees than the default tree. Because the data is already partitioned, validation partition is used for finding the objective value. The validate and out-of-bag error curves are exactly coinciding on one another.

27. Minimize the **Error Plot**. It would be interesting to note the optimal values of other hyperparameters.
28. Focus on the **Autotune Best Configuration** table.

Autotune Best Configuration	
Parameter	**Value**
Evaluation	51
Number of Trees	150
Number of Variables to Try	24
Bootstrap	0.1000
Maximum Tree Levels	30
Number of Bins	60
Kolmogorov-Smirnov	0.4688

The resulting values of parameters that were autotuned are displayed. By default, the genetic algorithm method is used as the autotune search method that uses an initial Latin Hypercube sample that seeds a genetic algorithm to generate a new population of alternative configurations at each iteration.

The autotuned best configuration suggests a forest with 150 trees, 24 inputs considered per split, and in-bag sampling of around 10% with a maximum tree depth of 30. While searching over several iterations, this configuration resulted in a KS of 0.4688.

29. Close the **Results** of the Forest Autotuned node.
30. Open the **Results** of the Model Comparison node.

Model Comparison				
Champion	Name	Algorithm Name	KS (Youden)	Misclassification Rate
⊡	Forest Autotuned	Forest	0.4684	0.2585
	Forest Tuned	Forest	0.4649	0.2673
	Forest Default	Forest	0.4576	0.2574

Interestingly, all three modes are quite different in terms of their tunable properties, yet they have comparable performance on several assessment statistics. Autotuned forest, however, comes out as the champion on KS (Youden) criteria.

	Forest Default	Forest Tuned	Forest Autotuned
Number of trees	100	200	150
Splitting criteria	IGR	Entropy	IGR
In-bag sample proportion	0.60	0.40	0.10
Number of inputs per split	7	5	24
Minimum leaf size	5	20	5
Maximum depth	20	25	29

Like the other two models, the autotuned forest does not have many trees but has smaller in-bag samples. A large value of number of inputs were considered per split that had increased the variance of the ensemble but reduced the bias of individual trees and perhaps that profited the forest model.

All three forest models have smaller minimum leaf size. A smaller leaf usually makes the model more prone to capturing noise in the training data. This property is directly associated with the maximum depth. However, you might want to try multiple leaf sizes to find the most optimum for your use case. The minimum leaf size is autotunable.

Therefore, there is no single recipe that would guarantee an accurate and efficient forest model. It is highly data dependent and experimentation is the key!

31. Close the **Results** of the Model Comparison node.

End of Demonstration

Quiz

1. A single decision tree poses a threat to the accuracy of the model as well as its stability.
   a.   True
   b.   False

2. Which of the following statements are TRUE? (select all that apply).
   a.   ensemble models can be created on the same sample.
   b.   bagging and boosting models are created on the same sample.
   c.   bagging is an approximation of the parallel bootstrap aggregation on different samples.
   d.   boosting is an approximation of the sequential bootstrap aggregation on different samples.

3. In a forest model, you can generate many trees and then aggregate the results of these trees. Which of the following is TRUE about individual tree in a forest model? (Select all that apply).
   a.   individual trees are built on a subset of input variables
   b.   individual trees are built on all input variables
   c.   individual trees are built on a subset of observations
   d.   individual trees are built on full set of observations

4. The training data for an individual tree excludes some of the available data to assess the fit of the model in the Forest node. The data that is withheld from training is called:
   a.   the validation sample
   b.   the test sample
   c.   the out-of-bag sample
   d.   the holdout sample

5. Increasing the number of trees in a forest does not increase the risk of overfitting the data. However, if the predictions from different trees are correlated, then increasing the number of trees makes little or no improvement.
   a.   True
   b.   False

Answers

1. False	2. a, c and d	3. a and c	4. c	5. True

Chapter 7: Additional Forest Models

Open-Source Random Forest Models

In recent years, many data scientists have used SAS software, Python, R, and other open-source or vendor-specific tools to mix and match various tasks of the analytical life cycle. SAS can embrace and extend the capabilities of open source as part of an analytics ecosystem. You can get the most out of SAS and open source working together. The word "open" signifies the fact that the power of SAS to build and deploy analytics can be accessed via many programming languages—not just SAS, but also Python, R, Lua, Java, or RESTful APIs. Integrating SAS and open-source technologies is often advantageous in two main scenarios:

1. Programmatically accessing SAS using open-source software
2. Bringing open-source models into SAS for side-by-side comparison.

This leads us to the two types of openness in SAS Viya. The first type is SAS in open source, and the second type is open source in SAS. This is illustrated in Figure 7.1.

SAS to Open-Source Languages and Interfaces (SAS in Open Source)

You can manage data and create analytical models using a variety of open-source languages. Multiple SAS interfaces can be used to connect to CAS, including SAS Studio, Model Studio, SAS Visual Analytics, SAS Enterprise Guide, SAS Display Manager, and others. Apart from SAS, you can also use Java, Lua, Python, or R to connect to CAS.

Figure 7.1: SAS Viya Integration with Open-Source Software

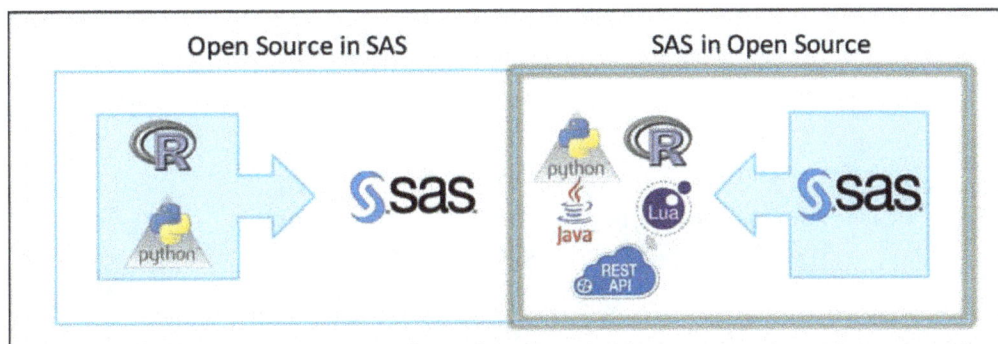

SAS in open source will be discussed in Chapter 9.

Open Source to SAS Language and Interfaces (Open-Source in SAS)

There might be times when it is beneficial to add components of open-source technologies into Model Studio. Open source in SAS Viya supports Python and R languages and requires Python or R and necessary packages to be installed on the same machine as the SAS Compute Server. It downloads data samples from SAS Cloud Analytic Services for use in Python or R code and transfers data by using a data frame or CSV file using the Base SAS Java Object.

Open-source in SAS is the primary focus in this chapter.

> "SAS® and Open Source: Two Integrated Worlds" is a great paper for more information. (https://www.sas.com/content/dam/SAS/support/en/sas-global-forum-proceedings/2019/3415-2019.pdf)

Open Source Code Node

The Open Source Code node enables you to import external code that is written in Python or R. The version of Python or R software does not matter to the node, so any version can be used as the code is passed along. The Python or R script executable must be in a system path on Linux, or the install directories can be specified with PYTHONHOME or RHOME on Windows. Prominent information about the node is listed in Figure 7.2.

The Open Source Code node enables the user to prototype machine learning algorithms that might exist in open-source languages but have not yet been vetted to be included directly as a node in Model Studio. This node can subsequently be moved to a Supervised Learning group if a Python or R model needs to be assessed and included to be part of model comparison. The node can execute Python or R software regardless of their versions.

Figure 7.2: Salient Features of Open Source Code Node

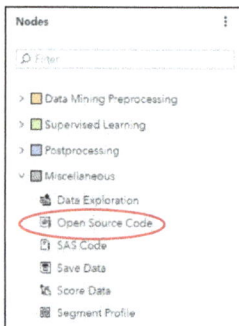

- The Open Source Code node is used to run Python or R code in a pipeline.
- Requires Python or R and necessary packages to be installed on the same machine as the SAS Compute Server.
- Cannot be part of an ensemble.
- Does not support registering, publishing, or downloading scoring code or scoring APIs.
- Enables the comparison of Python or R models within a Model Studio pipeline.

Figure 7.3: Code Editor

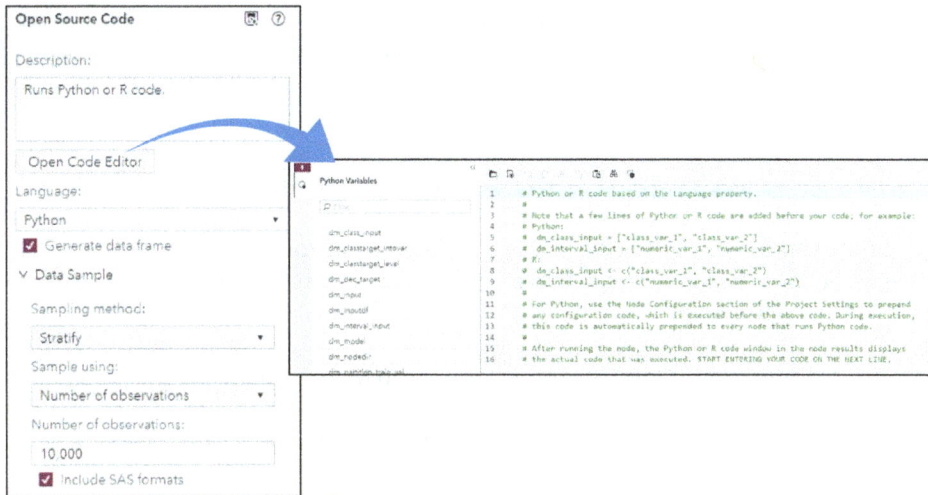

After selecting the language (Python or R) from properties, use the **Open Code Editor** button, shown in Figure 7.3, to enter respective code in the editor. Because this code is not executed in CAS, a data sample (10,000 observations by default) is created and downloaded to avoid movement of large data. Use Data Sample properties to control the sample size and method. Apply caution and do not specify full data or a huge sample when the input data is large. When performing model comparison with other Supervised Learning nodes in the pipeline, note that this node might not be using full data.

Input data can be accessed by the Python or R code via a CSV (comma-separated-value) file or as a data frame. When **Generate data frame** is selected, a data frame is generated from the CSV, and input data is available in **dm_inputdf**, which is a Pandas DataFrame in Python or an R data frame. When data are partitioned, an additional data frame, **dm_traindf**, is also available in the editor. That frame contains training data. If a Python or R model is built and needs to be assessed, corresponding predictions or posterior probabilities should be made available in the **dm_scoreddf** data frame. To do so, right-click and select **Move ⇨ Supervised Learning** to indicate that model predictions should be merged with input data and model assessment should be performed. Note that the number of observations in **dm_inputdf** and **dm_scoreddf** should be equal for successful merge to occur.

Note that this node cannot support operations such as **Download score code**, **Register models**, **Publish models**, and **Score holdout data** from the Pipeline Comparison tab because it does not generate SAS score code.

The Code Editor window enables the user to view a list of R variables or Python variables, depending on what open-source language is being used, that are available to the editor session.

Further information about the Open Source Code node in Model Studio, including a short video illustrating use of the node, can be found here: https://communities.sas.com/t5/SAS-Communities-Library/How-to-execute-Python-or-R-models-using-the-Open-Source-Code/ta-p/499463

Demo 7.1: Executing Open-Source Models in SAS Viya

In this demonstration, you use the Open Source Code node available for Python and R scripts that help you explore data or build models within a pipeline. You create random forest models in R and Python on the **insurance** data.

1. Ensure that the **Insurance_ClassTree** project is open in Model Studio and that you are on the **Forest Models** pipeline.
2. Right-click the **Data** node and select **Add child node** ⇨ **Data Mining Preprocessing** ⇨ **Imputation**.
   Leave the settings of the Imputation node at the defaults. Many open-source packages do not like missing values.

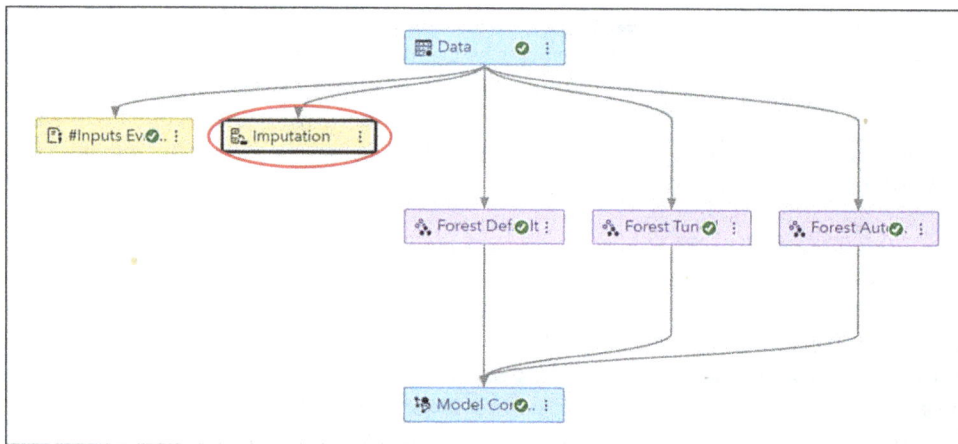

Note: Both Python and R packages sometimes do not support missing values in data. It is your responsibility to prepare the data as necessary for these packages. It is highly recommended that you add an Imputation node before the Open Source Code node to handle missing values. If the training data does not contain missing values but if either the validation or test data does contain missing values, consider enabling the **Impute nonmissing variables** property in the Imputation node.

3. Right-click the **Imputation** node and select **Add child node** ⇨ **Miscellaneous** ⇨ **Open Source Code**.
4. Right-click **Open Source Code** node and rename the node **Python Forest**. Your pipeline should look like the one below.

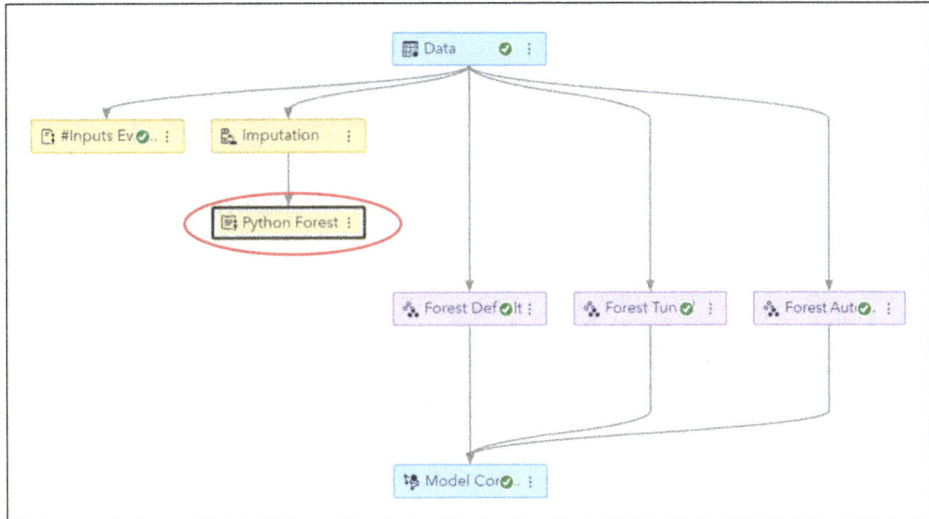

5. In the properties panel of the Open Source Code node (now renamed **Python Forest**), verify that the language is set to **Python**.

6. Expand the **Data Sample** properties. Clear the check box for **Include SAS formats**. This property controls whether the downloaded data sent to the Python or R software should keep SAS formats. It is strongly recommended that you keep SAS formats, and this should work in most cases. However, some numeric formats such as DOLLAR*w.d* add a dollar sign and change the data type of the variable when exporting to CSV. In such cases, these formats must be removed.

Note: We are sampling to use 10,000 observations, but you have the option to use all the data. The data sample is downloaded from CAS as a CSV file (**node_data. csv**). The default is stratified sampling, and stratification is done by partition variable or class target when applicable. This node uses a sample when performing model comparison.

Note: The **Include SAS formats** property either keeps or removes SAS formats for all variables in the data set. If input or target variables have SAS or user-defined formats that significantly modify the data, it is not recommended that you deselect this option because the model that is built might not be comparable to other models.

7. Click the **Open Code Editor** button (the one on the extreme left). When the Code Editor opens, select **Clear all code** icon and then select **Load source code file** icon (the one on the extreme right). Navigate to the data folder.

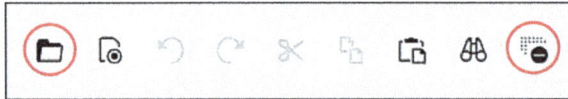

8. Select the **Python_Forest.py** file from the data folder.

```
1    from sklearn import ensemble
2
3    # Get full data with inputs + partition indicator
4    dm_input.insert(0, dm_partitionvar)
5    fullX = dm_inputdf.loc[:, dm_input]
6
7    # Dummy encode class variables
8    fullX_enc = pd.get_dummies(fullX, columns=dm_class_input, drop_first=True)
9
10   # Create X (features/inputs); drop partition indicator
11   X_enc = fullX_enc[fullX_enc[dm_partitionvar] == dm_partition_train_val]
12   X_enc = X_enc.drop(dm_partitionvar, 1)
13
14   # Create y (labels)
15   y = dm_traindf[dm_dec_target]
16
17   # Fit RandomForest model w/ training data
18   params = {'n_estimators': 100, 'max_depth': 20, 'min_samples_leaf': 5}
19   dm_model = ensemble.RandomForestClassifier(**params)
20   dm_model.fit(X_enc, y)
21   print(dm_model)
22
23   # Save VariableImportance to CSV
24   varimp = pd.DataFrame(list(zip(X_enc, dm_model.feature_importances_)), columns=['Variable Name', 'Importance'])
25   varimp.to_csv(dm_nodedir + '/rpt_var_imp.csv', index=False)
26
27   # Score full data
28   fullX_enc = fullX_enc.drop(dm_partitionvar, 1)
29   dm_scoreddf = pd.DataFrame(dm_model.predict_proba(fullX_enc), columns=['P_INS0', 'P_INS1'])
30
```

This fits a random forest classifier model in Python. The default values for the parameters that control the size of the trees (for example, **max_depth (default=none)**, **min_samples_leaf (default=1)**) lead to fully grown and unpruned trees, which can be exceptionally large data sets. To reduce memory consumption, the complexity and size of the trees are controlled by setting parameter values like the ones in the code above. The code that needs to be changed for different data sets is line 29, how your predictions are named using the **P_ + "*target*"** naming convention.

Note: Remember that we are just modeling the data here. Currently, there is not a way to do data preparation within the Open Source Code node so that a subsequent node will recognize it. If this is necessary, either prepare data before Model Studio or perform both of the following: (1) open-source data preparation with the Open Source Code node (in preprocessing group), and (2) modeling with the Open Source Code node (in supervised learning group).

9. In the upper right corner of the window, click the **Save** icon to save the Python code and then click the **Close** button to close the Code Editor window.

10. Run the **Python Forest** node.
11. Repeat the previous steps for fitting a forest model in R. Right-click the **Imputation** node and select **Add child node** ⇨ **Miscellaneous** ⇨ **Open Source Code**.
12. Right-click the **Open Source Code** node and rename it **R Forest**. Your pipeline should look like the one below.

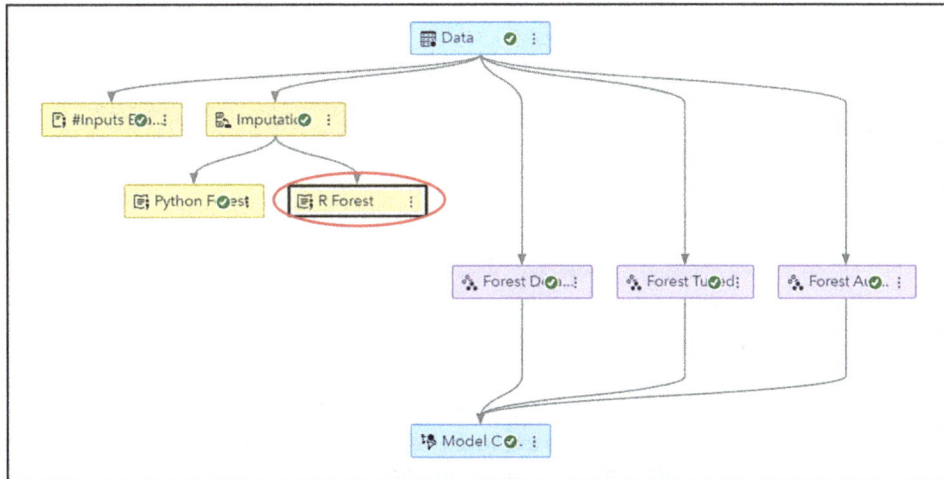

13. In the properties panel of the Open Source Code node (now renamed **R Forest**), set the language to **R** and clear the check box for **Include SAS formats**.

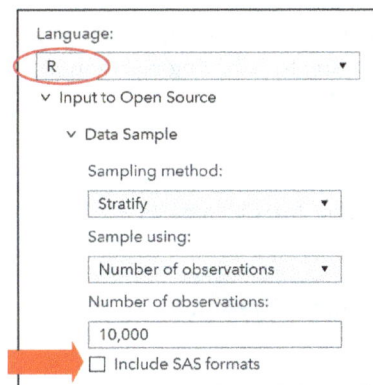

14. Click the **Open Code Editor** button. When the Code Editor opens, click the **Clear all code** icon and then click the **Load source code file** icon. Navigate to the data folder.

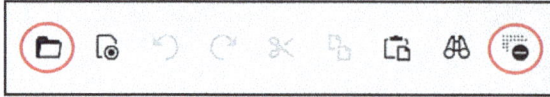

15. Select the **R_Forest.r** file from the data folder.

```
1    library(randomForest)
2
3    # RandomForest
4    dm_model <- randomForest(dm_model_formula, ntree=100, mtry=5, data=dm_traindf, importance=TRUE)
5
6    # Score
7    pred <- predict(dm_model, dm_inputdf, type="prob")
8    dm_scoreddf <- data.frame(pred)
9    colnames(dm_scoreddf) <- c("P_INS0", "P_INS1")
10
11   # Print/plot model output
12   png("rpt_forestMsePlot.png")
13   plot(dm_model, main='randomForest MSE Plot')
14   dev.off()
15
16   write.csv(importance(dm_model), file="rpt_forestIMP.csv", row.names=TRUE)
17   |
```

This fits Breiman and Cutler's random forest classifier model in R.

The code that needs to be changed for different data sets is line 9, how your predictions are named using the **P_ + "*target*"** naming convention.

Note: It is a good practice to execute the node in an empty state to validate whether Python or R is correctly installed and configured. In addition, you can view the precursor code that is added as part of the executed code. The code that is added depends on the combination of properties selected. The precursor code is part of the node results.

Note: Model assessment is performed automatically if the following are true:

- Predictions are saved in the **dm_scoreddf** data frame or the **node_scored.csv** file.
- Prediction variables are named according to following convention:
  P_<*targetVarName*> for interval target
  P_< *targetVarName* ><*targetLevel*> for class targets (All target level probabilities should be computed.)

16. Click the **Save** icon to save the R code and then click the **Close** button to close the Code Editor window.
17. Run the **R Forest** node.
18. Open the **Results** of the **Python Forest** node (or the R Forest node).

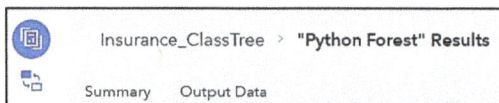

Why does the **Open Source Code** node does not have an **Assessment** tab even though it was successfully executed?

For model assessment, you need to move the nodes to the supervised learning group. Right-click the **Python Forest** node and select **Move ⇨ Supervised Learning**.

19. Repeat the same for the **R Forest** node.

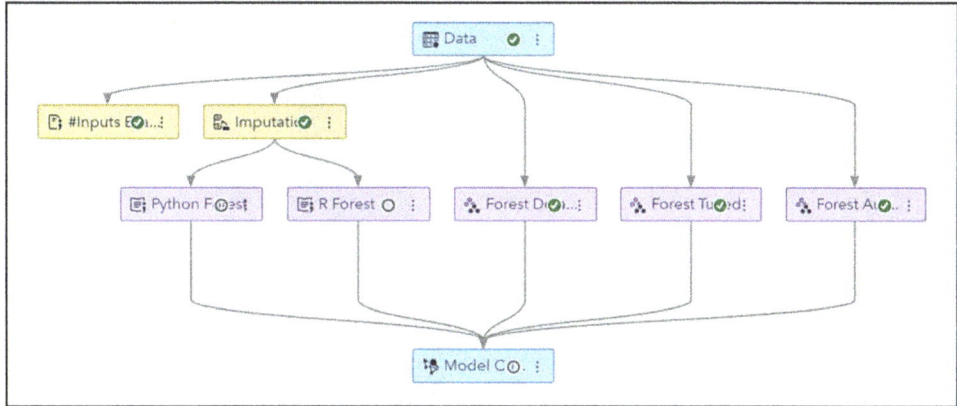

The color of both the nodes has changed to green, showing that these nodes have changed to the group of Supervised Learning nodes and they have been connected to the Model Comparison node. You can compare these open-source models with the models that you fit in Model Studio.

Notice that the nodes need to be rerun.

20. Click the **Run Pipeline** button.

21. Open the **Results** of the Python Forest node (or the R Forest node, or both).

The Python code (or R code) that was executed is displayed. The standard Python output (or R output) is displayed.

22. Scroll down and maximize the **Output** window.

When in Supervised Learning mode, the list of predicted variables expected by the project is displayed.

				Predicted Variables			
Obs	Predicted Variable Name	Variable Name	Level Frequency	Level Frequency Percent	Raw Numeric Value	Raw Character Value	Formatted Value
1	I_INS	INS	.	.			
2	P_INS0	INS	12652	65.3614	0		0
3	P_INS1	INS	6705	34.6386	1		1

In addition, if available, the top 10 observations from the input data and the scored data are displayed.

Obs	P_INS0	P_INS1
1	0.763909	0.236091
2	0.843076	0.156924
3	0.632841	0.367159
4	0.939175	0.0608251
5	0.739958	0.260042
6	0.862866	0.137134
7	0.623295	0.376705
8	0.784079	0.215921
9	0.80968	0.19032
10	0.222717	0.777283

Top 10 observations from the Scored Data (Generated in the Open Source Program)

Note: Results from Python (or R) execution can be viewed in the node when saved with the **rpt_** prefix.

- Files with the **.csv** extension are displayed as tables where the first row is the header (**rpt_VariableImportance.csv**).
- Files with a **.png**, **.jpeg/.jpg**, or **.gif** extension are displayed as images (**rpt_MeanSquareErrorPlot.png**).
- Files with the **.txt** extension are displayed as plain text (**rpt_GLMOutput.txt**).

Remember that the **rpt_** prefix is not case sensitive, and that the **rpt_** prefix and file extensions (**.csv**, **.txt**, **.png**, **.jpeg**, **.jpg**, **.gif**) are key in identifying which files to display.

23. Minimize the Output window and click the **Assessment** tab in the upper left corner.

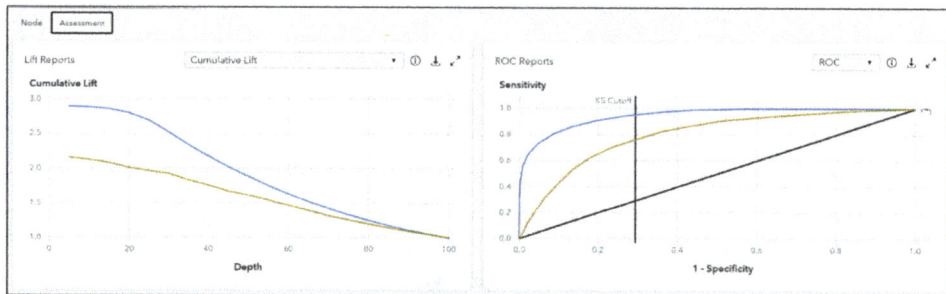

The usual assessment results are displayed.

24. Close the **Results** of the Python Forest node.
25. Open the **Results** of the Model Comparison node.

Model Comparison				± ↗
Champion	Name	Algorithm Name	KS (Youden)	Misclassification Rate
⊡	Forest Autotuned	Forest	0.4684	0.2585
	Python Forest	Open Source Code	0.4676	0.2511
	Forest Tuned	Forest	0.4649	0.2673
	R Forest	Open Source Code	0.4632	0.2530
	Forest Default	Forest	0.4576	0.2574

The open-source models and the SAS models can be compared here on different assessment measures. The SAS autotuned model comes out as a champion based on the KS (Youden) statistic.

26. Click the **Assessment** tab in the upper left corner and maximize the **Lift Reports** chart. Under **View chart**, select **Cumulative Lift** from the drop-down menu. In the left pane, click the **One partition** radio button under **Display** and then select **Validate** from the drop-down menu. Adjust the zoom-in view.

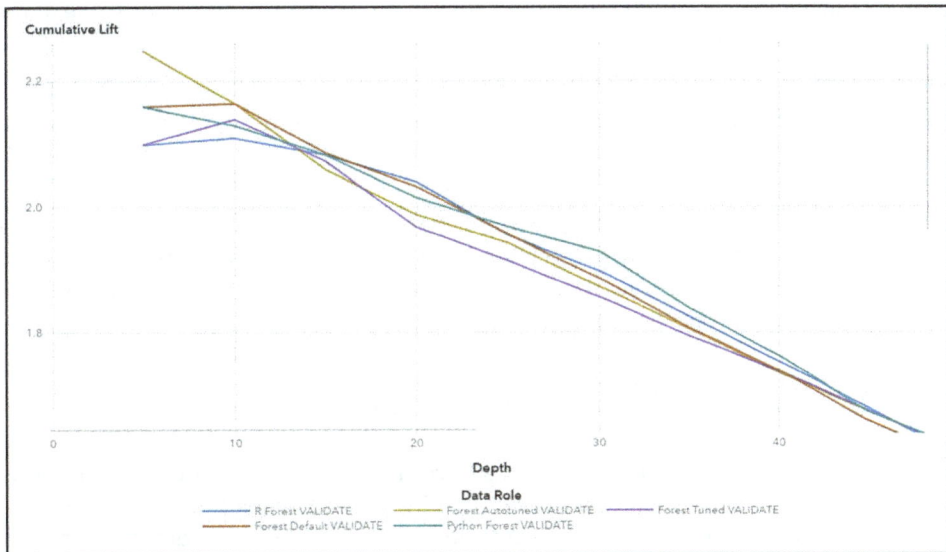

Infer the results as deemed fit.
27. Minimize the **Lift Reports** chart.
28. Close the **Results** of the Model Comparison node.

End of Demonstration

Figure 7.4: Common Examples of Anomaly Detection

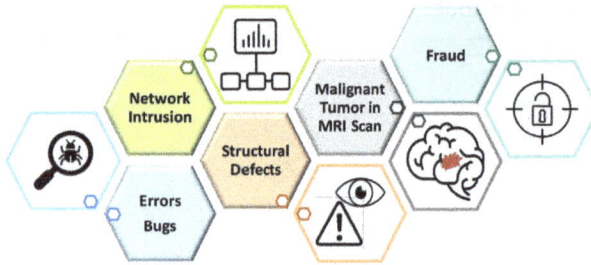

Isolation Forest Models

In data mining and machine learning, anomaly detection is the identification of rare items, events, or observations that raise suspicions by differing significantly from most of the items, events, or observations. Typically, these anomalous items have the potential of getting translated into problems such as network intrusion, structural defects, errors, event detection in sensor networks, detecting ecosystem disturbances, spotting a malignant tumor in an MRI (Magnetic Resonance Imaging) scan or fraud detection, and so on. Figure 7.4 represents some of the anomaly detection applications.

> Anomaly detection refers to the identification of items or events that do not conform to an expected pattern or to other items present in a data set.

Anomaly detection is often used in preprocessing to remove anomalous data from the data set. In Model Studio, the **Anomaly Detection** node is a Data Mining Preprocessing node that identifies and excludes anomalies (observations) using the Support Vector Data Description (SVDD). However, this is beyond the scope of this book.

Detecting Fraud Using Tree-Based Methods

Fraud is a growing concern for companies all over the globe. Fraud detection methods attempt to detect or impede illegal activity that involves financial transactions. Fraud is intentionally committing a misleading act for financial gain. Anomaly detection is one of the ways to detect fraud. You look to predict an event that occurs rarely and identify patterns in the data that do not conform to expected behavior, such as an abnormally high purchase made on a credit card.

When trying to identify fraud with machine learning, two approaches are commonly used. They are concisely described in Figure 7.5. The first approach is with methods associated with supervised machine learning. This method involves using historical data that contains examples of the type of fraud that the user is trying to find. The algorithm can then learn to detect the fraudulent event by training a model using the examples of fraudulent and non-fraudulent cases. Typical modeling methods used for this type of fraud detection are logistic regression,

Figure 7.5: Tree-Based Methods for Fraud Detection

- Supervised methods
  - decision tree, forest, gradient boosting
- Unsupervised methods
  - isolation forest
- Unsupervised methods are often preferred because
  - historical examples are unavailable
  - a new type of fraud is anticipated.

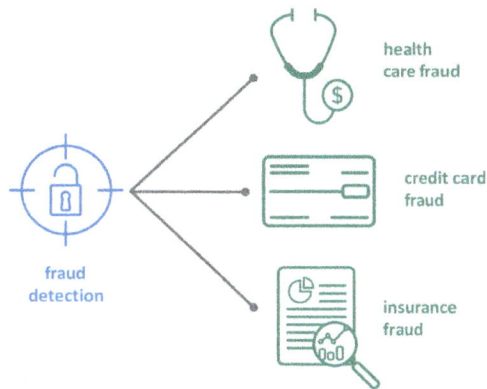

health care fraud

credit card fraud

fraud detection

insurance fraud

decision trees, random forests, gradient boosting, neural networks, and other types of classification models.

In addition to supervised methods, there are also unsupervised methods that are being used to detect fraudulent cases. These methods are referred to as unsupervised because there is no historical information about fraudulent cases that is used to train the model. Whereas a data set used to train a supervised model would have a variable indicating fraud/non-fraud that could be used to train the model, a data set used for unsupervised modeling does not contain a variable with this information. Instead, unsupervised methods are used to find anomalies by locating observations within the data set that are separated from other heavily populated areas of the data set. By analyzing customers or transactions relative to each other, we are able to spot unusual observations. These observations can potentially be indicative of fraud and, by identifying them, we are able to examine what is occurring and if it is of a fraudulent nature.

The assumption behind this is that fraudulent behavior can often appear as anomalous within a data set. It should be noted that just because an observation is anomalous, it does not mean it is fraudulent or of interest to the user. Similarly, fraudulent behavior can be disguised to be hidden within more regular types of behavior. However, without labeled training data, unsupervised learning is a good method to use to begin to identify deviant accounts or transactions.

There are several reasons why a user might want to use unsupervised methods instead of supervised methods. As described above, a user might not have historical examples of the type of fraud they are trying to detect. Without labeled observations, options for supervised learning are limited.

Another reason for using unsupervised methods instead of supervised methods, or in addition to them, is to try to find new types of fraud that might not have been captured within the historical data. Fraud patterns can evolve or change, so it is important to constantly be searching for ways to identify these new patterns as early as possible. If purely relying on supervised models built with historical data, these new patterns can be missed. However, since the unsupervised methods are not limited by the patterns present in the historical data, they can potentially identify these new patterns as they can represent behavior that is unusual or anomalous.

One unsupervised method that we will examine to identify anomalies is isolation forest.

What Is an Isolation Forest?

An Isolation forest is an algorithm that would identify the outliers in a multidimensional space, a method that in principle is like the forest algorithm. An isolation forest (Liu, Ting, and Zhou 2008) is a specially constructed forest that is used for anomaly detection instead of target prediction.

Isolation forests explicitly identify anomalies instead of profiling normal data points. An isolation forest, like any tree ensemble method, is built based on decision trees. For each split in an isolation forest, one input variable is randomly chosen. If the variable is an interval variable, then it is split at a random value between the maximum and minimum values of the observations in that node. If the variable is a nominal variable, then each level of the variable is assigned to a random branch. By constructing the forest in this way, anomalous observations are likely to have a shorter path from the root node to the leaf node than non-anomalous observations have.

How Do Isolation Forests Detect Anomalies?

Anomalous observations are less frequent than regular observations and are different from them in terms of values (they lie farther away from the regular observations in the feature space). That is why by using such random partitioning, they should be identified closer to the root of the tree (shorter average path length, that is, the number of edges an observation must pass in the tree going from the root to the terminal node), with fewer splits necessary. This is described in Figure 7.6.

The shortest path to isolating an observation can also be understood by visual examination provided by Gillespie 2019. In Figure 7.7, you see a two-dimensional data set with two anomalies

Figure 7.6: Isolating Anomalies in a Completely Random Tree

- Grow a random decision tree until each instance is in its own leaf.
- Observations that are more anomalous will have shorter paths from the root node of the tree to the leaf node.
- These observations will be isolated by fewer splits of the data.

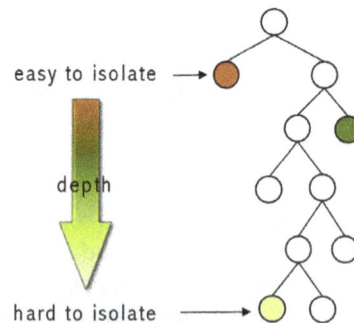

Figure 7.7: 2-D Space with Two Highlighted Observations

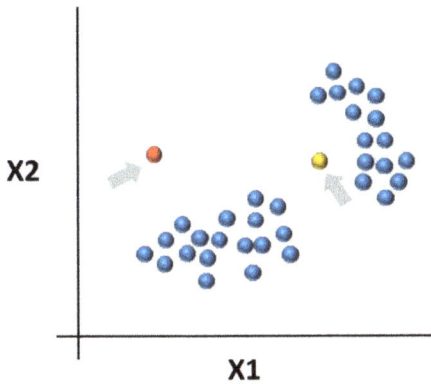

Figure 7.8: 2-D Space with the Red Observation Isolated

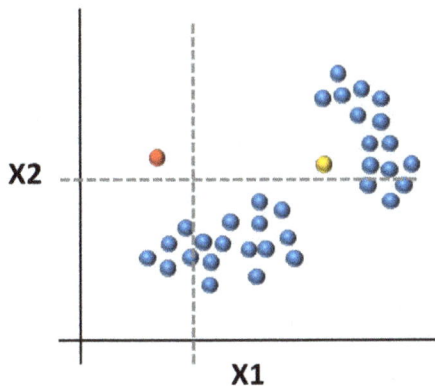

that are highlighted, one in red and one in orange. By examining how many divisions of the X1 and X2 variable we would require for isolating each observation, we can get a sense of how anomalous the observation is considered.

Looking at the observation that is colored red, you can see that it takes two divisions to isolate this record from the remaining records in the data set. See Figure 7.8.

However, when you try to isolate the orange observation, you can see that it takes three divisions. See Figure 7.9.

A non-anomalous observation requires more partitions to be identified than an anomalous observation. Based on the two divisions, we would say that the red colored observation would be more anomalous than the orange, which had three divisions. This is the central idea behind isolation forests. However, during the implementation of the algorithm, it is applied as an average of path lengths that make up the ensemble of decision trees used to construct the isolation forest as illustrated in Figure 7.10.

Figure 7.9: 2-D Space with the Orange Observation Isolated

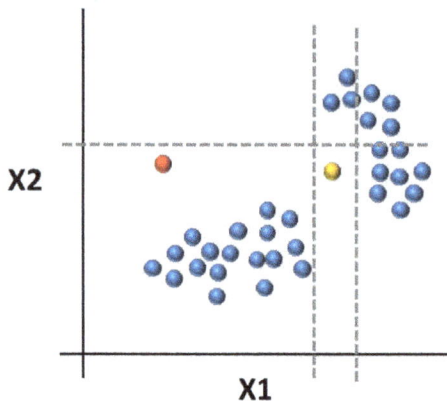

Figure 7.10: Creating a Forest

- Create several partitions (completely random trees) to form an isolation forest.

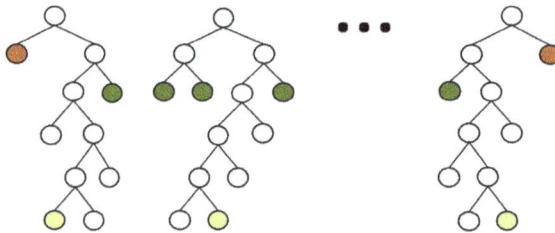

- By averaging the path lengths for each observation, we can find observations that are more distinct within the data set.

Measuring Anomalies

The algorithm will return an anomaly score for each observation based on the path lengths for that observation within the constructed trees. The anomaly score of an observation is calculated as:

$$\text{Anomaly Score} = 2^{-\left(\frac{\text{Average path length for observation}}{\text{Average length of all paths}}\right)}$$

The quantity in the superscript is the average, over all trees, of the length of the path from the root node to the leaf node that contains that observation, divided by the average length of all paths across all trees.

The anomaly score is always between 0 and 1, where values closer to 1 indicate a higher chance of the observation being an anomaly. With this anomaly score, you can then analyze the rare observations for their likelihood of anomalous (for example, fraud).

Figure 7.11: Detecting Anomalous Observations

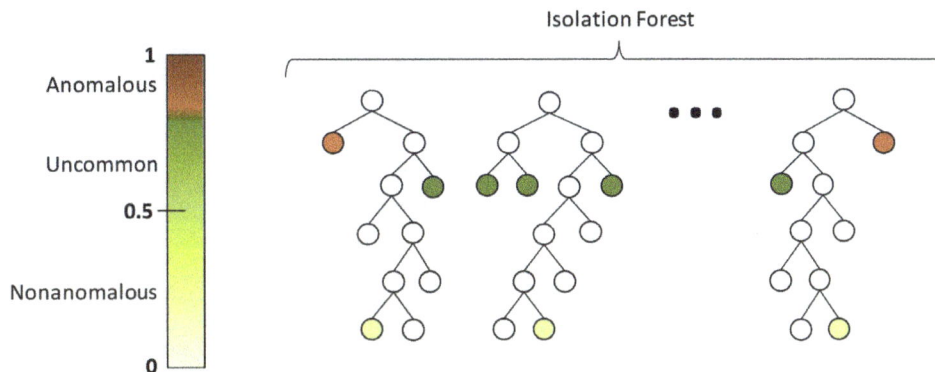

When the FOREST procedure creates an isolation forest, it writes anomaly scores in the scored data table.

A score close to 1 indicates anomalies. A score much smaller than 0.5 indicates normal observations. If all scores are close to 0.5, then the entire sample does not seem to have clearly distinct anomalies.

One way to measure the validity of the isolation forest in finding fraud events is to test it against a data set that does contain labeled fraudulent observations.

Isolation Forest Using the FOREST Procedure

ISOLATION is an option in the PROC FOREST statement in the FOREST procedure that generates an isolation forest for anomaly detection instead of a forest for target prediction.

```
ISOLATION < (SAMPLEN=number) >
```

ISOLATION < (SAMPLEN=*number*) > creates an isolation forest for anomaly detection instead of creating a forest for target prediction.

You can specify the following suboption:

SAMPLEN=*number* specifies the number of observations, sampled without replacement, to use in each tree of the isolation forest. By default, SAMPLEN=100.

When the ISOLATION option is specified, the GROW and TARGET statements are ignored and the BINMETHOD, INBAGFRACTION, RBAIMP, VARS_TO_TRY, and VII options in the PROC FOREST statement are ignored. Fit statistics are not computed.

Fraud Example

- The **PaySim** data set contains simulated mobile- based payment transactions for a variety of transactions, with 11 variables and 6,362,620 observations.

- The fraudulent behavior of the agents involves a misleading act for financial gain by taking control of customer accounts and trying to empty the funds by transferring to another account and then cashing out of the system.

PaySim is a simulated financial data set for fraud detection that contains mobile money transactions based on a sample of real transactions extracted from one month of financial logs from a mobile money service implemented in an African country. The original logs were provided by a multinational company, who is the provider of the mobile financial service that is currently running in more than 14 countries all around the world. This synthetic data set is scaled down to one-fourth of the original data set and consists of both numerical and categorical features like transaction type, amount transferred, account numbers of sender, and recipient accounts.

You will implement an isolation forest on this data in the next demonstration using the SAS Studio interface. As an alternative interface, you can also use a SAS Code node in the Model Studio pipeline.

Table 7.1: Data Dictionary of PaySim Data

Variable	Description	Role	Level
amount	Amount of the transaction (say, $)	Input	Interval
isFlaggedFraud	Flags illegal attempts when more than $200,000 transferred from one account to another in a single transaction	Rejected	Binary
isFraud	Indicates whether transactions were made by the fraudulent agents	Target	Binary
nameDest	Name of the recipient	Rejected	Nominal
nameOrig	Name of the sender	ID	Nominal
newbalanceDest	Recipient account balance after transaction	Input	Interval
newbalanceOrig	Sender account balance after transaction	Input	Interval
oldbalanceDest	Recipient account balance before transaction	Input	Interval
oldbalanceOrg	Sender account balance before transaction	Input	Interval
step	Maps a unit of time (1 step is 1 hour of time)	Rejected	Interval
type	Transaction type (cash-in / cash-out / debit / payment / transfer)	Rejected	Nominal

The data source is provided by E. A. Lopez-Rojas, A. Elmir, and S. Axelsson. "PaySim: A financial mobile money simulator for fraud detection." The 28th European Modeling and Simulation Symposium-EMSS, Larnaca, Cyprus, 2016.

Demo 7.2: Detecting Fraud Using an Isolation Forest in SAS Studio

In this demonstration, you create an isolation forest to detect fraud in a data set that contains synthetic transactions for mobile payments, known as **PaySim** data. For the purposes of our comparison, we will use the **isFraud** variable as a check for our unsupervised model's ability to detect the fraudulent observations.

1. Go to the **Applications** menu (☰) in the upper left side.
2. Click **Develop SAS Code** in the application shortcut area. This launches SAS Studio. You can use predefined tasks here to generate SAS code or write new ones.

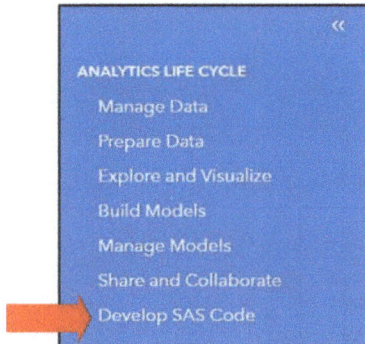

3. Click the **Snippets** icon in the left pane. Under **My Snippets** ⇨ **VBBF**, either double-click or drag and drop **Isolation Forest Code** onto the canvas.

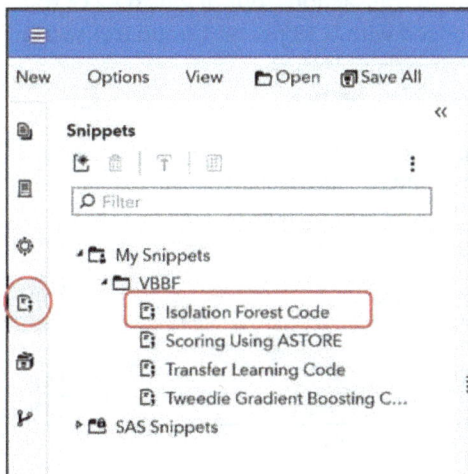

The program opens in the Code Editor window. First, a cas session is started and all the default libraries are assigned.

```
cas;

caslib _all_ assign;
```

4. Run this part of the program. See the log panel on your right.

```
NOTE: The session CASAUTO connected successfully to Cloud Analytic Services server.demo.sas.com using port 5570. The UUID is
      9ceb47cd-63c8-9d45-bd0a-054c66c0692a. The user is student and the active caslib is CASUSER(student).
NOTE: The SAS option SESSREF was updated with the value CASAUTO.
NOTE: The SAS macro _SESSREF_ was updated with the value CASAUTO.
```

5. Go back to the code panel.

The next part of the program loads the **PaySim** data CSV file into CAS and explores the data by the target variable.

```
proc casutil;
load file="/home/student/casuser/VBBF/PaySimData.csv"
importoptions=(filetype="csv")
outcaslib= "public" casout="PaySimData"; *promote;
run;

proc freq data=public.PaySimData;
tables isFraud isFraud*type;
run;
```

6. Run this part of the program. The **Results** tab opens.

The FREQ Procedure

isFraud	Frequency	Percent	Cumulative Frequency	Cumulative Percent
0	6354407	99.87	6354407	99.87
1	8213	0.13	6362620	100.00

The data set consists of around six million transactions, out of which 8213 transactions are labeled as fraud. It is highly imbalanced with 0.13 percent fraud transactions.

Table of isFraud by type

Frequency Percent Row Pct Col Pct	isFraud	CASH_IN	CASH_OUT	DEBIT	PAYMENT	TRANSFER	Total
	0	1399284	2233384	41432	2151495	528812	6354407
		21.99	35.10	0.65	33.81	8.31	99.87
		22.02	35.15	0.65	33.86	8.32	
		100.00	99.82	100.00	100.00	99.23	
	1	0	4116	0	0	4097	8213
		0.00	0.06	0.00	0.00	0.06	0.13
		0.00	50.12	0.00	0.00	49.88	
		0.00	0.18	0.00	0.00	0.77	
	Total	1399284	2237500	41432	2151495	532909	6362620
		21.99	35.17	0.65	33.81	8.38	100.00

The data set contains five categories of transactions labeled as CASH_IN, CASH_OUT, DEBIT, PAYMENT, and TRANSFER. Going forward, we examine only the CASH_OUT and TRANSFER transaction types, as these are the categories containing the fraudulent observations. Limiting the data set to these categories reduces the total number of observations to 2,770,409.

7. Go back to the code panel.

The next part of the program creates an isolation forest model using the ISOLATION option in a FOREST procedure and shows a sample of the scored data.

```
proc forest data=public.PaySimData (where=(type='CASH_OUT' or
type='TRANSFER'))
        isolation outmodel=public.PaySimIsoForest seed=8844
        ntrees=50 numbin=32 minleafsize=1 maxdepth=8 vars_to_try=3;
input amount oldbalanceOrg newbalanceOrig oldbalanceDest
newbalanceDest / level = interval;
target isFraud / level = nominal;
id nameOrig;
output out=public.PaySimScore copyvars=(_ALL_);
run;

proc print data=public.PaySimScore (obs=10);
var nameOrig amount oldbalanceOrg newbalanceOrig oldbalanceDest
newbalanceDest isFraud _Anomaly_;
run;
```

This piece of code runs the FOREST procedure to generate an isolation forest with 50 trees. The OUTPUT statement scores the training data and saves the results to a new table named PaySimScore, which contains the anomaly score for each observation. For the purposes of our comparison, we will use the **isFraud** variable as a check for our unsupervised model's ability to detect the fraudulent observations. However, as inputs to the model, we will use only the **amount, oldbalanceOrg, newbalanceOrig, oldbalanceDest**, and **newbalanceDest** variables.

Note: The FOREST procedure does require a target variable. The target is ignored with the ISOLATION option. The target can be a constant or otherwise random. All that matters is that one is specified, and it is not specified as an input.

8. Run this part of the program.

9. An isolation forest model is created with 50 trees.

The FOREST Procedure	
Model Information	
Number of Trees	50
Number of Variables Per Split	1
Seed	99
Bootstrap Percentage	100
Number of Bins	32
Maximum Number of Tree Nodes	197
Minimum Number of Tree Nodes	35
Maximum Number of Branches	2
Minimum Number of Branches	2
Maximum Depth	8
Minimum Depth	8
Maximum Number of Leaves	99
Minimum Number of Leaves	18
Maximum Leaf Size	2769993
Minimum Leaf Size	1
Average Number of Leaves	52.04

Note: You will not get the variable importance table in an isolation forest model. Isolation forest scores typically contain the anomaly scores (_Anomaly_) for each observation. A sample of the first 10 observations is shown.

Obs	nameOrig	amount	oldbalanceOrg	newbalanceOrig	oldbalanceDest	newbalanceDest	isFraud	_Anomaly_
1	C214996481	118206.10	0	0	1327819.45	1446025.55	0	0.4975253813
2	C137598890	331251.56	0	0	2525811.54	2857063.10	0	0.498853334
3	C2133482303	750229.56	0	0	2199564.71	2949794.27	0	0.5004683403
4	C1264432372	251758.89	0	0	400764.79	652523.68	0	0.4970489898
5	C1135956736	983131.48	0	0	2705946.75	3689078.22	0	0.5004683403
6	C1462702619	2329617.01	0	0	4895783.67	7225400.68	0	0.5082754194
7	C2054244450	1781661.06	0	0	2059437.71	3841098.77	0	0.502463254
8	C1554889164	284604.24	103246	0	994461.40	1279065.64	0	0.4991535766
9	C896130058	429459.01	247188	0	583091.22	1012550.23	0	0.4998129644
10	C452570998	82976.95	12430	0	3834.51	86811.45	0	0.4977269691

The higher the anomaly score, the more anomalous the observation is.

10. Go back to the code panel.

The last part of the program reduces the scored table for the purpose of quick evalua-tions, interpretable graphs that includes a scatter plot of anomaly scores on two of the input variables, and a histogram of anomaly scores grouped by actual fraud cases.

```
proc partition data=public.PaySimScore samppct=7 seed=8844 partind
nthreads=3;
   by isFraud;
   output out=public.SampledScore copyvars=(_ALL_);
run;

proc template;
   define statgraph anomalyPlot;
      begingraph;
         layout overlay;
         scatterplot y=amount x=newbalanceDest /
            name='color'
            markerattrs=(symbol=circlefilled)
            colormodel=(cyan ligr red)
            colorresponse=_Anomaly_;
            continuouslegend 'color'/ title='_Anomaly_';
         endlayout;
      endgraph;
   end;
 run;

ods graphics on;
proc sgrender data=public.SampledScore (where=(_PartInd_=1))
template=anomalyPlot;
run;

proc sgplot data=public.SampledScore (where=(_PartInd_=1));
      histogram _Anomaly_ / group=isFraud;
      yaxis grid;
run;
ods graphics / reset;
```

The graph template plots the data with a color ramp from cyan to red, based on the anomaly score. The most anomalous observations are colored red.

11. Run this last part of the program now.

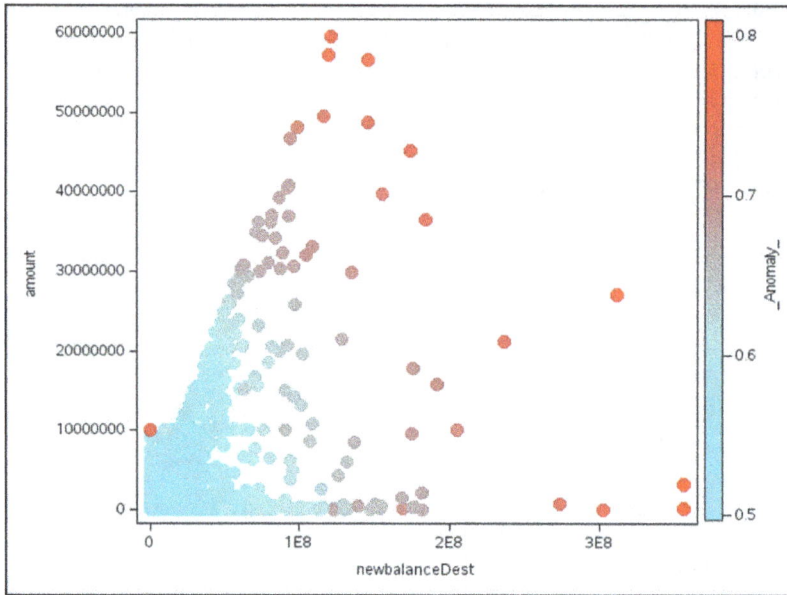

Most of the anomalous observations are the ones that have higher transaction amounts and higher recipients' account balance after transaction. Before generalizing this outcome, note that this is based on a sample of only 7% from the scored data. It would be interesting to see similar graphs on the other input variables also.

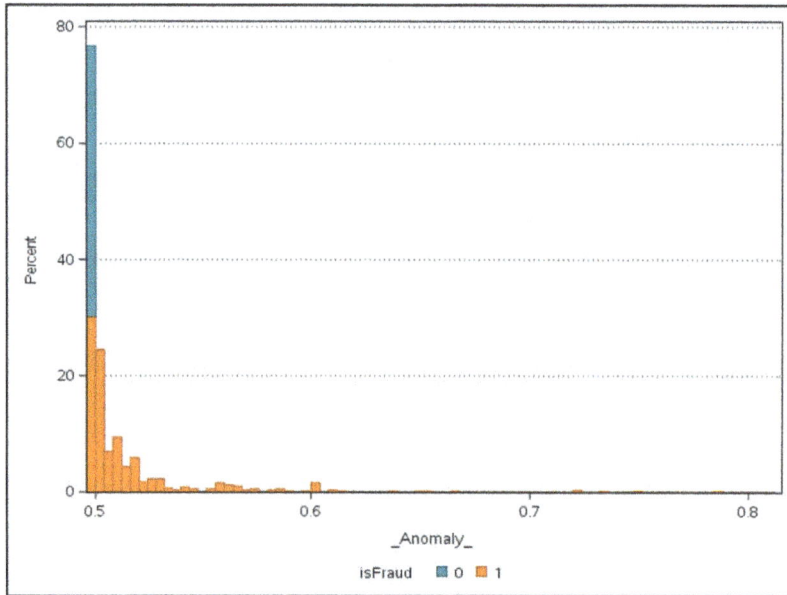

The histogram indicates that a substantial portion of observations that were fraud (orange) have higher anomaly scores (only anomaly scores higher than 0.5 are shown). However, in a practical situation you might not have target labels to compare with. In fraud or other rare event detection, a company might also be limited by the number of events it can investigate or assess. As such, the isolation forest is also evaluated by looking at the number of events captured within the 1,000 observations (for example) with the highest anomaly score. The thinking being that if the company had only the time to look at a set number of observations, they would prefer to maximize the number of true positives within that resource constrained span.

12. Select **Build Models** from the **Applications** menu (☰) to close SAS Studio and come back to the Model Studio.

End of Demonstration

Introduction to Deep Forest Models

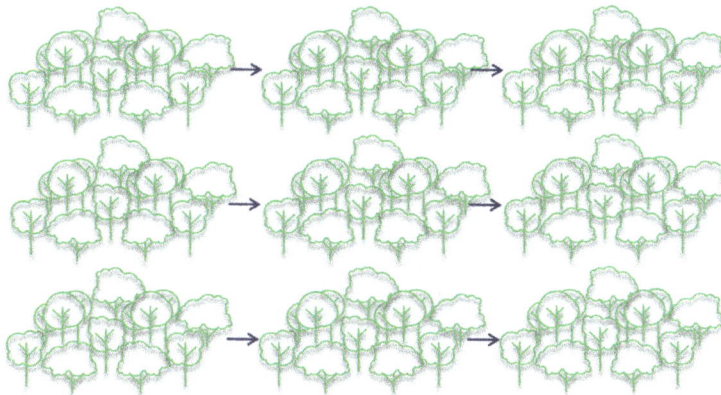

A new deep learning framework was introduced by Zhi-Hua Zhou and Ji Feng in 2017 as an alternative to deep neural networks (DNNs) in a seminal paper "Deep Forest: Towards an Alternative to Deep Neural Networks." Commonly called deep forest, also known as gcForest (multi-grained cascade forest), this framework combines several ensemble-based methods, including random forests and stacking, into a structure that is like a multi-layer neural network, but each layer in the deep forest contains random forests instead of neurons.

Current deep learning models are mostly built upon neural networks, that is, multiple layers of parameterized differentiable nonlinear modules that can be trained by back propagation. The deep forest algorithm builds deep models based on non-differentiable modules. Success of deep neural networks depends much on three characteristics: layer-by-layer processing, in-model feature transformation, and enough model complexity. Deep forest holds similar characteristics. This is a decision tree ensemble approach, with fewer hyper-parameters than deep neural networks, and its model complexity can be automatically determined in a data-dependent way.

Deep forest is simple for training due to a small number of hyper-parameters, it does not use backpropagation training, and it outperforms many well-known methods, including deep neural networks, when there is only small-scale training data.

Inspired by the deep neural networks, deep forest has a multi-layer cascade structure, but each layer contains many random forests instead of neurons in deep neural networks. Deep forest consists of two ensemble components as shown in Figure 7.12. The first one is multi-grained scanning, which adopts a sliding window structure to scan local context from high dimensionality to learn representations of input data according to different random forests. This is an optional ensemble component to be used when raw inputs have high dimensionality. The second one is the cascade forest, which learns more discriminative representations under supervision of input representations at each layer, thus giving more accurate predictions according to ensemble of random forests. If you do not want to do the multi-grained scanning, raw input features are directly used as input to the cascade forest instead of a feature vector coming out of a multi-grained scan.

Deep forests use traditional forests as a subroutine, adding depth and sliding-window application as the main innovative ideas.

As shown in Figure 7.13, deep forest uses a cascade structure, where each level of cascade receives feature information processed by its preceding level and sends its processing result to the next level. The trees in the first-layer forests have as their input either the feature vector or the raw data (if multi-grained scanning is not performed). The trees deeper within the model have as their input the raw data concatenated with the classifications according to the

Figure 7.12: Deep Forest: High-Level View

The deep forest has a main structure called the cascade forest and an additional component called the multi-grained scanning.

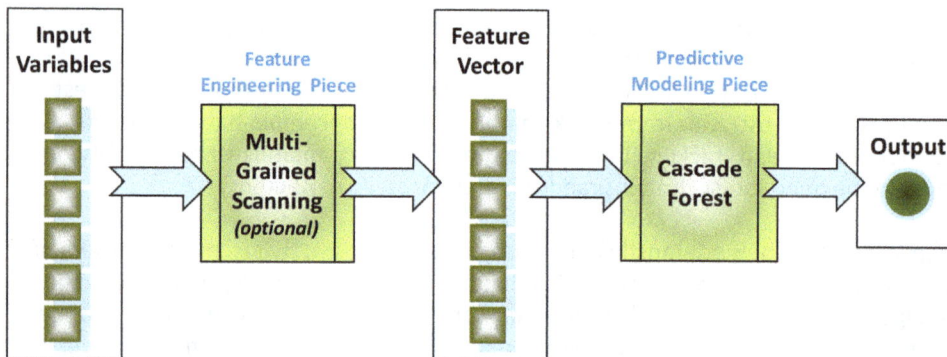

Figure 7.13: Cascade Structure

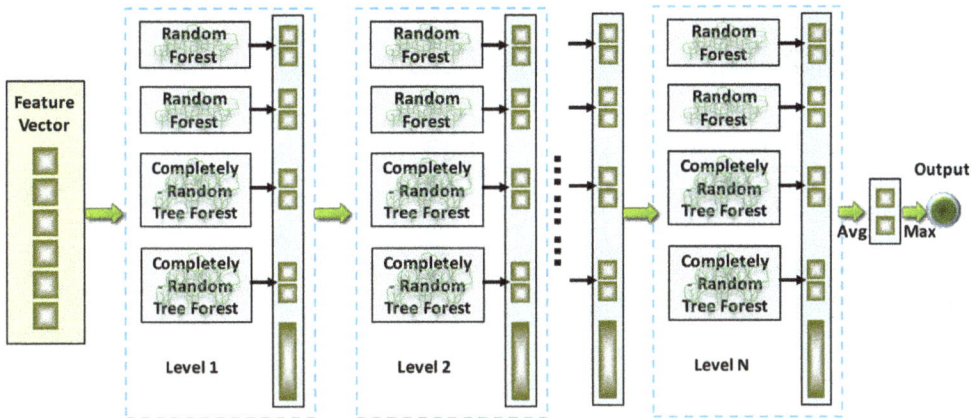

previous layer's forests. Each level is an ensemble of decision tree forests, that is, an ensemble of ensembles.

Diverse types of forests are included to encourage the diversity, because it is crucial for ensemble. For example, two completely random tree forests and two random forests are used. Completely random tree forest contains trees generated by randomly selecting a feature for split at each node of the tree. Similarly, each random forest contains trees by randomly selecting $\sqrt{d}$ number of features as candidate (d is the number of input features) and choosing the one with the best Gini value for split.

The forests are directly trained to classify whatever they are connected to regardless of their level of depth within the deep forest. However, only the ultimate layer's predictions are averaged as the overall prediction used in evaluation; this is unlike most deep neural networks where only the last layer is computing a classification and the others are hidden layers trained via backpropagation and therefore not trivially interpretable.

Because a key element in the feature representation at every cascade level of deep forest is a probability class vector of every random forest, which are computed by averaging the corresponding decision tree-class vectors across all trees in the random forest, you can then specify a loss function by controlling the class vectors by means of weighing trees and computing weighted averages of the tree-class vectors, where weights are trained using the same training data. These features are simply predictions from individual random forests, trained using the final class labels, which involves stacking.

Cascade forest structure representation learning in DNNs mostly relies on the layer-by-layer processing of raw features. Inspired by this recognition, deep forest takes advantage by a procedure of multi-grained scanning shown in Figure 7.14. Sliding windows are used to scan the raw features.

Figure 7.14: Multi-Grained Scanning

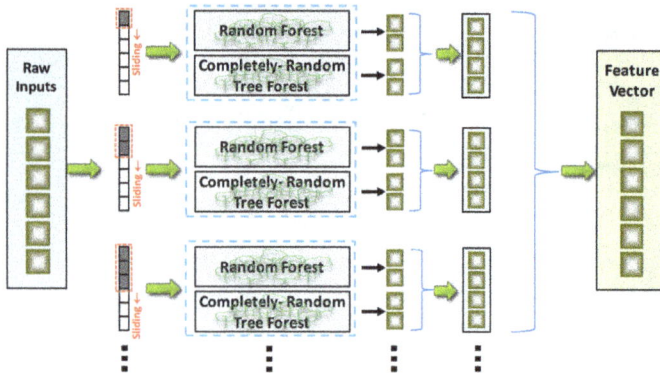

For example, suppose that the original input is of 400 raw features, and three window sizes of 100, 200, and 300 features are used for multi-grained scanning. For sequence data, a 100-dimensional feature vector is generated by sliding the window for one feature; in total 301 feature vectors are produced. For *n* training examples, a window with size of 100 features generates a data set of 301×*n* 100-dimensional training examples. These data will be used to train a completely random tree forest and a random forest, each containing say 500 trees.

In case of a binary target where two classes are to be predicted, a 1,204-dimensional feature vector [(301+301)+(301+301)] will be obtained. All feature vectors extracted from training examples are regarded as instances, which will then be used to generate class vectors. These class vectors are generated and concatenated as transformed features. The instances extracted from the sliding windows are simply assigned labels of the original training example.

The transformed training set will then be used to train the first grade of the cascade forest. Given a test instance, it will go through the multi-grained scanning procedure to get its corresponding transformed feature representation, and then go through the cascade until the last level. The final prediction will be obtained by aggregating the four three-dimensional class vectors at the last level and taking the class with the maximum aggregated value.

Deep Forest Actions in SAS Viya

SAS Viya comes with the capability of training and scoring the deep forest models. The deep forest algorithm is included in the **decisionTree** action set. See Figure 7.15. You can focus on two of the actions. The **deepForestTrain** action trains a deep forest model and the **deepForestScore** action scores a table using deep forest. Both actions are currently hidden. The **tuneDeepForest** is another action coming up soon to tune the hyperparameters of a deep forest model. In each layer, the number of forests and the number of trees in each forest are hyperparameters in practice.

Currently these actions are experimental and therefore hidden. Deep forest demonstration is not included here.

Figure 7.15: Deep Forest Actions in the decisionTree Action Set

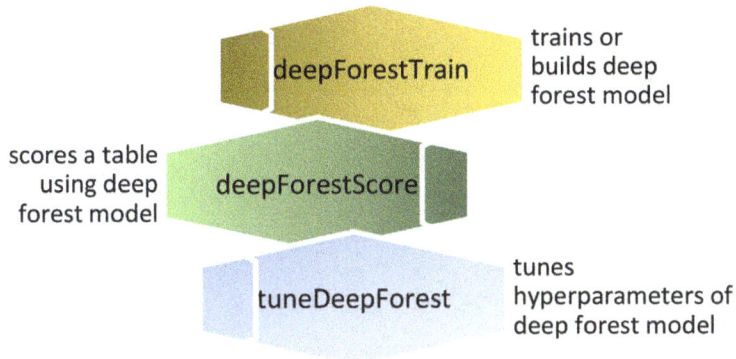

deepForestTrain — trains or builds deep forest model

deepForestScore — scores a table using deep forest model

tuneDeepForest — tunes hyperparameters of deep forest model

Quiz

1. Which of the following statements is TRUE regarding the isolation forest model?
   a. It works on the principle of isolating anomalies, instead of the most common techniques of profiling normal points.
   b. It is a supervised learning algorithm.
   c. It is used for feature extraction.
   d. All of the above

Answers

1. a

Chapter 8: Tree-Based Gradient Boosting Machines

Gradient Boosting Models

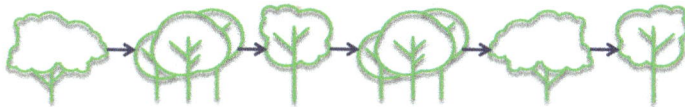

Gradient boosting is an iterative approach that creates multiple models, preferably decision trees, on the training data. A tree-based gradient boosting machine is essentially a boosting algorithm that creates a series of classification or regression trees. The boosting adds new decision trees to the ensemble sequentially. At each particular iteration, a new weak, base-learner model is trained with respect to the error of the whole ensemble that was trained so far.

The gradient boosting model improves its predictions by minimizing a specified loss function, such as average square error. The first step creates a baseline tree. Each subsequent tree is fit to the residuals of the previous tree, and the loss function is minimized. This process is repeated a specific number of times. The final model is a single function, which is an aggregation of the series of trees that can be used to predict the target value of a new observation.

Allowing trees to be greedily formed from subsamples of the training data set was a major breakthrough in bagging ensembles and forest models. This advantage can also be used to reduce the correlation between the trees in the sequence in a gradient boosting model. Stochastic gradient boosting is the name for this form of boosting.

The term *stochastic gradient boosting* refers to training each new tree based on a subsample of the data. This typically results in a better model. For gradient boosting models, each new observation is fed through a sequence of trees that are created to predict the target value of each new observation.

Figure 8.1: Stochastic Gradient Boosting

- At each iteration, an independent subsample of the training data is drawn at random (without replacement) from the full training data set to fit the base learner.
- This reduces the correlation of the predictions of the trees, which in turn improves the boosting model.

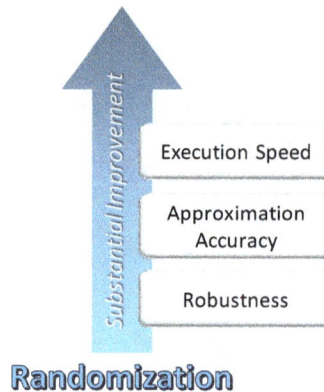

Key Components of Gradient Boosting

Gradient boosting typically involves four imperative components that are listed in Figure 8.2. First, a weak learner to make predictions (for example, a decision tree model). Second, a loss function to measure the difference between the observed target and the predicted value of the target (for example, average squared error). Third, a numerical optimization method to optimize the loss function (for example, gradient descent). Fourth, an additive ensemble model to add several weak learners (for example, boosting algorithm). The name *gradient* boosting comes from the last two elements, gradient descent and boosting.

Each of these four components and their associated properties in SAS Viya are discussed next.

Figure 8.2: Components of a Gradient Boosting Model

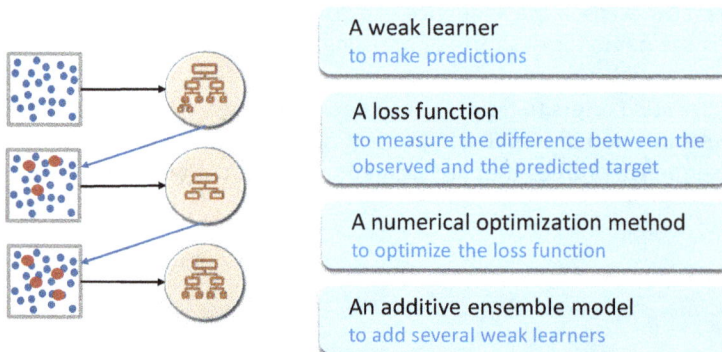

Weak Learner Model

The idea behind boosting is that each sequential model builds a simple weak model to improve the remaining errors. A weak model is one whose error rate is slightly better than conjecturing randomly. Gradient boosting nearly at all times uses decision trees as the weak learners because in many practical applications, small trees and tree-stumps (a special case of a decision tree with only one split, that is, a tree with two terminal nodes) provide considerably accurate results (Wenxin 2002).

Every tree is considered as a weak learner in that iteration and improved in the subsequent iteration. Many weak learners together form a strong predictive model. With regard to decision trees, shallow trees (with only one to six splits) represent a weak learner. It is common to constrain the decision tree weak learners in specific ways, such as number of branches, maximum depth, leaf sizes, and so on.

Tree-Splitting Options You Already Know

You are already familiar with the splitting options for individual trees that are used as weak learners in a gradient boosting model. They have been discussed in Chapter 3. Here is a quick discussion of such options in the context of gradient boosting models.

Tree Structure and Complexity

Gradient boosting in SAS Viya offers the options shown in Figure 8.3 for controlling the structure and complexity of a decision tree weak learner model.

Maximum number of branches specifies the maximum number of branches to consider for a training node split in the tree. Possible values range from 2 to 5. The default value is 2.

Maximum depth specifies the maximum number of generations of nodes. The original node, generation 0, is called the root node. The children of the root node are the first generation. Possible values range from 1 to 50. The default value is 4.

Figure 8.3: Branches, Depth, and Leaf Size Options

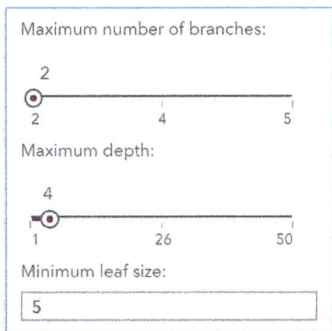

Minimum leaf size specifies the smallest number of training observations that a leaf can have. The default value is 5.

Missing Values

You already know that the tree-based models use observations that have missing input values. If the value of a target variable is missing, the observation is excluded from training and from evaluating the model. Gradient boosting in SAS Viya offers almost same strategies for handling missing values as offered for the forest models. Figure 8.4 represents the available options.

Missing values specifies the action to take when there are values missing in an input variable. You can specify to **Ignore, Use as machine smallest**, or **Use in search**. The default value is **Use in search**.

Minimum missing use in search specifies a threshold for using missing values in the split search. The default value is 1.

Interval Binning

You are already familiar with binning of the interval variables in tree-based models. Figure 8.5 represents these options.

Number of interval bins specifies the number of bins in which to bin the interval input variables. Bin size is (maximum value - minimum value)/(**Number of interval bins**). The default value is 50.

Figure 8.4: Missing Values Options

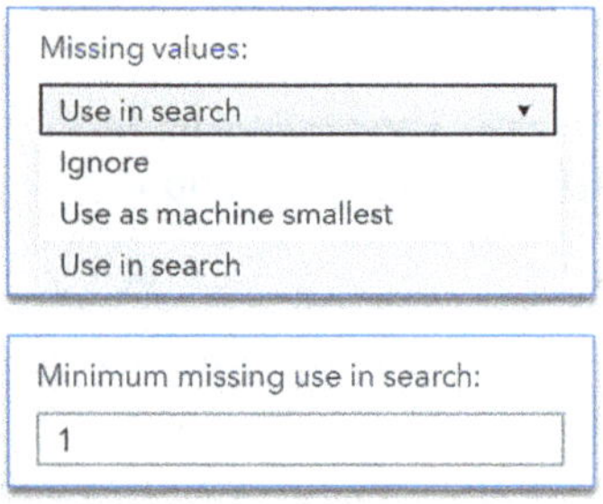

Figure 8.5: Interval Binning Options

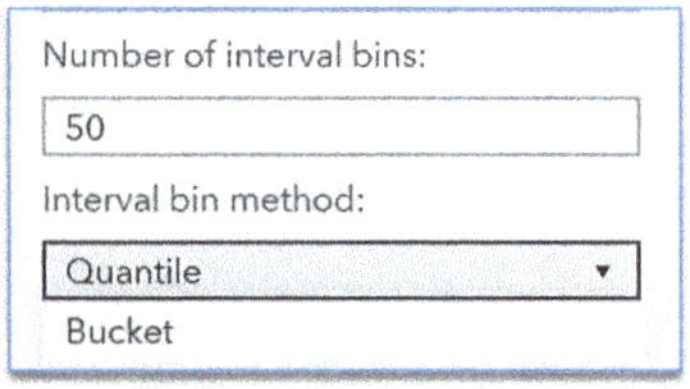

Interval bin method specifies the method used to bin the interval input variables. Select **Bucket** to divide input variables into evenly spaced intervals based on the difference between maximum and minimum values. Select **Quantile** to divide input variables into approximately equal sized groups. The default value is **Quantile**.

Number of Inputs for Split

Gradient boosting in SAS Viya trains a decision tree by splitting the subsampled data, and then splitting each resulting segment, and so on recursively until some constraint is met. Splitting involves selecting some candidate input variables independently in every node. Then a split search is performed on all the selected variables, and the best rule is kept splitting the node. Properties related to number of inputs for splitting are shown in Figure 8.6.

The split search seeks to maximize the reduction in the gain for a nominal target and the reduction in variance of an interval target.

Use default number of inputs to consider per split specifies whether to use the default number of inputs to consider per split. By default, this option is selected.

Number of inputs to consider per split specifies the number of input variables randomly sampled to use per split. This option is not available if **Use default number of inputs to consider per split** is selected. The default value is 100.

Loss Function

Loss function measures the difference between the observed target and the predicted value of the target. Loss function is a method of evaluating how well your algorithm models your data set. If your predictions are totally off, obviously, the loss function will result in a higher number. If they're pretty good, it will yield a lower number. The objective is to minimize this loss function in every iteration of the gradient boosting model.

The loss function used depends on the type of target. The gradient boosting in SAS Viya uses one of the following three loss functions:

- For the continuous target, the loss function is the L2 loss, also known as squared error loss, which is the sum of all the squared differences between the true value and the predicted value.

Figure 8.6: Number of Inputs Options

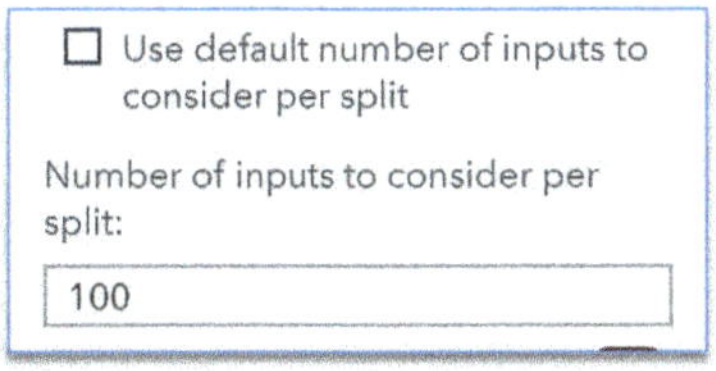

- For the binary target, the loss function is the logistic loss.
- For the nominal target, the loss function is the multiclass logistic loss.

You do not have a choice of changing them, and it is not required either.

Numerical Optimization Method

The whole point of gradient boosting is to find the function that best approximates the data, and therefore, it is primarily an optimization problem. Here the objective is to minimize the cost function, which is the average of the loss functions for all the training examples represented in Figure 8.7.

A cost function (or the loss function) evaluates the performance of your model. The loss function computes the error for a single training example, and the cost function is the average of the loss functions for all the training examples. Henceforth, these terms will be used interchangeably in the book.

A tree ensemble that seeks to predict an observed target, y_i, by using the input, x_i, can be written as

$$\hat{y}_i = F(x_i) = \sum_{m=1}^{M} f_m(x_i)$$

where f_m is the functional representation of the mth tree.

The Gradient Boosting node (or GRADBOOST procedure) seeks to minimize the regularized objective function shown above where $l(\hat{y}_i, y_i)$ is the loss function that measures the difference between the observed target, y_i, and the predicted value of the target, $\hat{y}_i$.

You use gradient descent to update the parameters of your model. The concept of gradients is more general and useful than the concept of residuals. This enables you to consider other loss functions and derive the corresponding algorithms in the same way.

Figure 8.7: Objective Function in Gradient Boosting

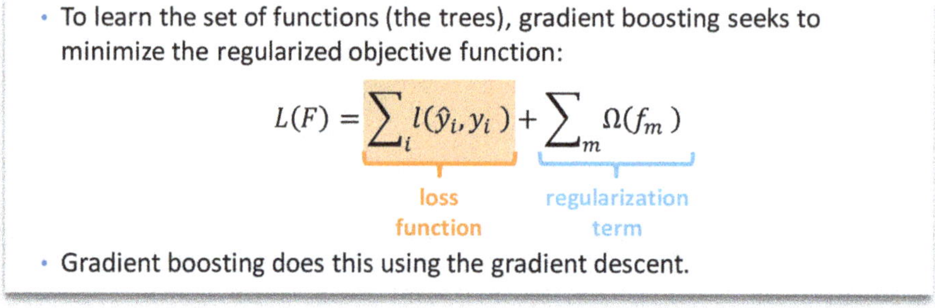

- To learn the set of functions (the trees), gradient boosting seeks to minimize the regularized objective function:

$$L(F) = \sum_i l(\hat{y}_i, y_i) + \sum_m \Omega(f_m)$$

loss function   regularization term

- Gradient boosting does this using the gradient descent.

Gradient Descent

Gradient descent is one of the most generic optimization algorithms capable of finding optimal solutions to a wide range of problems. The general idea of gradient descent is to modify parameters iteratively in order to minimize a cost function. It measures the local gradient of the error function with regard to the parameter vector ϑ, and it goes in the direction of descending gradient. Once the gradient is zero, you have reached a minimum! You start by filling ϑ with random values (this is called *random initialization*), and then you improve it gradually, taking one baby step at a time, each step attempting to decrease the cost function, until the algorithm *converges* to a minimum.

The gradient descent optimization method minimizes a cost function by iteratively moving in the direction of steepest descent as defined by the negative of the gradient. This is graphically illustrated in Figure 8.8.

Gradient means steepness of a slope (incline or decline) that is commonly measured as a ratio of the vertical distance to the actual distance along the slope, that is, sine of the slope angle or a ratio of the vertical distance (elevation) to the horizontal distance, that is, tangent of the slope angle. One of the most trivial examples includes steepness of a straight line, and therefore, in a simple linear regression case, slope of the line is going to be the gradient. Gradient boosting essentially involves minimizing some loss function such as a squared error loss function. This is directly related to gradient descent, which is an optimization method for minimizing an objective function that is written as a sum of differentiable functions. Gradient descent is based on the observation that if the multi-variable function $F(x_{ij})$ is defined and differentiable in a neighborhood of a point a, then $F(x_{ij})$ decreases *fastest* if one goes from a in the direction of the negative gradient of F at $(a, -\nabla F(a))$.

Suppose we have a loss function $L(y_i, F(x_{ij})) = (y_i - F(x_{ij}))^2/2$ that you want to minimize. Because $F(x_{ij})$ are only some numbers, we can treat $F(x_{ij})$ as parameters and take derivatives:

$$\frac{\partial J}{\partial F(x_{ij})} = \frac{\partial \sum_i L(y_i, F(x_{ij}))}{\partial F(x_{ij})} = \frac{\partial L(y_i, F(x_{ij}))}{\partial F(x_{ij})} = F(x_{ij}) - y_i$$

Figure 8.8: Gradient Descent Illustrated

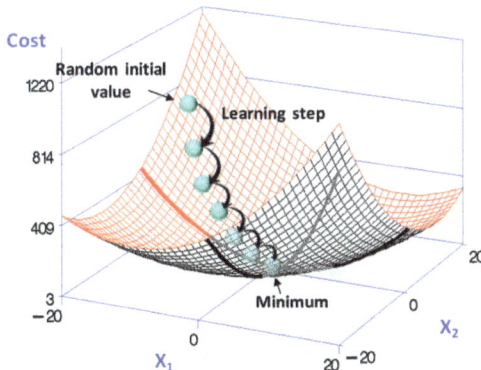

To find a local minimum of a function using gradient descent, one takes steps proportional to the *negative* of the gradient (or of the approximate gradient) of the function at the current point.

$$y_i - F(x_{ij}) = -\frac{\partial J}{\partial |F(x_{ij})}.$$

If $G(x_{ij})$ is an additional model that is added to compensate the shortcoming of the existing model, then

$$
\begin{aligned}
F(x_{ij}) :&= F(x_{ij}) + G(x_{ij}) \\
&= F(x_{ij}) + y_i - F(x_{ij}) \\
&= F(x_{ij}) - 1\frac{\partial J}{\partial F(x_{ij})}
\end{aligned}
$$

Therefore, residuals can be interpreted as negative gradients.

$$\theta_i := \theta_i - \rho\frac{\partial J}{\partial \theta_i}$$

A gradient is a vector pointing at the greatest increase of a function, and a negative gradient is a vector pointing at the greatest decrease of a function. Therefore, we can minimize a function by iteratively moving a little bit in the direction of negative gradient.

Gradient Descent Characteristics

Because gradient is a vector, it has characteristics like direction and magnitude. If you compute the derivative of a function, you know in which direction to proceed to minimize it. The gradient always points in the direction of steepest increase in the cost. The gradient descent algorithm takes a step in the direction of the negative gradient in order to rapidly reduce the loss. These features and associated properties are represented in Figure 8.9.

Figure 8.9: Optimization Properties

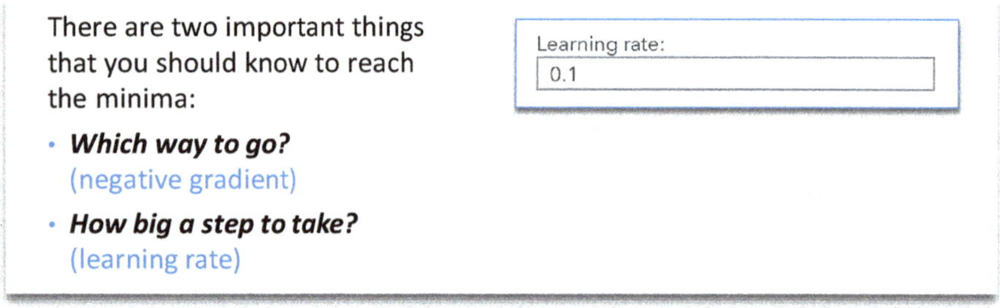

There are two important things that you should know to reach the minima:

Learning rate:
0.1

- *Which way to go?*
  (negative gradient)
- *How big a step to take?*
  (learning rate)

To determine the next point along the cost function curve, the gradient descent algorithm adds some fraction of the gradient's magnitude to the starting point and then repeats this process, edging ever closer to the minimum. The size of these steps is called the learning rate.

Learning rate specifies the step size for gradient descent optimization. The default value is 0.1.

Learning rate governs the contribution of each tree on the outcome and controls how quickly the algorithm proceeds down the gradient descent (learns). This hyperparameter is also called *shrinkage*.

If the learning rate is too small (left graph), then the algorithm has to go through many iterations to converge. You can confidently move in the direction of the negative gradient because you are recalculating it so frequently. A low learning rate is more precise, but calculating the gradient is time-consuming, so it will take you a long time to get to the bottom. (See the graph on the left in Figure 8.10.)

In contrast, if the learning rate is too high (right graph), you might jump across the valley and end up on the other side, possibly even higher up than you were before. This might make the algorithm diverge, with larger and larger values, failing to find a good solution. You can cover more ground with each step, but you risk overshooting the lowest point because the slope of the hill is constantly changing. (See the graph on right in Figure 8.10.)

Additive Ensemble Model

Gradient descent is used to minimize a set of parameters, such as the coefficients in a regression equation or weights in a neural network. After calculating error or loss, the weights are updated to minimize that error. Instead of parameters, we have weak learner decision trees. After calculating the loss, to perform the gradient descent procedure, the algorithm adds a tree to the ensemble model that reduces the loss. This is done by parameterizing the tree, and then

Figure 8.10: Learning Rate

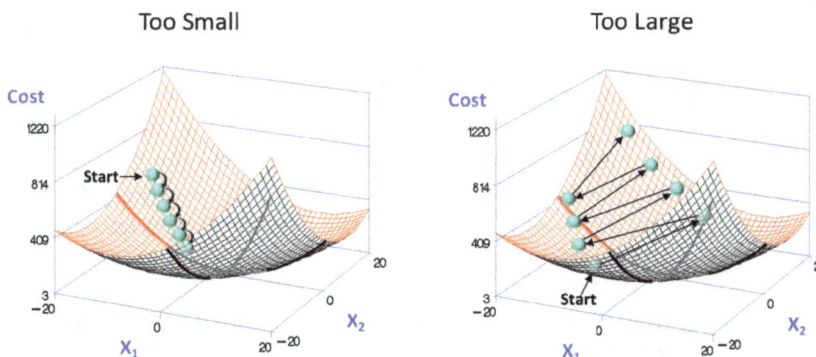

Figure 8.11: Boosting

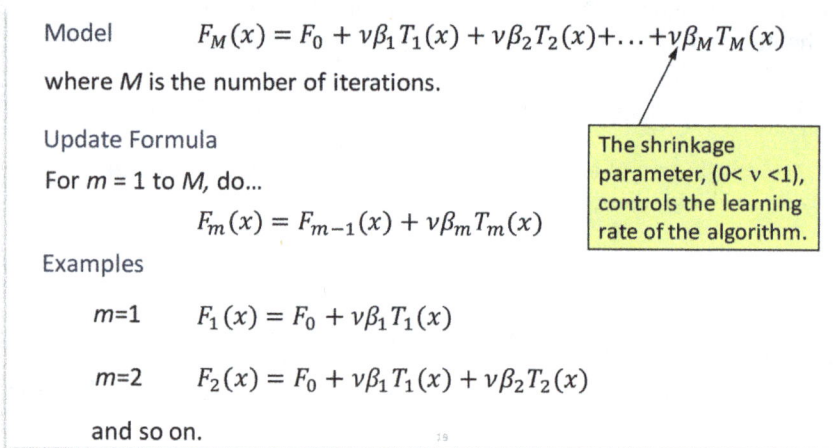

Model
$$F_M(x) = F_0 + v\beta_1 T_1(x) + v\beta_2 T_2(x) + \ldots + v\beta_M T_M(x)$$

where *M* is the number of iterations.

Update Formula

For *m* = 1 to *M*, do…

$$F_m(x) = F_{m-1}(x) + v\beta_m T_m(x)$$

> The shrinkage parameter, (0< v <1), controls the learning rate of the algorithm.

Examples

m=1 $\quad F_1(x) = F_0 + v\beta_1 T_1(x)$

m=2 $\quad F_2(x) = F_0 + v\beta_1 T_1(x) + v\beta_2 T_2(x)$

and so on.

modifying the parameters of the tree and moving in the right direction, that is, following the gradient. This approach is called functional gradient descent or gradient descent with functions.

The gradient boosting algorithm is an additive ensemble model that adds several weak learners. It is a weighted $(\beta_1\ldots\beta_M)$ linear combination of (usually) simple decision trees $(T_1\ldots T_M)$ (Friedman 2001). This is illustrated in Figure 8.11.

The Gradient Boosting node (or GRADBOOST procedure) in SAS Viya creates a series of trees that together form a single predictive model. Trees are added one at a time, and existing trees in the model are not changed. Each tree uses a subsample of the data. The sequence of trees and how each tree affects a subsequent tree are discussed in Hastie, Tibshirani, and Friedman (2001); Friedman (2001); and Chen and Guestrin (2016).

The Gradient Boosting node runs a stochastic gradient boosting that is like standard boosting, except that on each iteration, the target is the residual from the previous decision tree model. The residual is defined in terms of the derivative of a loss function. A gradient descent procedure is used to minimize the loss when adding trees. Each time that the data are used to grow a tree, the accuracy of the tree is computed. Successive samples are adjusted to accommodate previous inaccuracies. Each successive sample is weighted per the accuracy of the previous models.

Begin with an initial guess, F_0, and proceed in a stage-wise manner fitting subsequent (*m*) tree models to "pseudo" residuals $(\tilde{y}_{im})$. The residuals are computed from target values (y_i) and predictions from the function at the previous iteration $(F_{m-1}(x_i))$. The function $F_m(x)$ is updated by adding the fitted model, $v\beta_m T_m(x)$ to $F_{m-1}(x)$.

The shrinkage parameter, $v(0<v<1)$ controls the learning rate of the algorithm. Friedman (2001) found that small values (≤ 0.1) lead to better generalization.

In regression trees with interval targets and least-square loss criterion, the "pseudo" residual, $\tilde{y}_{im}$, and the "guess," F_0, are defined as follows:

$$\tilde{y}_{im} = y_i - F_{m-1}(x_i)$$

$$F_0 = \bar{y}$$

In classification trees with a binary target ($y \in \{-1,1\}$) and binomial log-likelihood loss criterion, the "pseudo" residuals and F_0 are as follows:

$$\tilde{y}_{im} = 2y_i / (1 + exp(2y_i F_{m-1}(x_i)))$$

$$F_0 = \frac{1}{2} log \left(\frac{1+\bar{y}}{1-\bar{y}}\right).$$

A gradient boosting tree fit to the Boston housing data with an interval target (median home value, **MEDV**) provides a simple illustration of the model is shown in Figure 8.12. Here the number of iterations is two (M=2) and maximum tree depth is one. The resulting model is a combination of two decision trees (T1 and T2) each with two leaves. The value of 22.275 is the mean **MEDV** in the data (F_0) and **P_MEDV** is the predicted value. For example, an observation with LSTAT=6 and RM=5 would have a **P_MEDV** value of 22.275 + 0.95 − 0.17.

Perturbing and Combining Boosted Trees

While perturbing and combining trees in a gradient boosting model, there are two important characteristics – the total number of trees to fit and not getting stuck in a local minimum or plateau of the loss function gradient by randomizing the samples. These features and associated properties are represented in Figure 8.13.

Gradient boosting often requires many trees. However, unlike forests, gradient boosting can overfit, so the goal is to find the optimal number of trees that minimize the loss function.

Figure 8.12: Simple Example of Gradient Boosting

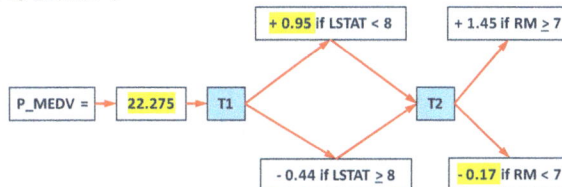

Figure 8.13: Perturb and Combine Properties

Two distinguishing characteristics
in a tree-based gradient boosting:

- *How many trees to boost
  the ensemble?*

Number of trees:
100

- *How much randomization
  to induce?*

Subsample rate:
0.5

The **Number of trees** property specifies the number of iterations in the boosting series. For interval and binary targets, the number of iterations equals the number of trees. For a nominal target, a separate tree is created for each target category in each iteration. The default value is 100.

Subsampling mostly improves the generalization properties of the model, at the same time reducing the required computation efforts (Sutton 2005). The idea is to introduce some randomness into the learning procedure. At each learning iteration, only a random part of the training data is used to fit a consecutive weak-learner decision tree. The training data is sampled without replacement and you can specify the proportion of random sampling.

The **Subsample rate** property specifies the proportion of training observations to train a tree with. A different training sample is taken in each iteration. Trees trained in the same iteration have the same training data. The default value is 0.5.

Another advantage of subsampling is that it naturally adapts to large data sets when there is no reason to use all the potentially enormous amounts of data at once.

To demonstrate gradient boosting, Natekin and Knoll (2013) used an artificially generated data set. The data set is sampled from a sinc(x) function with two sources of artificially simulated noise: the Gaussian noise component $\varepsilon \sim N(0, \sigma 2)$ and the impulsive noise component $\xi \sim$ Bern(p). For this experiment, they used the $L2$ loss. As the input dimension is equal to one, they chose to use the tree-stumps. The resulting fitted models are shown in Figure 8.14. To demonstrate the progress of the fitting procedure, the number of iterations *nTrees* was varied

Figure 8.14: Different Number of Boosts

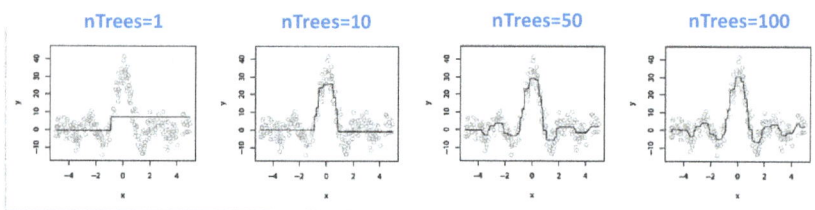

Image source:
Natekin and Knoll (2013). Gradient boosting machines, a tutorial. *Front. Neurorobot.* 7:21. doi: 10.3389/fnbot.2013.00021

from 1 to 500. The similar behavior of consecutive improvements in the fit accuracy, when the number of iterations *nTrees* increases, is apparent on these charts. However, the model can easily overfit the data with the increasing number of iterations.

Avoiding Overfitting in Gradient Boosting

One of the most important concerns about building a machine-learning model like gradient boosting is the resulting model's generalization capabilities. You can easily imagine a situation where new weak-learner trees are added to the ensemble until the data is completely overfitted.

Gradient boosting has quite a few hyperparameters that overlap in their purpose. Tree complexity can be controlled by maximum depth, minimum leaf size, or maximum number of branches. Any combination of these might be optimal for some problem. Overfitting can also be combatted with the learning rate versus the number of trees (and early stopping), subsample rate, and either of the regularization penalties.

All these different approaches help to constrain the fitting procedure and thus balance the predictive performance of the resulting model (Sutton 2005, Zhang and Yu 2005, Zou and Hastie 2005).

Regularization and early stopping are two important post-pruning methods in SAS Viya and are illustrated in Figure 8.15.

Regularization limits how extreme the weights (or influence) of the leaves in a tree can become.

Another method to combat overtraining in gradient boosting is early stopping, which stops the gradient boosting process when some error criterion on the validation data is met. An additional advantage to early stopping is reduced training time in cases where the stopping criterion is met well before the specified maximum number of iterations occur.

Figure 8.15: Post-Pruning Ways to Avoid Overfitting

Apart from many pre-pruning ways, there are two distinguishing post-pruning ways to avoid overfitting in a tree-based gradient boosting.

- ***How extreme the influence of the leaves in a tree can be?***
(regularization)

- ***When to stop the training?***
(early stopping)

Regularization

Regularization actually serves to reduce the depth of trees in gradient boosting because it is applied to *leaf scores* rather than directly to features as in a regression model. Consequently, this tends to reduce the impact of less-predictive features, but it is not so dramatic as removing a feature in the regression model.

The term $\Omega(f_m)$ in the objective function is the regularization term, which acts as a penalty on the model complexity. The decay parameter Ω, which can range from 0 to 1, controls the relative importance of the penalty term. The regularization term and associated properties are shown in Figure 8.16.

You can specify two of the regularization parameters. L1 regularization is like L2 regularization in that both methods penalize the objective function for large ensemble weights. The two regularizations differ in that L2 regularization penalizes the objective function by an amount proportional to the weight size, whereas L1's penalty is constant relative to changes in the weights. This means that L2 is likely to penalize larger weights to a greater degree than L1, but to a lesser degree than L1 when the weights are small.

L1 regularization specifies the L1 regularization parameter (LASSO= option in GRADBOOST procedure). The default value is 0.

L2 regularization specifies the L2 regularization parameter (RIDGE= option in GRADBOOST procedure). The default value is 1.

Early Stopping

When you build a gradient boosting model, a single model is estimated by adding several weak learners sequentially. This model is a weighted linear combination of simple decision trees. The magnitude of the weight coefficients allows the complexity of that model to change.

Figure 8.16: Regularization Options

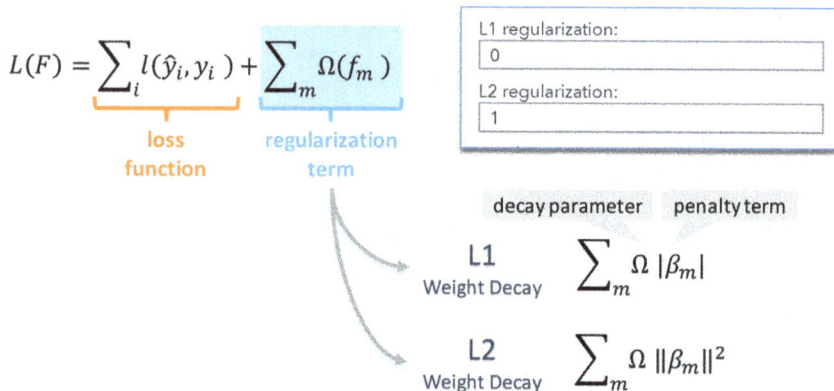

$$L(F) = \sum_i l(\hat{y}_i, y_i) + \sum_m \Omega(f_m)$$

loss function

regularization term

L1 regularization:
0

L2 regularization:
1

decay parameter   penalty term

L1 Weight Decay $\sum_m \Omega \, |\beta_m|$

L2 Weight Decay $\sum_m \Omega \, \|\beta_m\|^2$

Early stopping takes advantage of the fact that boosting is an iterative process. This means that the prediction error can be measured on a validation data set at each iteration of the process. When the prediction error on the validation data meets specified criterion, the gradient boosting process stops, yielding a model that is less overtrained than if the boosting process could continue until completion.

The prediction error is average square error, misclassification rate, and log-loss prediction error respectively for an interval, binary, and nominal target. The error is measured on validation data. Without validation data, early stopping is usually ineffective.

Figure 8.17 plots the value of the objective function (here, the average squared error) on the Y axis, and the iteration step on the X axis (from 0 to 77). In most of the early iterations, the objective function decreases for both the training and the validation data. On the training data, notice that the ASE decreases for all of the iterations. However, the same is not true of the ASE on the validation data set. At approximately iteration 12, the ASE continues to grow as the weight estimates are updated. At the 60th iteration, the weight estimates appear to be nearly stable, and the model performance on the training data stops improving.

The philosophy of early stopping says that the optimal model comes from an earlier iteration where the objective function on the validation data has a smaller value than it does for the final iteration.

Iteration 12 provides the gradient boosting where the complexity is optimized. Based on this gradient boosting model's performance on the validation data, it generalizes better than the model at iteration 77.

Figure 8.17: Early Stopping and Related Options

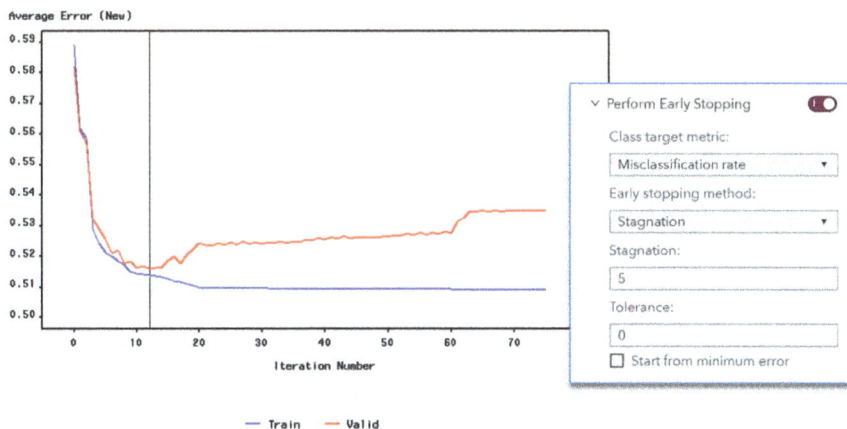

The **Perform Early Stopping** property specifies whether to stop training when the model begins to overfit. The Gradient Boosting node (or GRADBOOST procedure) uses two approaches to early stopping:

1. Stagnation method stops when the relative change in the validation error in consecutive iterations satisfies criteria based on the **Stagnation, Tolerance,** and **Start from minimum error** properties.
2. Threshold method stops when the validation error exceeds a threshold based on the **Threshold** and **Threshold iterations** properties.

The training stops if the relative error in each of the *N* consecutive iterations (**Stagnation**) is less than a threshold (**Tolerance**). Early stopping cannot be used if there is no validation partition. By default, this option is selected. The following options are available:

The **Early stopping method** property specifies the method to use for early stopping.

The **Stagnation** property specifies the number of consecutive iterations (*N*) to consider when using the **Stagnation** early stopping method. The default value is 5.

The **Tolerance** property specifies the threshold for the relative change in validation error when using the **Stagnation** early stopping method. The default value is 0.

The **Start from minimum error** property specifies whether to count iterations starting from the iteration that has the smallest validation error when using the **Stagnation** early stopping method.

The **Threshold** property specifies the threshold value when using the **Threshold** early stopping method. Training is stopped when the validation error equals or exceeds this value.

The **Threshold iterations** property specifies the minimum number of training iterations to run before the validation error is compared to the specified **Threshold**.

> Early stopping is not performed if there is no validation partition. When early stopping is enabled and a validation partition exists, autotuning **Number of Trees** is not performed if selected.

Demo 8.1: Building a Default Gradient Boosting Model in Model Studio and Scoring Using ASTORE in SAS Studio

In this demonstration, you create a stochastic gradient boosting model using the Gradient Boosting node in Model Studio. You will continue working on the **insurance_part** data set and observe some of the default options in the Gradient Boosting node.

1. Ensure that the **Insurance_ClassTree** project is open in Model Studio.
2. Click the **Pipelines** tab and press **+** to add a new pipeline.
3. In the New Pipeline window, name the pipeline as **GradBoost**. Ensure that **Blank Template** is selected under **Template**.
4. Click **OK**.
5. Right-click the **Data** node and select **Add child node** ⇨ **Supervised Learning** ⇨ **Gradient Boosting**.
6. Click the **Run Pipeline** button to produce a stochastic gradient boosting model with default settings.
7. Open the **Results** of the Gradient Boosting node. Maximize the **Error Plot** window. A gradient boosting model with 100 trees is created.

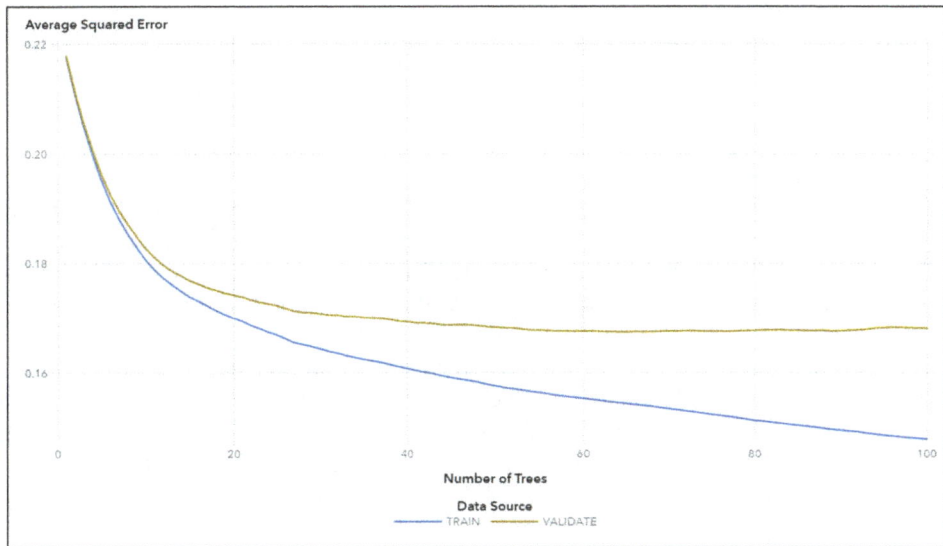

The error plot displays a graph of the average squared error as a function of the number of trees. The average squared error is given for each of the data roles. To examine the misclassification rate as a function of the number of trees, use the drop-down menu in the upper right corner.

Because the trees are not independent (unlike forest), a gradient boosting model beyond 60 trees is expected to overfit the data. You would probably want to verify this by running a separate gradient boosting model with the number of trees set to 60. The default settings show the number of trees was set to 100 and early stopping enabled. However, a gradient boosting model with 100 trees is created. Does that mean no early stopping is applied?

8. Minimize the error plot and return to the Results of the Gradient Boosting node.
9. Scroll down in the Results window and maximize the **Output** window.

10. Observe the **Fit Statistics** table.

Number of Trees	Training Average Square Error	Validation Average Square Error	Training Misclassification Rate	Validation Misclassification Rate	Training Log Loss	Validation Log Loss
1	0.218	0.218	0.346	0.346	0.626	0.627
2	0.211	0.211	0.346	0.346	0.611	0.612
3	0.205	0.205	0.346	0.346	0.598	0.599
4	0.200	0.201	0.337	0.337	0.587	0.589
5	0.195	0.196	0.319	0.327	0.577	0.579
6	0.191	0.192	0.296	0.303	0.568	0.570
7	0.188	0.190	0.285	0.291	0.561	0.563
8	0.185	0.187	0.277	0.286	0.554	0.558
9	0.183	0.185	0.271	0.281	0.548	0.552
10	0.181	0.183	0.267	0.279	0.543	0.547
11	0.179	0.181	0.264	0.279	0.539	0.543
12	0.177	0.180	0.261	0.273	0.535	0.539

The gradient boosting model builds trees sequentially. With early stopping, when the stopping criterion is met, it stops building more trees. By default, for a nominal target, it uses the validation misclassification rate as the stopping criterion. However, the GRADBOOST procedure supports the suboption METRIC=LOGLOSS also for early stopping.

The default settings show that **Perform Early Stopping** was enabled and **Stagnation** set to 5 and **Tolerance** at 0, which means the gradient boosting will stop building more trees when the model doesn't improve in five consecutive trees. From the results, you can see that the first three trees have the same MCR (misclassification rate); thus, it continued building more trees as 3 is less than the stagnation value of 5. Similarly, a pattern of 2, 3, and 4 stagnant MCRs, but not 5, are observed until 100 trees.

11. Minimize the **Output** window and close the **Results**.
12. Change the **Stagnation** value from 5 to **2**.

13. Rerun the **Gradient Boosting** node.
14. Open the **Results** of the Gradient Boosting node.
15. Maximize the **Error Plot** window and change **View chart** from Average Squared error to **Misclassification Rate**. Also see the **Fit Statistic** table from the **Output** window.

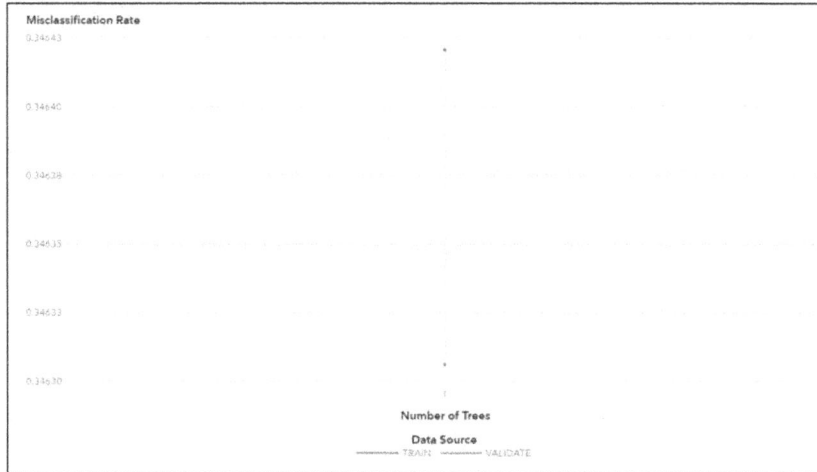

Fit Statistics						
Number of Trees	Training Average Square Error	Validation Average Square Error	Training Misclassification Rate	Validation Misclassification Rate	Training Log Loss	Validation Log Loss
1	0.218	0.218	0.346	0.346	0.626	0.627

Gradient boosting returned the model with only one tree, which makes sense because it is the simplest model and with the same MCR for three consecutive iterations. If you set a larger value for stagnation, you might be able to build a model with more trees if MCR does improve later.

16. Minimize the **Output** window and close the **Results** of the Gradient Boosting node.
17. Change the **Stagnation** value to **5**, which was the default.
18. Click the **Run Pipeline** button.
19. Rename the Gradient Boosting node to **GB Default**.

20. Open the **Results** of the GB Default node and maximize the **Variable Importance** table.

Variable Label	Role	Variable Name	Training Importance	Importance Standard Devi...	Relative Importance
Saving Balance	INPUT	SAVBAL	20.5710	55.8641	1
Branch of Bank	INPUT	BRANCH	12.3211	3.2895	0.5990
Checking Balance	INPUT	DDABAL	10.0438	15.7753	0.4883
Money Market	INPUT	MM	6.3871	41.7518	0.3105
Checking Account	INPUT	DDA	4.7701	21.0140	0.2319
Age of Oldest Account	INPUT	ACCTAGE	4.3234	2.1745	0.2102
Certificate of Deposit	INPUT	CD	2.9896	20.5234	0.1453
Credit Score	INPUT	CRSCORE	2.7607	1.0870	0.1342
ATM Withdrawal Amount	INPUT	ATMAMT	2.7326	2.4261	0.1328
CD Balance	INPUT	CDBAL	2.5509	8.5064	0.1240
Number of Checks	INPUT	CHECKS	2.3261	3.0320	0.1131
Income	INPUT	INCOME	2.0818	1.1893	0.1012
Age	INPUT	AGE	2.0729	1.2740	0.1008
Amount Deposited	INPUT	DEPAMT	1.8058	1.5795	0.0878

The table displays importance for each variable. These statistics include train importance, importance standard deviation, and relative importance.

21. Minimize the **Variable Importance** table.
22. Scroll down in the Results window and maximize the **Properties** table.
    Observe the default values of Number of trees (100), Learning rate (0.1000), Sampling rate (0.5000), L1 (0), L2 (1), and Use default number of inputs to consider per split (True).

Property Name	Property Value
train	true
ntrees	100
learningRate	0.1000
subsampleRate	0.5000
lasso	0
ridge	1
distribution	GAUSSIAN
power	1.5000
seed	12,345
maxBranch	2
maxDepth	4
minLeafSize	5
missingValue	USEINSEARCH
minUseInSearch	1
intervalBins	50
intBinMethod	QUANTILE
defaultVarsPerTree	true
varsToTry	100
earlyStop	true

Because **Use default number of inputs to consider per split** is by default selected (defaultVarsPerTree=true), all input variables will be considered per split and **Number of inputs to consider per split** (varsToTry=100) is not applicable. This value can be autotuned.

In the next demo, you modify some of these properties and try improving the gradient boosting model.

23. Minimize the **Properties** table and close the **Results**.
24. Open the **Results** of the Model Comparison node. Click the **Assessment** tab and maximize the **Fit Statistics** table.

Statistics Label	Train: GB Default	Validate: GB Default
Area Under ROC	0.8483	0.8032
Average Squared Error	0.1480	0.1681
Divisor for ASE	13,550	5,807
Formatted Partition	1	0
Gamma	0.7256	0.6367
Gini Coefficient	0.6967	0.6064
KS (Youden)	0.5351	0.4772
KS Cutoff	0.3500	0.3000
Misclassification at Cutoff	0.2142	0.2500
Misclassification Rate	0.2142	0.2500
Misclassification Rate (Event)	0.2142	0.2500
Multi-Class Log Loss	0.4568	0.5083

The default gradient boosting model has a ROC Index of 0.8032, an ASE of 0.168, and a MISC of 0.2500.

25. Close the **Results** of the Model Comparison node.
26. Right-click the **GB Default** node and click **Download Score Code**.

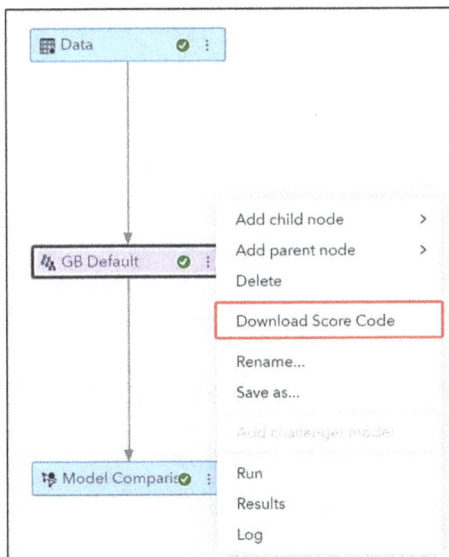

A ZIP file that contains a SAS program called **dmcas_epscorecode.sas** is saved onto the server that contains the SAS Viya installation. For more information about this process, see "Export Models for Production" in *SAS Visual Data Mining and Machine Learning: User's Guide*.

Note: There are two types of score code that Model Studio nodes can create: DATA step or analytic store (also known as ASTORE). The Decision Tree node produces DATA step score code, whereas the Forest and Gradient Boosting nodes produce ASTORE.

27. Open the **score_code_GB Default.zip** file. Double-click **dmcas_epscorecode.sas** to open it. By default, this opens in the SAS Windowing Environment.

```
/*
 * This score code file references one or more analytic stores that are located in the caslib "Models".
 * This score code file references the following analytic-store tables:
 *    12RS1ETX25AOERXGTK267TPYB_ast
 */
data sasep.out;
    dcl package score _12RS1ETX25AOERXGTK267TPYB();
    dcl double "P_INS1" having label n'Predicted: INS=1';
    dcl double "P_INS0" having label n'Predicted: INS=0';
    dcl nchar(32) "I_INS" having label n'Into: INS';
    dcl nchar(4) "_WARN_" having label n'Warnings';
    dcl double EM_EVENTPROBABILITY;
    dcl nchar(8) EM_CLASSIFICATION;
    dcl double EM_PROBABILITY;
```

28. Copy the name of the ASTORE file (highlighted above) on the clipboard and minimize the SAS Windowing Environment window and Windows Explorer window, if required.

29. Return to SAS Viya. Go to the **Applications** menu (≡) in the upper left side. Click **Develop SAS Code** to launch SAS Studio.

30. Click the **Snippets** icon in the left pane. Under **My Snippets** ⇨ **VBBF**, either double-click or drag and drop **Scoring Using ASTORE** onto the canvas.

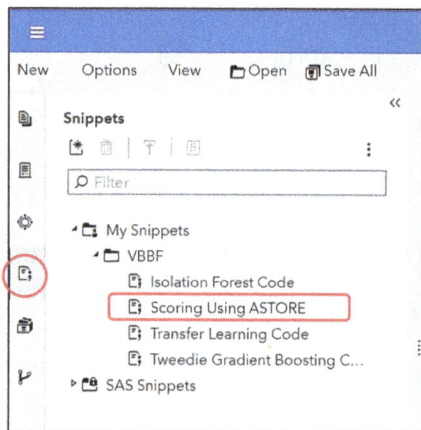

31. The first piece of code starts a CAS session and assigns the library names.
```
options cashost="server" casport=5570;
cas mysess;
caslib _all_ assign;
```

The next piece of the code loads the analytic store table for scoring.

```
proc casutil;
load casdata="_12RS1ETX25AOERXGTK267TPYB_ast.sashdat"
     incaslib="Models" casout="GBDefault" outcaslib="casuser";
quit;
```

The last piece of the code uses the ASTORE procedure to run the EP score code against the data.

```
proc astore;
score data= public.score_insurance
out=casuser.ins_scored rstore=casuser.GBDefault;
run;
```

32. Replace the name with the one (highlighted above) you copied on clipboard. Run the code using the running man icon.

33. After it runs successfully, see the Results.

The ASTORE Procedure

Output CAS Tables

CAS Library	Name	Number of Rows	Number of Columns
CASUSER(student)	INS_SCORED	7000	5

An output scored table is created.

35. Click the **Output Data** tab to see the scored table and scores therein.

Log	Results	Output Data			
CASUSER.INS_SCORED ▾		Columns: 5 of 5	Total rows: 7000	Rows 1 to 200	

	⊛ P_INS1	⊛ P_INS0	⚠ I_INS	⚠ _WARN_	⊛ IDNUM
1	0.3809236617	0.6190763383	0		20002
2	0.1165661465	0.8834338535	0		20004
3	0.1894258015	0.8105741985	0		20006
4	0.1614032258	0.8385967742	0	M	20007
5	0.0741842863	0.9258157137	0	M	20009
6	0.1028959183	0.8971040817	0		20017
7	0.1117378593	0.8882621407	0		20019
8	0.5934280879	0.4065719121	1		20020
9	0.7276948911	0.2723051089	1	M	20024
10	0.1053303658	0.8946696342	0		20026

36. Again, go to the **Applications** menu (≡) in the upper left side. Click **Build Models** to come back to the Model Studio.

End of Demonstration

Comparison of Gradient Boosted Decision Trees with Other Tree-Based Models

The gradient boosting algorithm is similar to standard boosting, except at each iteration, the target is the residual from the previous decision tree model.

At each iteration, the accuracy of the tree is computed.

Successive samples are adjusted to accommodate previous inaccuracies.

The model is a weighted linear combination of (usually) simple trees.

The principal idea is to construct the new base-learners to be maximally correlated with the negative gradient of the loss function, associated with the whole ensemble.

Gradient boosting can be thought of as:

Gradient Boosting = Gradient Descent + Boosting

Table 8.1 is a quick comparison of three tree-based models discussed in this book. Every model has some advantages and disadvantages. Overall, ensemble learning is very powerful. No model is universally best.

Tuning a Gradient Boosting Model

You are required to work on more than a few parameters to tune your gradient boosting model well.

Table 8.1: Comparison of Tree-Based Models

Models	Advantages	Disadvantages
Decision Tree	Easy to understand, interpret, and implement, extremely fast to train	Highly unstable, not powerful enough to handle complex data
Forest	More stable and robust because of "wisdom of the crowd" approach, fast to train, easier to tune, highly generalizable model, less prone to overfit	Very difficult in model interpretability and explanability
Gradient Boosting	High performing, effectively capture complex nonlinear function dependencies, solves interval target distributions other than normal	Slow to train due to sequential nature, a small change in the feature set or training set can create drastic changes in the model, not easy to understand predictions

Figure 8.18: Tree-Splitting Options for Tuning a Gradient Boosting Model

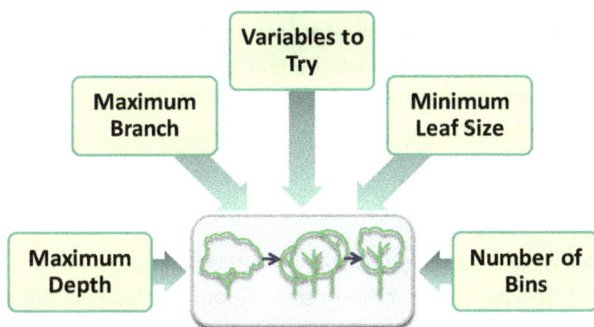

There are a few tree-specific parameters that affect each individual tree in the model and are represented in Figure 8.18. This includes tree-splitting options like maximum number of branches, maximum depth, minimum leaf size, number of inputs to consider per split, and so on. You are already familiar with all of them. (See the discussion in the previous chapters.)

For example, a larger number of inputs to consider per split can lead to overfitting. Few other options are related to tree complexity, which controls whether interactions are fitted: a tree complexity of one (single decision stump; two terminal nodes) fits an additive model, a tree complexity of two fits a model with up to two-way interactions, and so on.

The sequential model-fitting process builds on trees fitted previously, and increasingly focuses on the hardest observations to predict. Accordingly, there are several boosting parameters that affect the boosting operation in a gradient boosting model such as the number of trees, the learning rate, the subsample rate, L1 and L2 regularizations, and so on. They are represented in Figure 8.19.

Our primary focus of discussion in this section revolves around the boosting hyperparameters.

Figure 8.19: Boosting Options for Tuning a Gradient Boosting Model

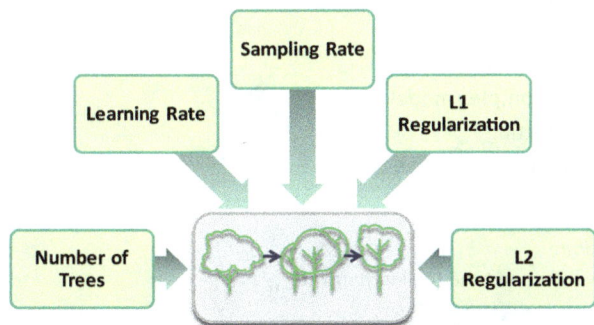

Tuning Parameters Manually

Hyperparameter tuning is choosing a set of optimal hyperparameters for a learning algorithm. A *hyperparameter* is a model argument whose value is set before the learning process begins. Experienced modelers prefer to tune these hyperparameters manually. Another option is to use the SAS Viya autotuning feature that is discussed later.

Number of Trees

The number of trees represents the number of sequential trees to be modeled in gradient boosting. Most implementations of gradient boosting are configured by default with a relatively moderate number of trees, such as hundreds or thousands, although this number is data dependent. Usually, the higher the number of trees, the better to learn the data. However, adding a lot of trees can slow down the training process considerably.

Gradient boosting is fairly robust at a higher number of trees, but it can still overfit at a point. Recall that each new tree attempts to model and correct for the errors made by the sequence of previous trees, and therefore, the model reaches a point of diminishing returns. Hence, this should be tuned using a validation data for a particular learning rate (discussed ahead). In order to find the optimal number of trees, you can also use early stopping.

Parameter tuning of the number of trees in a forest is summarized in Figure 8.20.

Learning Rate

The learning rate scales the contribution of each tree. Gradient boosting starts with an initial estimate, which is updated using the output of each subsequent tree. The learning parameter controls the magnitude of this change in the estimates.

Lower values of the learning rate generally make the model robust to the specific characteristics of the tree and thus allow it to generalize well. A low learning rate would require more trees to

Figure 8.20: Tuning the Number of Trees

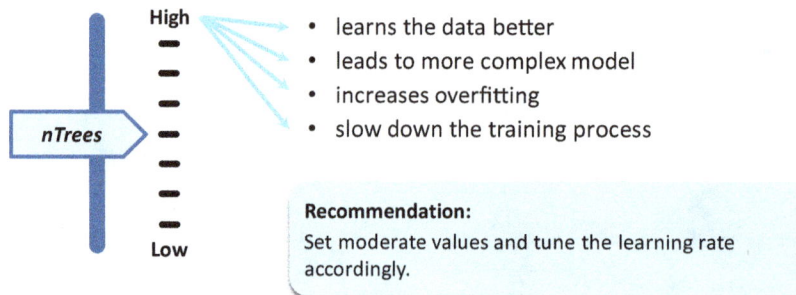

High
- learns the data better
- leads to more complex model
- increases overfitting
- slow down the training process

nTrees

Low

Recommendation:
Set moderate values and tune the learning rate accordingly.

Figure 8.21: Tuning the Learning Rate

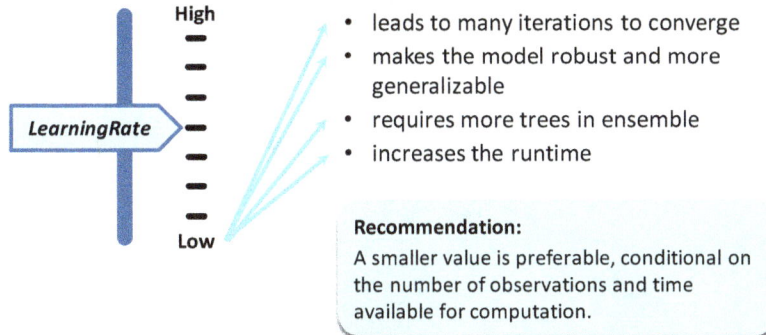

- leads to many iterations to converge
- makes the model robust and more generalizable
- requires more trees in ensemble
- increases the runtime

Recommendation:
A smaller value is preferable, conditional on the number of observations and time available for computation.

Figure 8.22: A Good Learning Rate!

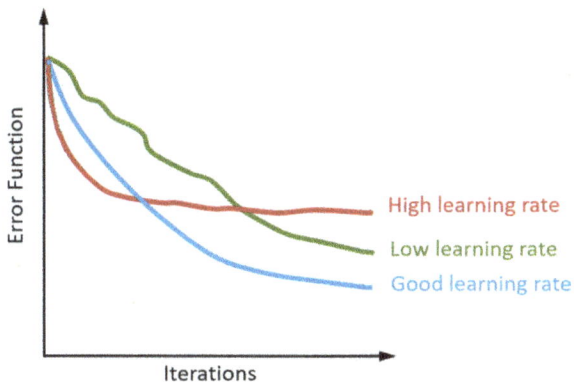

model all the relations. A low learning rate is more precise, but calculating the gradient is time-consuming, so it is computationally expensive.

Learning rate tuning in a gradient boosting model is summarized in Figure 8.21. Figure 8.22 further explains the trade-off between setting a high and low learning rate.

Setting the learning rate too high yields great progress in decreasing the error function in the first few iterations, but at a risk of diverging later in the training process. On the other hand, setting the learning rate too small can result in very long training times. A "good" learning rate is a balance between the two and is problem specific.

Trade-off Between Learning Rate and Number of Trees

By now you are aware that the learning rate shrinks the contribution of each tree as it is added to the model. Decreasing (slowing) the learning rate increases the number of trees required, and in general a smaller learning rate (and larger number of tress) is therefore preferred. The usual

Figure 8.23: Learning Rate Versus Number of Trees

Learning rate and number of trees have an inverse relationship.

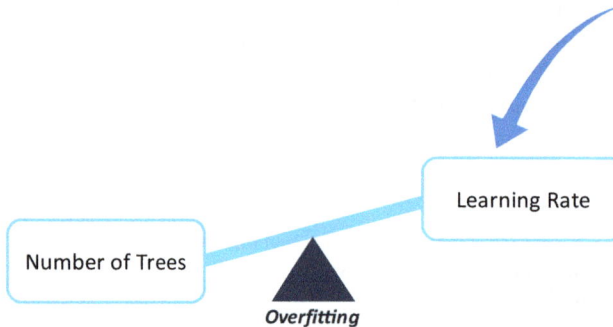

approach is to estimate the optimal number of trees and the learning rate with an independent test set or with cross validation, using reduction in the loss function as the measure of success. The trade-off between learning rate and number of trees is illustrated in Figure 8.23.

Friedman (1999) commented on the trade-off between the number of trees (M) and the learning rate (λ):

> *"The λ-M trade-off is clearly evident; smaller values of λ give rise to larger optimal M-values. They also provide higher accuracy, with a diminishing return for $\lambda < 0.125$. The misclassification error rate is very flat for M > 200, so that optimal M-values for it are unstable. ... the qualitative nature of these results is fairly universal."*

– Friedman 1999

He suggests to first set a large value for the number of trees, and then tune the shrinkage parameter to achieve the best results. Studies in the paper preferred a shrinkage value of 0.1, a number of trees in the range 100 to 500, and the number of terminal nodes in a tree between 2 and 8.

How does the shrinkage (learning rate) affect overfitting?

Presented in Figure 8.24 are the learning error curves for gradient boosting models with different shrinkage (λ) parameters using the *sinc*(x) data referred previously. The left graph shows the performance on the training data, whereas the right graph shows the performance on the validation data.

Observe that the training set error is substantially dropping, but the speed of this improvement heavily depends on the shrinkage parameter λ. The validation-error hyperparameter M, corresponding to the error minima of each of the models, is highlighted with circles. You can see

Figure 8.24: Effect of Learning Rate on Overfitting

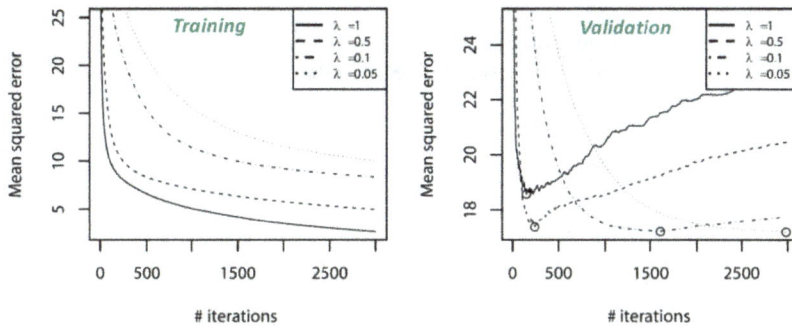

Image source:
Natekin and Knoll (2013). Gradient boosting machines, a tutorial. *Front. Neurorobot.* 7:21. doi:
10.3389/fnbot.2013.00021

that increasing the shrinkage leads to both finding a better hyperparameter *M* minima and to improving the generalization of the model. The latter corresponds to the fact that shrunk models have a flatter plateau beyond their error minima, and it takes them many more iterations to initiate overfitting. Yet, it also means that these models naturally take longer to learn. Hence, a high learning rate commonly results in overfitting (Natekin and Knoll 2013).

Subsample Rate

Subsample rate sets the fraction of samples to be used for fitting the individual base learners by applying random sampling. Friedman (1999) found that almost all subsampling percentages are better than so-called deterministic boosting (subsample rate=1):

> *"… the best value of the sampling fraction … is approximately 40% (f=0.4) … However, sampling only 30% or even 20% of the data at each iteration gives considerable improvement over no sampling at all, with a corresponding computational speed-up by factors of 3 and 5 respectively."*
>
> *– Friedman 1999*

When the amount of data, measured by the number of data points *N* is not of practical concern, setting the default value of 0.5 gives a reasonable result. If an optimal subsample rate is of interest, you can simply estimate it by comparing predictive performance under different parameter values.

Smaller subsample rates generally result in more diverse trees and consequently a more robust and generalized ensemble model. However, you should also consider the effect of reducing the sample size on the model estimates. If the subsample rate becomes too low, you might get a poorly fit model due to the lack of degrees of freedom.

Figure 8.25: Tuning Subsample Rate

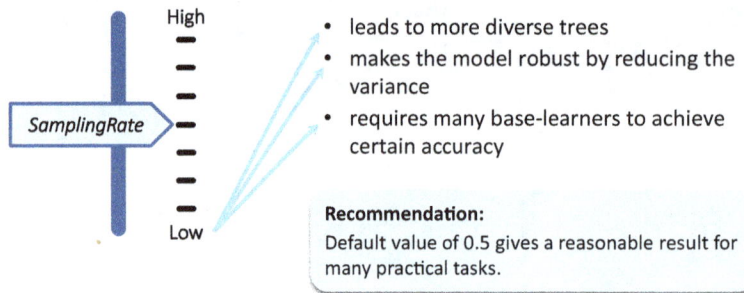

In general, the more data that is available for fitting a base-learner, the more accurate the estimate will be, if enough data was used. Hence, when there are large amounts of data available, you might want to consider a trade-off between the number of observations used for fitting each of the base-learners and their respective accuracy improvement.

At times, it is more efficient to have a larger number of base-learners, learned with the smaller subsample rate to reach the desired accuracy, than the one with a smaller amount of more carefully fitted base-learners with a higher subsample rate.

Parameter tuning of the subsample rate in gradient boosting is summarized in Figure 8.25.

L1 and L2 Regularizations

The decay parameter controls the relative importance of the penalty term. Specifying too large of a penalty term risks the model under-fitting the data.

The two regularization terms have different effects on the weights. L2 regularization encourages the weights to be small, whereas L1 regularization encourages sparsity, so it encourages weights to go to 0. This is generally helpful in regression models where you want some feature selection, but in decision trees, you have already selected features while trees are being built, so zeroing their weights might not be helpful. For this reason, you can set a high L2 and a low (or 0) L1. You might think of L1 regularization as more aggressive against less-predictive features than L2 regularization. However, it might make sense to use both: some L1 to penalize the less-predictive features, but then also some L2 to further penalize large leaf scores without being so harsh on the less-predictive features.

Parameter tuning of L1 and L2 regularizations in gradient boosting is summarized in Figure 8.26.

Figure 8.26: Tuning L1 And L2 Regularizations

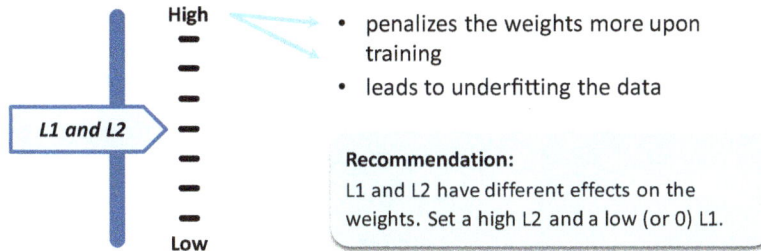

- penalizes the weights more upon training
- leads to underfitting the data

Recommendation:
L1 and L2 have different effects on the weights. Set a high L2 and a low (or 0) L1.

Autotuning Hyperparameters

Autotuning specifies whether to perform autotuning of any gradient boosting parameters. Autotuning will automatically choose the optimal values for the hyperparameters. You can determine which hyperparameters to be tuned. Figure 8.27 lists gradient boosting hyperparameters that can be autotuned.

If **Perform Autotuning** is selected, the following options are available:

L1 Regularization specifies whether to autotune L1 Regularization.

L2 Regularization specifies whether to autotune L2 Regularization.

Learning Rate specifies whether to autotune the learning rate.

Maximum Depth specifies whether to autotune the maximum number of generations of nodes.

Minimum Leaf Size specifies whether to autotune the minimum leaf size.

Figure 8.27: Autotunable Parameters in Gradient Boosting

- L1 Regularization
- L2 Regularization
- Learning Rate
- Maximum Depth
- Minimum Leaf Size
- Number of Interval Bins
- Number of Inputs per Split
- Number of Trees
- Subsample Rate

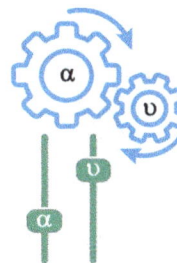

Number of Interval Bins specifies whether to autotune the number of interval bins.

Number of Inputs per Split specifies whether to autotune the number of inputs per split.

Number of Trees specifies whether to autotune the number of trees in a boosting series.

> When early stopping is enabled and a validation partition exists, autotuning **Number of Trees** is not performed if selected.

Subsample Rate specifies whether to autotune the subsample rate.

If any of the above options are selected, the following option is available for each one of them:

Initial value specifies the initial value for autotuning that hyperparameter. The default value is set for that hyperparameter. Use the **From** and **To** options to specify the range. The default **From** value and the default **To** value are set for that hyperparameter.

> You can perform cross validation in the **General Options** of **Perform Autotuning** in the Gradient Boosting node. However, if your data is already partitioned, then that partition is used and all the other options: Validation method, Validation data proportion, and cross validation number of folds are all ignored. If not, then in the Autotuning options, you can select **K-fold cross validation** and then specify the number of partition folds in the cross validation process.

Demo 8.2: Modifying and Autotuning a Gradient Boosting Model

In this demonstration, you try to improve the previously created default gradient boosting model by tuning certain parameters. You continue working on the **insurance_part** data set and experiment with some of the options in the Gradient Boosting node.

1. Ensure that the **Insurance_ClassTree** project is open in Model Studio and that you are on the **GradBoost** pipeline.

2. Right-click the **Data** node and select **Add child node** ⇨ **Supervised Learning** ⇨ **Gradient Boosting**. Rename this Gradient Boosting node to **GB Tuned**. Your pipeline should look like the one below.

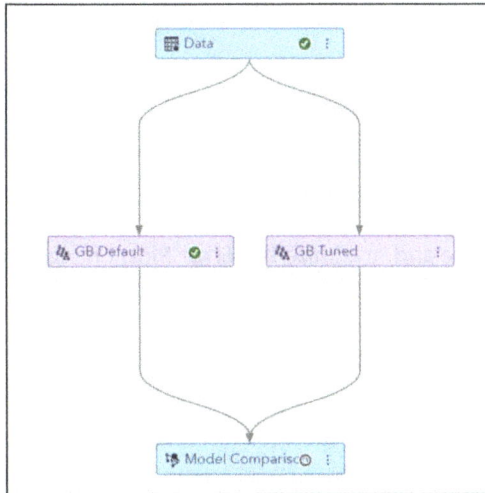

3. You will modify some of the boosting options as below. Change the **Number of trees** from 100 to **25**, **Learning rate** from 0.1 to **0.3**, and **Subsample rate** from 0.5 to **1**.

The idea is to have a fewer number of trees and larger step sizes in which all the trees to be fitted use the entire data, precisely a deterministic gradient boosting model.

4. Expand **Tree-splitting Options** and change **Maximum depth** from 4 to **6** and **Number of interval bins** from 50 to **200**.

5. Turn off **Perform Early Stopping** using the slider button.

This is reasonable considering the meager number of tress.

6. Click the **Run Pipeline** button.

7. Open the **Results** of the GB Tuned node and maximize the **Error Plot**. Change **View Chart** to **Misclassification Rate**.

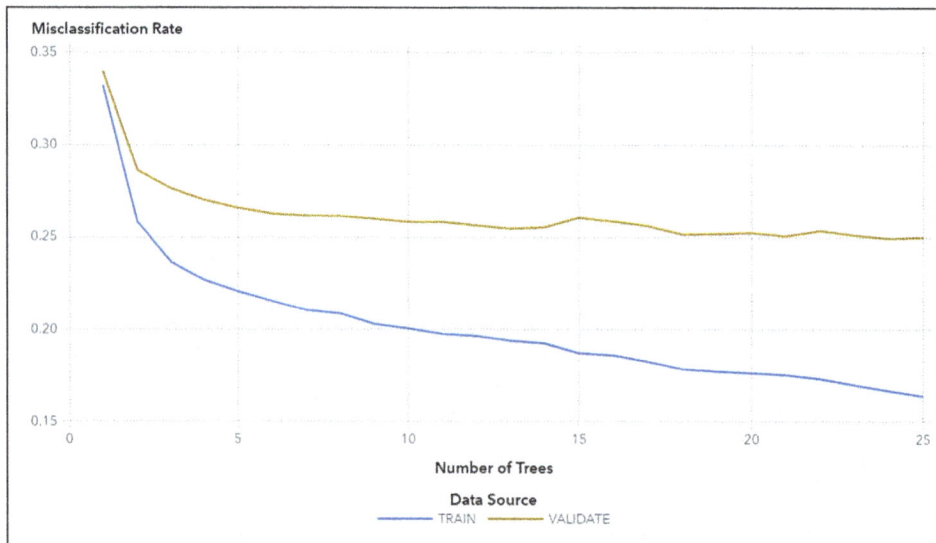

The misclassification rate is falling for every addition of a decision tree. Therefore, you can try to increase the number of trees a bit.

8. Close the **Results** of the GB Tuned node.

9. Open the **Results** of the Model Comparison node.

Model Comparison				
Champion	Name	Algorithm Name	KS (Youden)	Misclassification Rate
▣	GB Default	Gradient Boosting	0.4772	0.2500
	GB Tuned	Gradient Boosting	0.4549	0.2502

Both models have comparable performance. However, the GB Default model has a slightly better KS (Youden) statistic than the GB Tuned model. It would be interesting to see the other fit statistic indicators like ROC, lift, ASE, and so on. Creating fewer but deeper trees using all the data seems to be a hopeful choice.

You can further modify certain properties and search for more performance gain. Instead, we will use the autotuning functionality in SAS Viya.

10. Right-click the **Data** node and select **Add child node** ⇨ **Supervised Learning** ⇨ **Gradient Boosting**. Rename this Gradient Boosting node to **GB Autotuned**. Your pipeline should look like the one below.

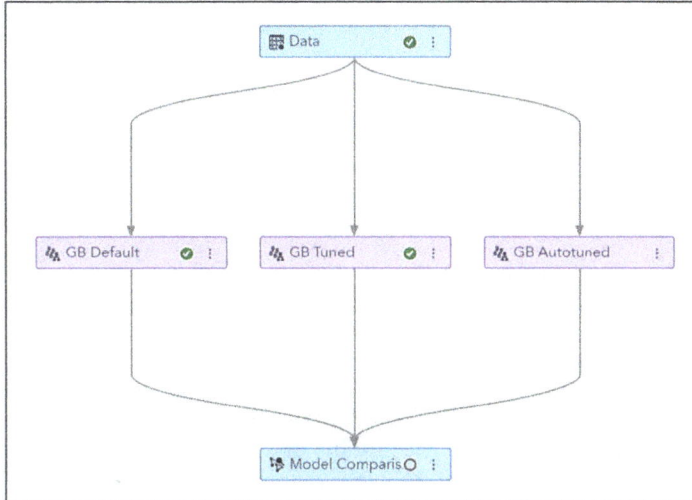

11. Expand the **Perform Autotuning** properties and set it to **On** using the slider button.

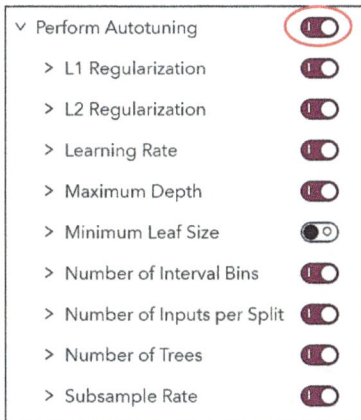

Note that the L1 and L2 regularizations, learning rate, maximum depth, minimum leaf size, number of interval bins, number of inputs per split, the number of trees, and the subsample rate are autotunable. By default, all of them are set to On except the minimum leaf size. You can expand each one of them and change their autotuning properties. However, currently no change is required.

12. Click the **Run Pipeline** button.

13. Open the **Results** of the Forest Autotuned node and maximize the **Error Plot** with **Misclassification Rate** chosen in the **View chart**.

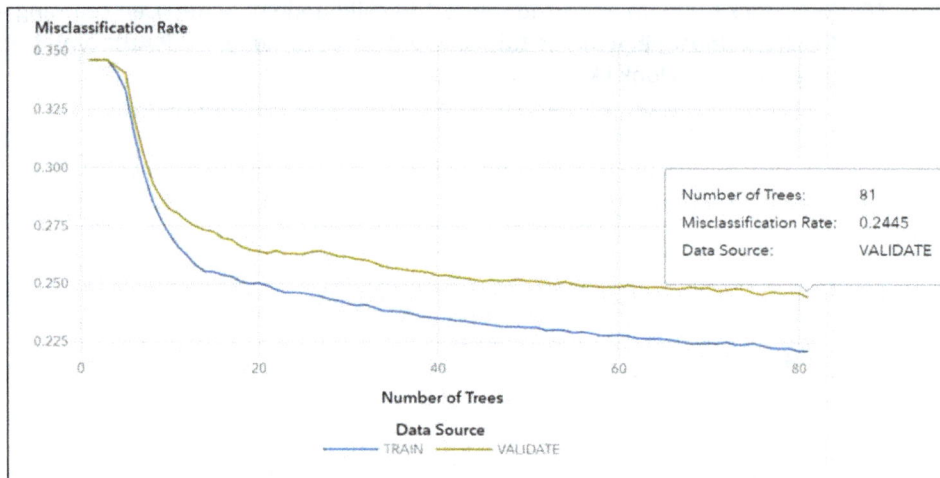

Autotuning gives a gradient boosting model with 81 trees, considerably more than the manually tuned model with 25 trees. The number of trees is determined by the early stopping criterion. Note that the early stopping was enabled. Therefore, autotuning of the number of trees is not performed even if you had selected that in the autotuning window.

14. Minimize the **Error Plot**. It would be interesting to note the optimal values of other hyperparameters.

15. Focus on the **Autotune Best Configuration** table.

Parameter	Value
Evaluation	52
Number of Variables to Try	17
Learning Rate	0.1031
Sampling Rate	0.8335
Lasso	7.6401
Ridge	1.7069
Number of Bins	54
Maximum Tree Levels	6
Kolmogorov-Smirnov	0.4866

The resulting values of parameters that were autotuned is displayed. By default, the genetic algorithm method is used as the autotune search method.

The autotuned best configuration suggests a gradient boosting model with 17 inputs considered per split, a stochastic (non-deterministic) gradient boosting model with 83% sampling rate and a learning rate of 0.1031, and L1 and L2 regularizations as 7.6401 and 1.7069 respectively. While searching over several iterations, this configuration resulted in a KS of 0.4866.

16. Close the **Results** of the GB Autotuned node.
17. Open the **Results** of the Model Comparison node.

Champion	Name	Algorithm Name	KS (Youden)	Misclassification Rate
⊡	GB Autotuned	Gradient Boosting	0.4866	0.2445
	GB Default	Gradient Boosting	0.4772	0.2500
	GB Tuned	Gradient Boosting	0.4549	0.2502

Interestingly, all three models are very different in terms of their tunable properties, and still their performance is not that different. The autotuned gradient boosting model, however, comes out as the champion on almost all the model performance statistics including KS (Youden) and MISC.

	GB Default	GB Tuned	GB Autotuned
Number of trees	100	25	81
Early stopping	Yes	No	Yes
Subsample rate	0.50	1	0.8335
Learning Rate	0.1	0.3	0.1031
L1 regularization	0	0	7.6401
L2 regularization	1	1	1.7069
Number of inputs per split	47	47	17
Maximum depth	4	6	5
Minimum leaf size	5	5	5
Number of interval bins	50	200	54

The autotuned gradient boosting model has fewer trees than the default gradient boosting model but many more trees than the manually trained model. The sub-sample rate in the autotuned model is higher than the default model, whereas we did not apply any sample in the manually trained model. A small value of number of inputs were considered per split in the autotuned model.

All the three gradient boosting models have a smaller minimum leaf size. A smaller leaf usually makes the model more prone to capturing noise in the training data. This property is directly associated with the maximum depth. However, you might want to try multiple leaf sizes or autotune to find the most optimum for your use case. Therefore, there is no single recipe that would guarantee an accurate and efficient gradient boosting model. It is highly data dependent and experimentation is the key!

18. Close the **Results** of the Model Comparison node.

End of Demonstration

Quiz

1. What is the first step in the gradient boosting model algorithm?
   a. Fit each tree to the residuals of the previous tree, and the loss function is minimized.
   b. Create a baseline tree.
   c. Repeat the process of fitting several trees a specific number of times.
   d. Aggregate the series of trees to build a model, which is a single function to predict the target value of a new observation.

2. For a nominal target with three categories, how many trees get created in the gradient boosting model if you have iterated it 100 times (Number of trees=100)?
   a. 100
   b. 200
   c. 300
   d. 400

3. Boosting is less prone to overfit the data than a single decision tree, and if a decision tree fits the data fairly well, then boosting often improves the fit.
   a. True
   b. False

4. In a Gradient Boosting node, the Perform Early Stopping property specifies whether to stop training when the model begins to overfit. The training stops if the relative error in each of the N consecutive iterations (Stagnation) is less than a threshold (Tolerance). Early stopping cannot be used if there is no validation partition.
   a. True
   b. False

5. Which of the following statements is FALSE?
   a. In SAS Viya, boosting involves fitting models to a random sample of data with replacement.
   b. Boosting machines are applicable not only to decision trees but any predictive model.
   c. The gradient boosting algorithm is essentially a weighted linear combination of decision trees.
   d. None of the above

6. See the two error plots of gradient boosting using two different learning rates on the same training data comprising the raw features.

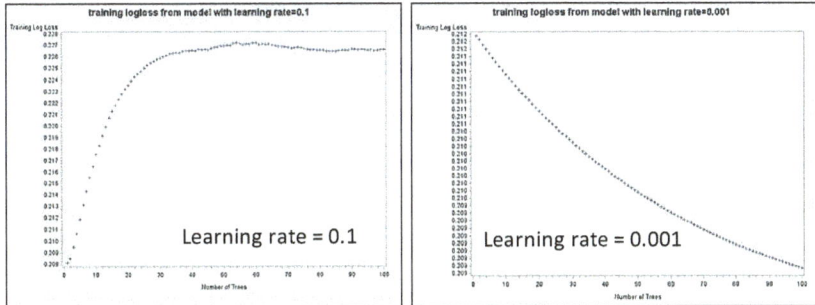

Which of the following statements are TRUE? (Select all that apply.)
a. Completely diverse (increasing and decreasing) log loss patterns are not possible.
b. It is better to use validation data to choose a good model.
c. Decreasing the learning rate is usually a good starting point.
d. Performing some feature engineering and then building a gradient boosting model is a good idea.

7. Higher values are better for which of the following hyperparameters of a gradient boosting algorithm?
a. maximum depth
b. number of inputs per split
c. minimum leaf size
d. number of trees
e. the optimal values of the hyperparameters are problem specific

8. Stochastic gradient boosting is different than gradient boosting in which of the following ways?
a. Different samples are taken for bootstrap aggregation.
b. The same training data is used at each iteration.
c. At each iteration, a subsample of the training data is drawn at random to fit the base learner and compute the model update for the current iteration.
d. None of the above

Answers

8. c   7. e   6. b, c, and d   5. a   4. True   3. True   2. c   1. b

Chapter 9: Additional Gradient Boosting Models

Gradient Boosting for Transfer Learning

Some data mining and machine learning applications have only a scarce amount of training data available, pictorially represented in Figure 9.1. In addition, we make a basic assumption that the training and the test data have the same distribution. Nevertheless, in many cases, this *identical distribution* assumption does not hold. The assumption might be violated when a task from one new domain comes, while there are only labeled data from a similar old domain. Labeling the new data can be costly, and it would also be a waste to throw away all the old data.

For example, insurance providers do not label many claims as fraudulent or non-fraudulent. Web mining is another example. Web data used in training a web page classification model can become easily outdated when applied to the web sometime later because the topics on the web change frequently. For more discussion, see Dai et al. (2007).

One method of building models in situations that have small training sets is to supplement the scarce training data with more abundant auxiliary data. The auxiliary data should have the same target and input variables but can come from a different context than the training data. You can use the auxiliary data to increase the number of observations for training the model.

Figure 9.1: Scarce Training Data!

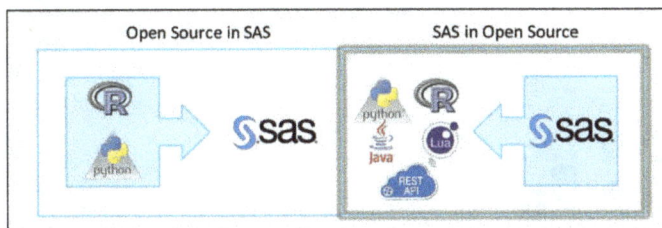

Consider a case where you have a limited amount of labeled data from the target population. However, you have ample labeled data available from a similar population or old domain, represented in Figure 9.2. Obviously, the inputs and target definitions are same in both the data.

Transfer learning augments the training data with auxiliary data and attempts to down-weight the influence of observations that are not representative of the original training data. The original training data typically come from a target population from which data are difficult to get.

Some of the observations in the auxiliary population would be like the target population and therefore can easily be used for training. However, the ones that are different than the target population are arbitrary to be used in modeling, and therefore it is important to identify them and reduce their influence.

For simplicity, you might refer the auxiliary observations that are like the target population as friends, and the other auxiliary observations as aliens. Figure 9.3 characterizes them as friends and aliens.

Gradient boosting tries to identify and down-weight the aliens. A part of the original training data is kept as test data for honest assessment.

Figure 9.2: Transfer Learning

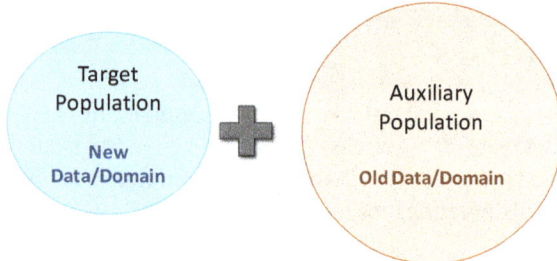

Figure 9.3: Friends and Aliens

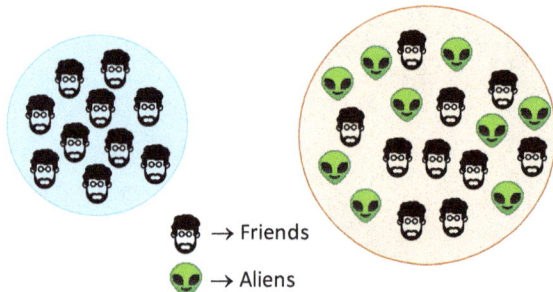

Figure 9.4: Identify and Down-Weight Aliens

Transfer Learning Using the GRADBOOST Procedure

PROC GRADBOOST tries to identify and down-weight the aliens. The TRANSFERLEARN statement enables you to train the gradient boosting model by using auxiliary data that are added to your training data. To prevent the model from being overly biased toward the auxiliary data, the GRADBOOST procedure down-weights auxiliary observations that are dissimilar from the training observations.

```
PROC GRADBOOST < options >;
CROSSVALIDATION < options >;
ID variables;
INPUT variables< / options >;
OUTPUT OUT=CAS-libref.data-table< option>;
PARTITION partition-option;
TARGET variable< / LEVEL=NOMINAL | INTERVAL >;
TRANSFERLEARN variable< / options >;
```

The PROC GRADBOOST statement invokes the procedure.

The CROSSVALIDATION statement performs k-fold cross validation to find the average estimated validation error. You cannot specify this statement if you specify either the AUTOTUNE statement or the PARTITION statement.

The ID statement lists one or more variables that are to be copied from the input data table to the output data tables that are specified in the OUT= option in the OUTPUT statement and the RSTORE= option in the SAVESTATE statement.

The INPUT statement names input variables that share common options. You can specify the INPUT statement multiple times.

The OUTPUT statement creates an output data table that contains the results of running PROC GRADBOOST.

The PARTITION statement specifies how observations in the input data set are logically partitioned into disjoint subsets for model training, validation, and testing.

The TARGET statement names the variable whose values PROC GRADBOOST predicts.

The TRANSFERLEARN statement names the variable whose value indicates whether the observation belongs to the training data or auxiliary data. A value between 0 and 1 indicates training data, and a value greater than or equal to 1 indicates auxiliary data. Observations that have missing values or negative values are ignored.

You can specify the following options:

- **BURN=***number*
  specifies the number of trees to create before downweighting any observation in the auxiliary data. By default, BURN=0.
- **SHRINKAGE=***number*
  specifies the number to apply as the weighting factor for downweighting auxiliary data, where the number must be between 0 and 1, exclusive. By default, SHRINKAGE=0.9.
- **TRIMMING=***number*
  specifies the number to use as a fraction of the distribution of gradients on the training data beyond which auxiliary observations are downweighted, where the number must be greater than 0 and less than or equal to 1/2. By default, TRIMMING=0.01.

> To use transfer learning in the GRADBOOST procedure, you must concatenate your training data and your auxiliary data and specify this combined data table in the DATA= option in the PROC GRADBOOST statement. To distinguish the two types of data, you must create an indicator variable.

Demonstration Scenario

Simulated data are used to show the transfer learning application using the GRADBOOST procedure. The DATA step code generated three data sets: one for training that includes the friends and the aliens, one without the aliens, and a third data set with test observations held out from training. All three data sets, **Train**, **Test**, and **NoAlien**, have 6000, 1000, and 4000 observations, and all of them have the following columns:

- **X1**   (normally distributed input variable)
- **X2**   (normally distributed input variable)
- **Y**   (binary target variable coded as 1 or -1)
- **constantZero** (univariate variable with all 0 values)
- **dataRole**   (data role indicator coded as -1, 0, 1 or 2 that represents test data, target population, friends, and aliens respectively)

The idea is to run gradient boosting three times. Recall that the TRANSFERLEARN statement specifies the variable that identifies the auxiliary observations.

The three data sets are pictorially represented in Figure 9.5.

The first gradient boosting model is run on the train data that includes training data from the target population as well as auxiliary data (friends and aliens both), illustrated in Figure 9.6. The TRANSFERLEARN statement names **dataRole** as the variable whose value indicates whether the observation belongs to the training data (**dataRole=0**) or auxiliary data (**dataRole=1,2**).

The second gradient boosting model is run on the training data that includes the target population as well as auxiliary data (friends and aliens both), illustrated in Figure 9.7. The TRANSFERLEARN statement names **constantZero** as the variable. When that variable is zero for all observations, transfer learning is not done. The **dataRole** variable is not used.

Figure 9.5: Simulated Data

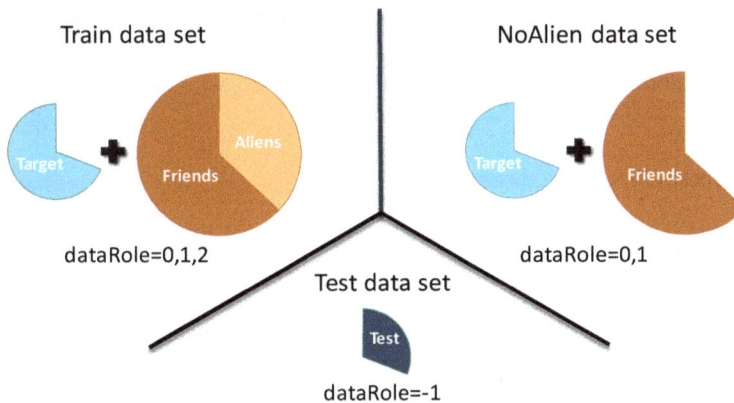

Figure 9.6: Model 1: With Transfer Learning

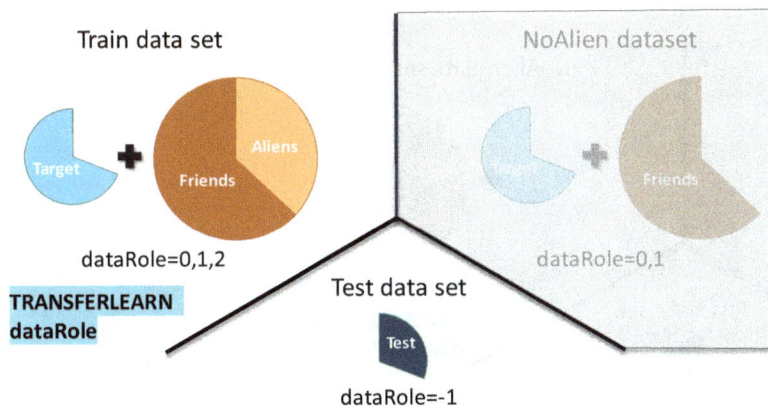

Figure 9.7: Model 2: Without Transfer Learning

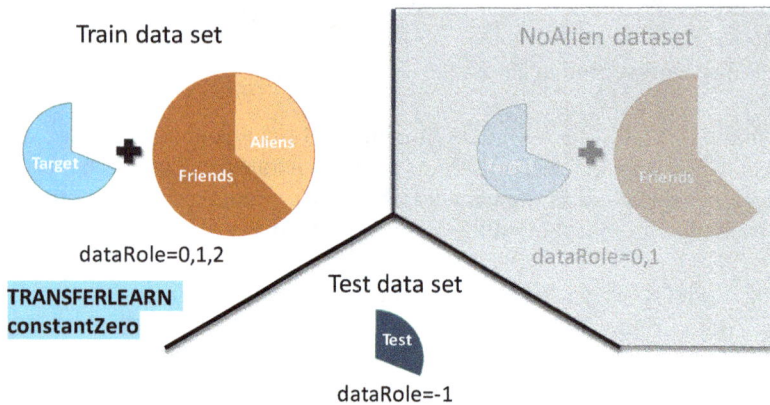

The third gradient boosting model is run on the **NoAlien** data that includes the target population and only friends from the auxiliary data, illustrated in Figure 9.8. The TRANSFERLEARN statement names **constantZero** as the variable whose value (**constantZero=0**) does not indicate whether the observation belongs to the training data or auxiliary data. The **dataRole** variable is also not used in this run.

The upcoming demonstration implements these three models.

Demo 9.1: Transfer Learning Using Gradient Boosting

This demonstration runs PROC GRADBOOST twice without using transfer learning: once with all the data, and once without the aliens. All models are evaluated with data from the target population that are not part of the training data. The model incorporating transfer learning

Figure 9.8: Model 3: Without Aliens

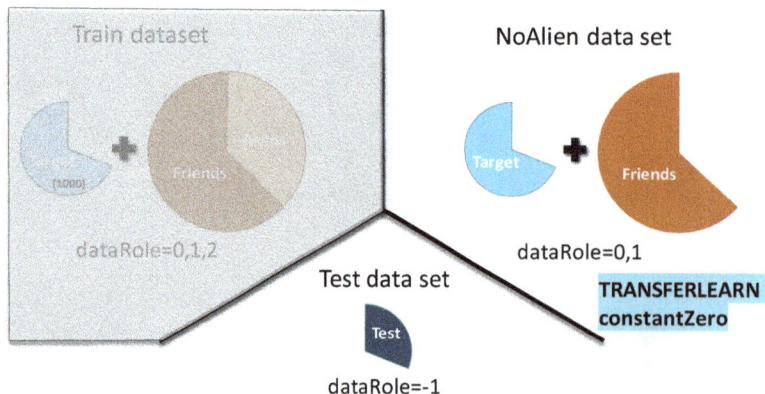

should provide a better fit than one without it, though not as good as the one in which the alien population is removed.

1. Go to the **Applications** menu (≡) in the upper left side.
2. Click **Develop SAS Code** in the application shortcut area to launch SAS Studio.
3. Click the **Snippets** icon in the left pane. Under **My Snippets** ⇨ **VBBF**, either double-click or drag and drop **Transfer Learning Code** onto the canvas.

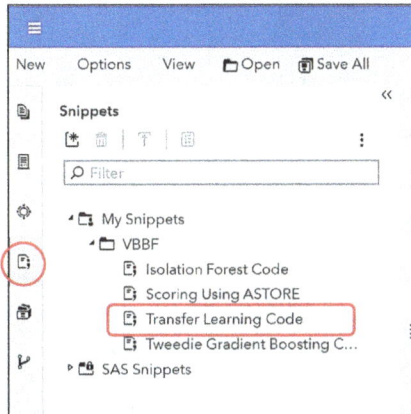

The program opens in the code editor window.

First, a caslib named **mycas** is created. Next, a local library named **local** is created. Then three DATA steps are used to create new data sets that are saved in memory.

```
libname mycas cas;
libname local '/home/student/casuser/VBBF';

data mycas.TL_train;
set local.TL_train;
run;

data mycas.TL_noAlien;
set local.TL_noAlien;
run;

data mycas.TL_test;
      set local.TL_test;
run;
```

4. Run this part of the program.
5. Go back to the **Code** tab.

The next part of the program uses a macro and invokes the GRADBOOST procedure twice: first to train the model and then again to apply the model to the test data and output fits statistics.

```
%macro runProc(data=, auxVar=, outfit=);
proc gradboost data=&data. outmodel=mycas.model seed=3331333;
```

```
input x: /level=interval;
target y /level=nominal;
transferLearn &auxVar. / burn=10;
run;

proc gradboost data=mycas.TL_test inmodel=mycas.model;
output out=mycas.score;
ods output FitStatistics=&outfit.;
run;

%mend;
%runProc(data=mycas.TL_train, auxVar=datarole, outfit=fit_train);

%runProc(data=mycas.TL_train, auxVar=constantZero,
outfit=fit_noTransfer);

%runProc(data=mycas.TL_noAlien, auxVar=constantZero,
outfit=fit_noAlien);
```

The factor that is applied to downweight observations is controlled by the SHRINKAGE= option in the TRANSFERLEARN statement. By default, SHRINKAGE=0.9 is used. To determine whether an observation is dissimilar, the GRADBOOST procedure computes a gradient for each observation. An observation in the auxiliary data that is dissimilar to the training data will have a gradient that is in either a low quantile or a high quantile of the gradients of the training data.

The TRIMMING= option in the TRANSFERLEARN statement controls the fraction of training gradients beyond which auxiliary observations are downweighted. By default, TRIMMING=0.01 is used.

The GRADBOOST procedure begins downweighting auxiliary observations only after 10 trees have been grown, as specified in the BURN= option in the TRANSFERLEARN statement.

6. Run this part of the program. The **Results** tab opens.

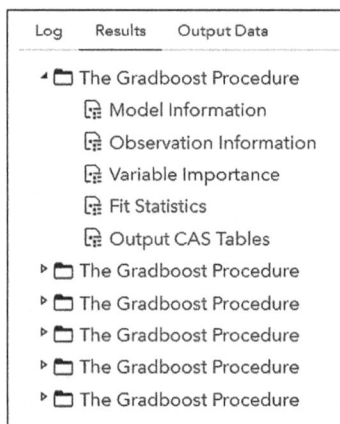

Six runs of gradient boosting results are generated: one with transfer learning, one without transfer learning, and one without aliens. Each one of them also run on test data for honest assessment.

7. Go back to the **Code** tab.

   The next part of the program combines average square error from the three models into a single table and then plots the ASEs for each model by the number of trees in the model.

```
%macro extractFit(data=, infit=);
data &data.;
set &infit.;
keep trees ase;
rename ase = &data._ase;
run;

%mend;
%extractFit(data=train, infit=fit_train);
%extractFit(data=noTransfer, infit=fit_noTransfer);
%extractFit(data=noAlien, infit=fit_noAlien);
data result;
merge train noTransfer noAlien;
by trees;
run;

proc template;
define statgraph transferLearning;
begingraph;
layout overlay;
scatterplot y=train_ase
x=trees / markerattrs=(color=blue)

name='with'
legendlabel="With Transfer Learning";
scatterplot y=noTransfer_ase
x=trees / markerattrs=(color=red)

name='without'
legendlabel="Without Transfer Learning";
scatterplot y=noAlien_ase
x=trees / markerattrs=(color=brown)

name='noAliens'
legendlabel="Without Aliens";
discretelegend 'without' 'with' 'noAliens';
endlayout;
endgraph;
end;
run;

proc sgrender data=result template=transferLearning;
run;
```

8. Run this last part of the program.

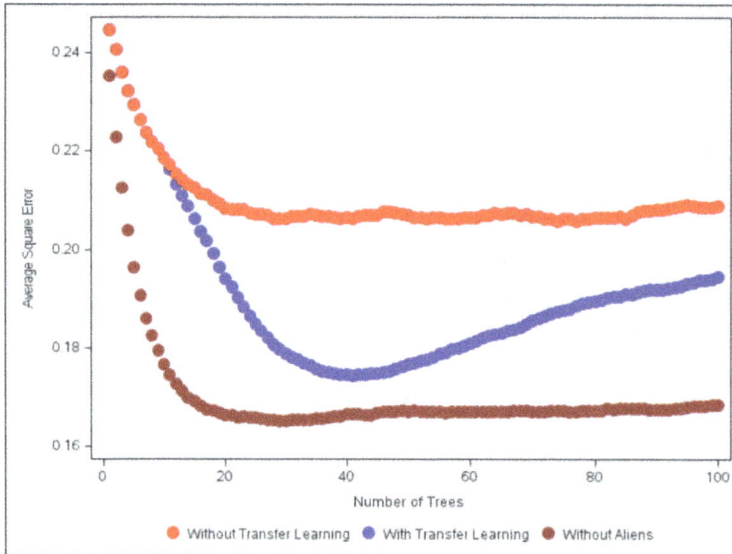

The plot of ASE versus Number of Trees for three models shows that the fit with transfer learning is better than without it, though not as good as when all the alien observations are removed from the data. The fit with transfer learning is identical to the fit without it for the first 10 trees because downweighting does not begin until tree 11 in this example.

9. Close the **Transfer Learning Code** window.
10. Do not close SAS Studio as you will continue using it in the next demonstration.

End of Demonstration

Transfer Learning and Autotuning

You can autotune a pre-trained gradient boosting model.

This sample program uses the home equity data set **HMEQ**, which is available in the **Sampsio** library that SAS provides. The data set contains observations for 5,960 mortgage applicants. A variable named **Bad** indicates whether the customer has paid on his or her loan or has defaulted on it.

By default, autotune starts with model train action default settings for hyperparameters, unless they are explicitly set. Then those values are used as the base model and basis for comparison. This can be done by setting values of those model parameters directly, or by setting the tuningParameters initValue option when calling the tune action directly.

```
proc cas noqueue;
    autotune.tuneGradientBoostTree result=r /
        trainOptions={
            table={name='HMEQ', vars={'CLAGE', 'DEBTINC', 'LOAN',
                    'MORTDUE', 'VALUE', 'REASON', 'JOB', 'bad'}},
            inputs={'CLAGE', 'DEBTINC', 'LOAN', 'MORTDUE',
                    'VALUE', 'REASON', 'JOB'},
            target='bad',
            nominals={'REASON', 'JOB', 'bad'},
            casOut={name='gbt_hmeq_model', replace=true}
            nbins=55
        }
        tunerOptions={seed=3791538}
    ;
    print r;
run;
quit;
```

Another option:

```
proc cas noqueue;
autotune.tuneGradientBoostTree result=r /
        trainOptions={
            table={name='HMEQ', vars={'CLAGE', 'DEBTINC', 'LOAN',
                    'MORTDUE', 'VALUE', 'REASON', 'JOB', 'bad'}},
            inputs={'CLAGE', 'DEBTINC', 'LOAN', 'MORTDUE', 'VALUE',
                    'REASON', 'JOB'},
            target='bad',
            nominals={'REASON', 'JOB', 'bad'},
            casOut={name='gbt_hmeq_model', replace=true}
        }
        tunerOptions={seed=3791538}
        tuningParameters = {
            {name="nBins", initValue=55}
        }
    ;
    print r;
run;
quit;
```

There are also historyTable and lookupTable options:

historyTable={*history_table*} specifies the CAS table that is created by the autotune action that contains all evaluation history data points.

lookupTable={*lookup_table*} specifies the CAS table used by the autotune action for evaluation lookup.

The lookupTable ensures that no configurations are repeated, but also that the best model from a previous tuning job can be selected from the lookupTable as the initial model and basis for comparison (and the model is not retrained).

Gradient Boosted Poisson and Tweedie Regression Trees

In the SAS Viya gradient boosting algorithm, the default distributions of the objective function are BINARY for a binary target, MULTINOMIAL for a nominal target, and GAUSSIAN for an interval target. However, in practice, not all interval targets follow the normal distribution. When the distribution of the observations comes from the family of exponential distributions, you can always build traditional generalized linear models with the appropriate distribution function. In addition, you can also create a gradient boosting model for regression problems using distributions other than normal. Figure 9.9 illustrates some of the most common distributions other than normal.

The **Interval target distribution** property (or DISTRIBUTION= option in PROC GRADBOOST statement) determines the objective measure that is optimized when searching for a split of a node. Possible values are Normal, Poisson, and Tweedie. For a binary or nominal target, the distribution is binary or multinomial, respectively. Gradient boosting can be viewed as an extended model of the generalized linear model with the same underlying distributions, but it can achieve better prediction. For more details, see Friedman, 2001.

The Poisson distribution is often used to analyze count data, where the response variable has nonnegative integer values (0, 1, 2, 3, and so on). This distribution is the benchmark distribution for count data in much the same way that the normal distribution is the benchmark for

Figure 9.9: Objective Function Distribution

- Not all interval targets follow normal distribution.
- The Poisson distribution is appropriate for count data.
- The Tweedie distribution is useful for modeling total losses in insurance.

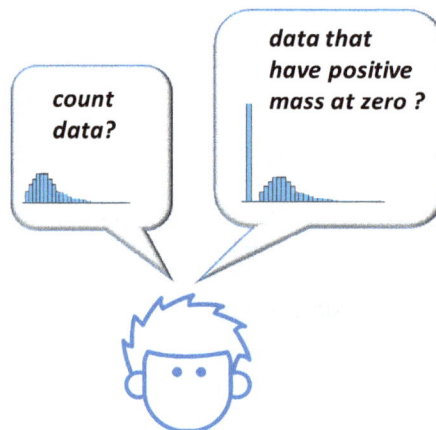

continuous data. The Poisson distribution can also be used to model the rate or incidence of an event. This type of outcome is widely seen in the medical sciences, biological sciences, social sciences, agriculture, engineering, and business. The OFFSET= option can be used for modeling rates adjusted for some measure of exposure. Insurance modelers commonly build separate frequency and severity models that are then combined. Many times, count data have too many zeros to use the standard Poisson distribution, and a zero-inflated Poisson distribution or Tweedie distribution must be used.

The family of Tweedie distributions are not as commonly known as some of distributions in the exponential family, although they are commonly used in many areas including insurance, microbiology, and genomics. Tweedie distributions are a special case of the exponential dispersion family, characterized by a variance equal to $\varphi\mu^p$, where μ is the mean, φ is a dispersion parameter, and p is an index parameter discovered by Tweedie (1984). The distribution is defined for all p except where p is between 0 and 1. The Tweedie simplifies to many other well-known distributions. For example, if p *is*

- 0, then the distribution is normal
- 1, then the distribution is Poisson
- 2, then the distribution is gamma
- 3, then the distribution is inverse Gaussian
- between 1 and 2, then the distribution is a compound Poisson-gamma mixture distribution, and a gamma scale parameter can be defined
- >1, then the mean is positive.

The most interesting range is from 1 to 2 in which the Tweedie distribution gradually loses its mass at 0 as it shifts from a Poisson distribution to a Gamma distribution. In this case, the Tweedie random variable can be generated from a compound Poisson distribution (Smyth 1996). This is the TWEEDIE distribution in PROC GRADBOOST; the default value of POWER=p is 1.5. In Model Studio, the **Tweedie power parameter** specifies the power parameter to use when **Tweedie** is selected as the **Interval target distribution**.

> Gradient boosting is a primary predictive modeling algorithm for tree-based models. If you want to specify some uncommon distributions to build a single regression tree, you can build a gradient boosting model with **ntree**=1 and DISTRIBUTION= option.

Demo 9.2: Building a Gradient Tree-Boosted Tweedie Compound Poisson Model

This demonstration runs PROC GRADBOOST with the DISTRIBUTION option to build a gradient tree-boosted Tweedie compound Poisson model on **bigPVA** data that contains data of charitable donations made to an American veterans' association. The data represent the results of a mail campaign to solicit donations.

Note: You can also run this demo in Model Studio using a Gradient Boosting node (Interval target distribution property) and a GLM node (Target probability distribution and Link function properties).

1. Ensure that you are in the **SAS Studio – Develop SAS Code** application.
2. Click the **Snippets** icon in the left pane. Under **My Snippets** ⇨ **VBBF**, either double-click or drag and drop **Tweedie Gradient Boosting Code** onto the canvas.

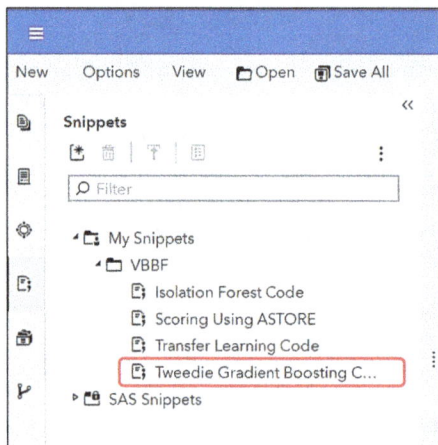

The program opens in the Code Editor window.

First, a caslib named **mycas** is created. Next, a local library named **local** is created. Then a DATA step is used to create new data set that is saved in memory.

```
/* Defining libraries and loading data */
libname mycas cas;
libname local '/home/student/casuser/VBBF';

data mycas.bigpva;
    set local.bigpva;
run;
```

3. Run this part of the program.
4. Stay on the **Code** tab to continue executing the code.

The next part of the program lists all the variables in the data, creates macro variables for interval and nominal inputs, and explores the data by plotting the target distributions.

```
/* Listing the variables */
proc contents data=mycas.bigpva out=contents(keep=name type);
run;

/* Creating macro variable for interval inputs */
proc sql; select name into: interval separated by " "
from work.contents
where type=1 and name not in("TARGET_B" "TARGET_D" "TARGET_D_with0"
"StatusCatStarAll");
quit; %put &interval;
```

```
/* Creating macro variable for nominal inputs */
proc sql; select name into: nominal separated by " "
from work.contents
where (type=2 or name = "StatusCatStarAll") and name ne "ID";
quit;
%put &nominal;
```

```
/* Exploring data */
proc univariate data=mycas.bigpva;
        var Target_D Target_D_with0;
        histogram Target_D Target_D_with0;
run;
```

5. Run this part of the program. The **Results** tab opens.

#	Variable	Type	Len	Format	Label
24	DemAge	Num	8		Age
23	DemCluster	Char	2		Demographic Cluster
25	DemGender	Char	3		Gender
26	DemHomeOwner	Char	3		Home Owner
27	DemMedHomeValue	Num	8	DOLLAR11.	Median Home Value Region
29	DemMedIncome	Num	8	DOLLAR11.	Median Income Region
28	DemPctVeterans	Num	8		Percent Veterans Region
10	GiftAvg36	Num	8	DOLLAR9.2	Gift Amount Average 36 Months
11	GiftAvgAll	Num	8	DOLLAR9.2	Gift Amount Average All Months
12	GiftAvgCard36	Num	8	DOLLAR9.2	Gift Amount Average Card 36 Months
9	GiftAvgLast	Num	8	DOLLAR9.2	Gift Amount Last
5	GiftCnt36	Num	8		Gift Count 36 Months
6	GiftCntAll	Num	8		Gift Count All Months
7	GiftCntCard36	Num	8		Gift Count Card 36 Months
8	GiftCntCardAll	Num	8		Gift Count Card All Months
14	GiftTimeFirst	Num	8		Time Since First Gift
13	GiftTimeLast	Num	8		Time Since Last Gift
1	ID	Char	8		Control Number
15	PromCnt12	Num	8		Promotion Count 12 Months
16	PromCnt36	Num	8		Promotion Count 36 Months
17	PromCntAll	Num	8		Promotion Count All Months
18	PromCntCard12	Num	8		Promotion Count Card 12 Months
19	PromCntCard36	Num	8		Promotion Count Card 36 Months
20	PromCntCardAll	Num	8		Promotion Count Card All Months
21	StatusCat96NK	Char	5		Status Category 96NK
22	StatusCatStarAll	Num	8		Status Category Star All Months
3	TARGET_B	Num	8		Target Gift Flag
4	TARGET_D	Num	8	DOLLAR9.2	Target Gift Amount
2	TARGET_D_with0	Num	8	DOLLAR9.2	Target Gift Amount with Zero

The data table has more than 100,000 rows and 29 columns. The table contains three possible target variables, one binary and two interval, an ID variable, and 25 potential input variables with respondents' PVA promotions, giving history and demographic data of the respondents. **TARGET_B** is a binary flag to indicate respondents to the appeal and the dollar amount of their donations is represented by **TARGET_D**. The difference between **TARGET_D** and **TARGET_Dwith0** is that the positive values of these two inputs are equal. In **TARGET_D**, the zero (**0**) values are replaced with missing (**.**) and you use this variable for the demo. The two candidate inputs have missing values. However, tree-based models can handle missing values very well.

Note: It would be interesting to try building a gradient boosting model on **GiftCntAll** variable using DISTRIBUTION=POISSON.

The UNIVARIATE Procedure			
Variable: TARGET_D (Target Gift Amount)			
Moments			
N	53273	Sum Weights	53273
Mean	15.6243444	Sum Observations	832355.7
Std Deviation	12.443969	Variance	154.852364
Skewness	5.16803399	Kurtosis	52.8001929
Uncorrected SS	21254307.3	Corrected SS	8249295.14
Coeff Variation	79.6447432	Std Error Mean	0.05391447

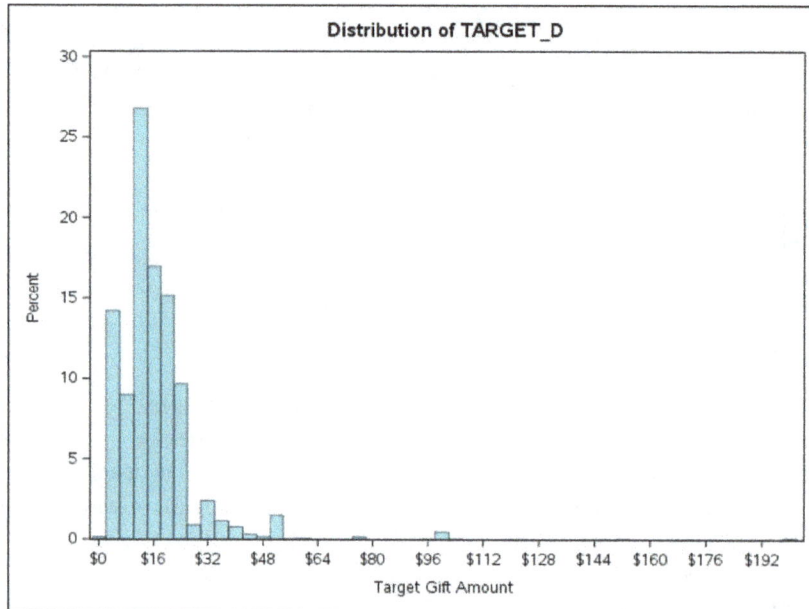

Distribution of TARGET_D

Target_D has no zeros, and they are coded as missing. The data is skewed to the right. In traditional regression modeling, a log transformation is commonly used to regularize skewed data.

The UNIVARIATE Procedure			
Variable: TARGET_D_with0 (Target Gift Amount with Zero)			
Moments			
N	106546	Sum Weights	106546
Mean	7.81217221	Sum Observations	832355.7
Std Deviation	11.7667354	Variance	138.456063
Skewness	4.17009664	Kurtosis	41.6872166
Uncorrected SS	21254307.3	Corrected SS	14751801.2
Coeff Variation	150.620533	Std Error Mean	0.03604852

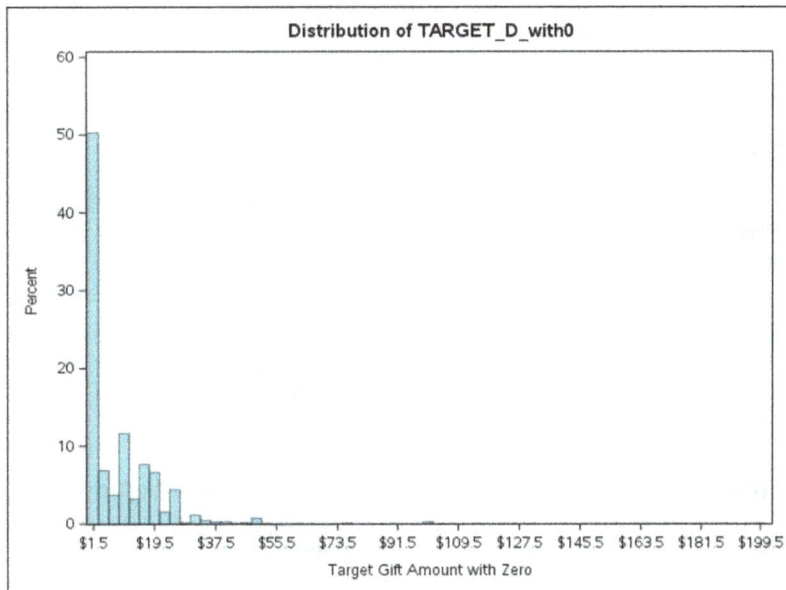

Distribution of TARGET_D_with0

Target_D_with0 has zeros instead of missing values. The distribution is more skewed and might be zero inflated. Gradient boosting models with a Gaussian distribution objective measure will not perform well on data distributed as above while searching for a split of a node.

6. Go back to the **Code** tab.

The next part of the program runs a gradient boosting model with the Tweedie distribution and a more traditional generalized linear model with the Tweedie distribution.

```
/* Gradient Boosting with Tweedie Distribution */
ods noproctitle;
proc gradboost data=mycas.bigpva distribution=tweedie;
        target Target_D_with0 / level=interval;
        input &interval / level=interval;
```

```
      input &nominal / level=nominal;
      ods output FitStatistics=Work._Gradboost_FitStats_;
      score out=mycas.GB_Tweedie copyvars=(_all_);
run;


/* Generalized Linear Model with Tweedie Distribution */
ods noproctitle;
ods graphics / imagemap=on;

proc genmod data=MYCAS.BIGPVA;
      class &nominal / param=glm;
      model Target_D_with0= &interval &nominal /
            dist=tweedie(p=1.5) link=log;
      ods output FitStatistics=Work._GenMod_FitStats_;
      output out=work.Genmod_Tweedie pred=pred_ resraw=r_;
run;

/* Making the GenMod residuals comparable to GradBoost residuals*/
data GenMod_Tweedie;
set Genmod_Tweedie;
r_=-r_;
run;
```

The GENMOD procedure fits a variety of statistical models. A generalized linear model with the Tweedie distribution and the log link function is used to compare the gradient boosting model. The Tweedie power parameter p is being set at 1.5 to make the model comparable. The Tweedie distribution has nonnegative support and can have a discrete mass at zero, making it useful to model responses that are a mixture of zeros and positive values. The Tweedie distribution belongs to the exponential family, so it conveniently fits in the generalized linear model framework.

7. Run this part of the program. Do not look at the Results.
8. Stay on the **Code** tab to run the program.

The next part of the program creates the plots of residuals by actuals for the two models.

```
/* Creating plots of residuals by actuals for the two models */
ods graphics / reset width=6.4in height=4.8in imagemap;

title 'Plot of Residuals by Actuals: Gradient Boosting Distribution=Tweedie';
proc sgplot data=mycas.GB_tweedie;
      scatter x=Target_D_with0 y=_Residual_ /;
      xaxis grid;
      yaxis grid;
run;


title 'Plot of Residuals by Actuals: Generalized Linear Model Distribution=Tweedie';
proc sgplot data=work.GenMod_Tweedie;
      scatter x=Target_D_with0 y=r_;
      xaxis grid;
      yaxis grid;
run;
```

9. Run this part of the program. The **Results** tab opens.

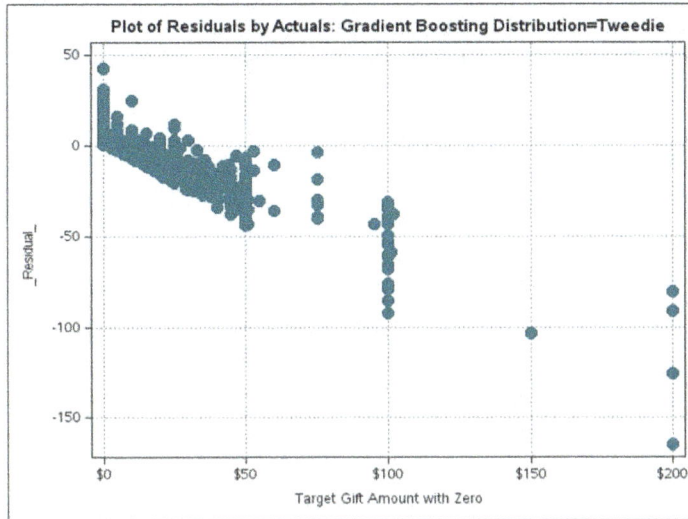

Plot of Residuals by Actuals: Gradient Boosting Distribution=Tweedie

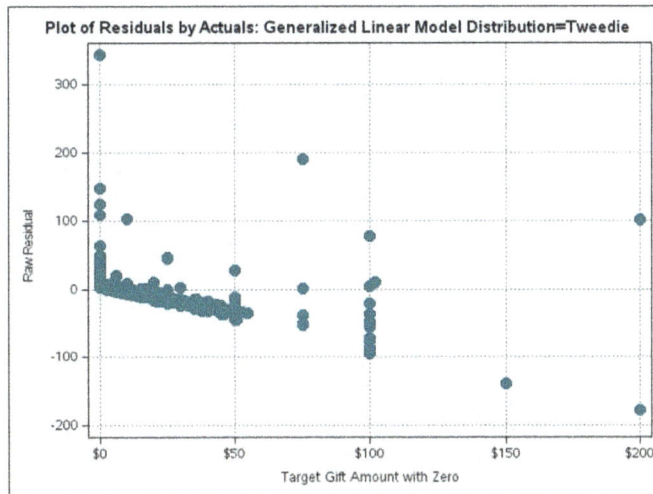

Plot of Residuals by Actuals: Generalized Linear Model Distribution=Tweedie

Plot of residuals by actuals have been generated for the two models. There are a few exceptionally large residuals in the Tweedie regression model, especially when the donation amount is small and close to zero. Both the models underpredicted the target, especially for larger donation amounts.

10. Click the **Code** tab to go back to the program.

The next part of the program creates the plots of predicted by actuals for the two models.

```
/* Creating plots of predicted by actuals for the two models */
```
title 'Plot of Predicted by Actuals: Gradient Boosting Distribution=Tweedie';
proc sgplot data=mycas.GB_tweedie;
      scatter x=Target_D_with0 y=P_Target_D_with0 /;
      series x=Target_D_with0 y=Target_D_with0 /
          lineattrs=(color=red); *reference line;
      xaxis grid;
      yaxis grid;
run;

title 'Plot of Predicted by Actuals: Generalized Linear Model Distribution=Tweedie';
proc sgplot data=work.GenMod_Tweedie;
      where pred_ between 0 and 200;
      scatter x=Target_D_with0 y=pred_ /;
      series x=Target_D_with0 y=Target_D_with0 /
          lineattrs=(color=red); *reference line;
      xaxis grid;
      yaxis grid;
run;
ods graphics / reset;
```

11. Run this part of the program. The **Results** tab opens.

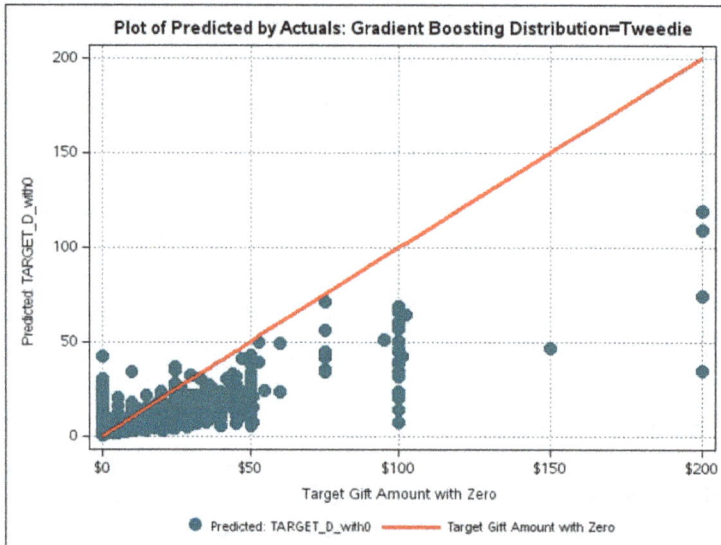

Plot of Predicted by Actuals: Gradient Boosting Distribution=Tweedie
```

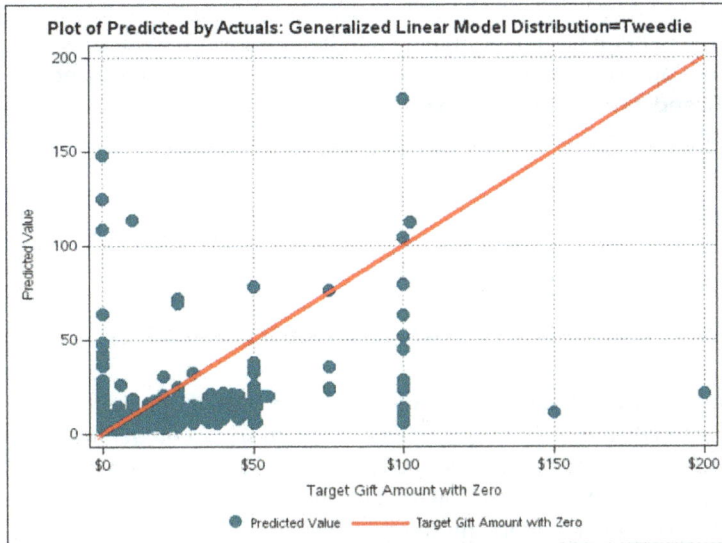

Plot of Predicted by Actuals: Generalized Linear Model Distribution=Tweedie

The gradient boosting model shows a better fit than the generalized linear model. This might be due to the obvious advantages that tree-based models have over the parametric models.

The gradient boosting residuals are smaller when donations are low. However, they are more dispersed when donations are high. On the other hand, generalized linear model residuals are higher when donations are low and high.

12. Click the **Code** tab to go back to the last part of the program that calculates the average square errors for model comparison.

```
/* Assessing Model Fit */
title "ASE";
proc sql;
select sum(_Residual_**2)/count(*) as GradBoost_Tweedie
          from mycas.GB_Tweedie;
       select sum(r_**2)/count(*) as GenMod_Tweedie
          from work.GenMod_Tweedie;
quit;
title;
```

13. Run this part of the program. The **Results** tab opens.

ASE
GradBoost_Tweedie
85.14791

ASE
GenMod_Tweedie
90.47644

The gradient boosting model with the Tweedie distribution has better average squared error than the generalized linear model with the Tweedie distribution. Modifying the power parameter of Tweedie can further improve the model.

14. Close the **Tweedie Gradient Boosting Code** tab.
15. Select **Build Models** from the **Application Menu** to close SAS Studio and come back to Model Studio.

End of Demonstration

SAS Gradient Boosting Using Open Source

Earlier in the book, you saw Python and R integration with SAS Viya in Chapter 7. The word "open" in Figure 9.10 signifies the fact that the power of SAS to build and deploy analytics can be accessed via many programming languages—not just SAS, but also Python, R, Lua, Java, or RESTful APIs. This integration enables analytical teams with varied backgrounds and experiences to come together and solve complex problems in new ways.

This section discusses an example of being able to access and execute SAS Analytics programmatically from open-source languages. For consistency, the primary focus is on calling SAS from Python, a popular general-purpose scripting language, via APIs. SAS provides a foundational open-source package called SWAT (Scripting Wrapper for Analytics Transfer) for doing this, available on GitHub. The SWAT package interfaces with SAS Viya.

Base SAS offers a Java object to incorporate a variety of external languages, including Python. This enables SAS procedures to be called from open-source tools. The Jupyter kernel for SAS brings the power of SAS data manipulation and analytics capabilities to the Jupyter notebook. Using SAS in open source can ease the transition for those who have not used SAS. Calling SAS via stored processes or APIs from other programming interfaces is a straightforward way for open-source programmers to access SAS.

Figure 9.10: SAS Viya Integration with Open-Source Software

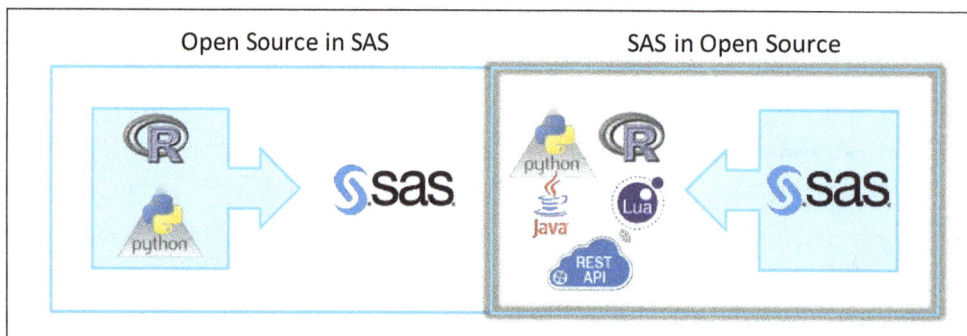

This openness enables data scientists to code in their language and interface of choice, and it allows SAS to extend open-source applications with productivity and the ability to scale to any data volume.

Demo 9.3: Training and Scoring a SAS Gradient Boosting Model Using Python

This demonstration runs a SAS gradient boosting model using Python script in a Jupyter Notebook. The **gbtreeTrain** and **gbTreeScore** actions in the decision tree action set are used to train and score a gradient boosting model on **insurance** data.

1. Minimize the **SAS Viya** tab.
2. From Windows OS: **Start** ⇨ **Anaconda3 (64-bit)** ⇨ **Jupyter Notebook (Anaconda3)**.
3. The Jupyter Notebook launches. Ensure that you are on the **Files** tab. Notice that a notebook document file is already uploaded.

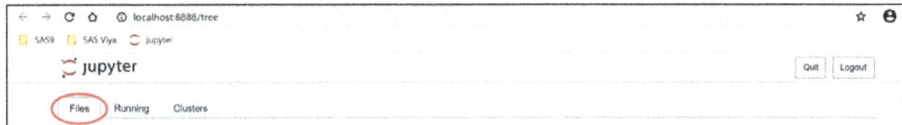

4. Double-click **SAS_GradBoost_Py_Jupyter.ipynb** to open the Python script for executing gradient boosting analysis in SAS Viya.

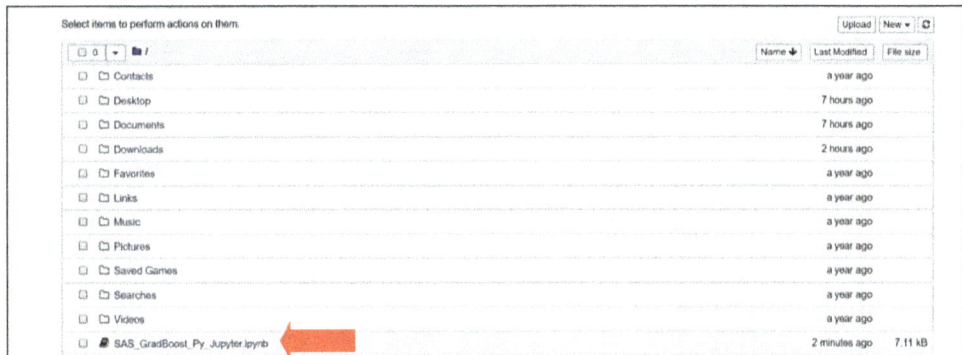

Alternatively, select **New** and then select **Python 3**.

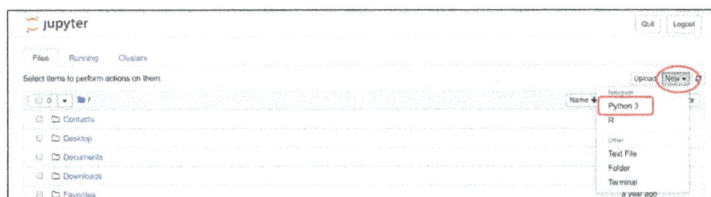

Copy the code from the **SAS_GradBoost Python Script.txt** file provided in the data folder.

```
In [ ]:
```

5. Run the Python code below.
You can run the code cell by cell, or you can run it in one go using the run the whole notebook option.

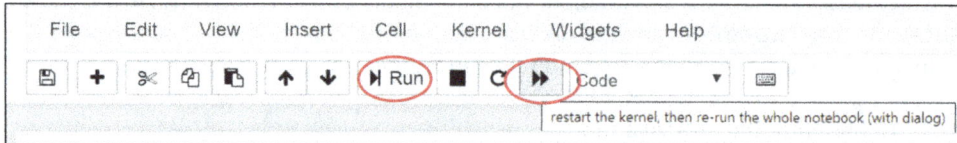

Python Code for SAS Gradient Boosting
Imports the SAS Scripting Wrapper (SWAT) for Analytics Transfer library for use in this example:

```
In [1]:  import swat
         import numpy as np
         import pandas as pd
         from matplotlib import pyplot as plt
         %matplotlib inline
         swat.options.cas.print_messages = True
```

Specifies the CAS host name and CAS port number to connect to a CAS session:

```
In [2]:  conn = swat.CAS("server", 8777, "student", "Metadata0", protocol="http")
```

Note that 8777 is the required port by the CAS Server Monitor process in SAS Viya and is used by clients to make REST HTTP calls to CAS, as with the Python REST interface. See *SAS Viya 3.5 for Linux: Deployment Guide* documentation for more details.
Loads analysis data into CAS:

```
In [3]:  uptab = conn.upload("D:/Workshop/winsas/VBBF/insurance.sas7bdat",
                             casout = dict(name="insurance", replace=True))
         indata = 'insurance'
         castbl = uptab.casTable

         NOTE: Cloud Analytic Services made the uploaded file available as table INSURAN
         CE in caslib CASUSER(student).
         NOTE: The table INSURANCE has been created in caslib CASUSER(student) from bina
         ry data uploaded to Cloud Analytic Services.
```

Returns the table information:

```
In [4]:  castbl.tableInfo()

Out[4]:  § TableInfo
```

	Name	Rows	Columns	IndexedColumns	Encoding	CreateTimeFormatted	ModTimeForma
0	INSURANCE	19357	50	0	utf-8	2020-01-29T05:13:25-05:00	2020 29T05:13:25-0

1 rows × 23 columns

elapsed 0.000729s · user 0.000706s · mem 0.7MB

The table has 19,357 observations and 50 variables.

Note: Unlike **insurance_part** data, this is unpartitioned data that will be partitioned later in the code. This is just to showcase another approach.

Returns column heads of the table:

```
In [5]:  castbl.head()

Out[5]:
```

Selected Rows from Table INSURANCE

	ACCTAGE	DDA	DDABAL	DEP	DEPAMT	CASHBK	CHECKS	DIRDEP	NSF	NSFAMT	...	HM
0	0.3	1.0	419.27	2.0	1170.06	0.0	0.0	0.0	0.0	0.0	...	
1	0.5	1.0	1594.84	1.0	1144.24	0.0	1.0	0.0	0.0	0.0	...	1
2	6.7	1.0	2813.45	2.0	1208.94	0.0	2.0	0.0	0.0	0.0	...	1
3	8.8	1.0	1437.57	2.0	2237.69	0.0	12.0	1.0	0.0	0.0	...	1
4	3.5	1.0	32812.98	2.0	1984.75	0.0	2.0	0.0	0.0	0.0	...	

5 rows × 50 columns

You have used this data many times and are familiar with these columns.

Loads a simple action set that will be used for summarizing the data:

```
In [6]:  conn.loadActionSet('simple')
         actions = conn.builtins.help(actionSet='simple')

         NOTE: Added action set 'simple'.
```

Computes the distinct number of values and missing values of the potential input variables. Only important variables are listed:

```
In [7]:  conn.simple.distinct(
            table = indata,
            inputs = ['SAVBAL', 'MM', 'CDBAL', 'DDABAL', 'DDA', 'BRANCH', 'ACCTAGE']
         )
```

Out[7]: **§ Distinct**

Distinct Counts for INSURANCE

	Column	NDistinct	NMiss	Trunc
0	SAVBAL	8552.0	0.0	0.0
1	MM	2.0	0.0	0.0
2	CDBAL	554.0	0.0	0.0
3	DDABAL	15269.0	0.0	0.0
4	DDA	2.0	0.0	0.0
5	BRANCH	19.0	0.0	0.0
6	ACCTAGE	414.0	1241.0	0.0

Assigns the target and input variables and specifies the continuous and nominal variables:

```
In [8]:  target = 'INS'
         cont = ['SAVBAL', 'CDBAL', 'DDABAL', 'ACCTAGE']
         inputs = ['SAVBAL', 'MM', 'CDBAL', 'DDABAL', 'DDA', 'BRANCH', 'ACCTAGE']
         nominals = [target] + ['MM', 'DDA', 'BRANCH']
```

Loads the "sampling" action set that will be used for data partition:

```
In [9]:  conn.loadActionSet('sampling')
         actions = conn.builtins.help(actionSet='sampling')

         NOTE: Added action set 'sampling'.
```

Samples a proportion of data from the input table and partitions the data into two parts:

```
In [10]:  conn.sampling.srs(
             table = indata,
             samppct = 70,
             seed = 919,
             partind = True,
             output = dict(casOut = dict(name = indata, replace = True),
                           copyVars = 'ALL')
          )
```

Out[10]: § OutputCasTables

	casLib	Name	Label	Rows	Columns	casTable
0	CASUSER(student)	insurance		19357	51	CASTable('insurance', caslib='CASUSER(student)')

§ SRSFreq

Frequencies

	NObs	NSamp
0	19357	13550

Verifies the sample proportion for training data:

```
In [11]:  castbl = conn.CASTable(name=indata)
          castbl['_PartInd_'].mean()
Out[11]:  0.70000516608978
```

Loads the "decisionTree" action set that will be used to train a gradient boosting model:

```
In [12]:  conn.loadActionSet('decisionTree')
          actions = conn.builtins.help(actionSet='decisionTree')

          NOTE: Added action set 'decisionTree'.
```

Calls the **gbtreeTrain** action of the decisionTree action set using the variable and table names that you want to use:

```
In [13]:  conn.decisionTree.gbtreeTrain(
              table = dict(name = indata, where = '_PartInd_ = 1'),
              target = target,
              inputs = inputs,
              nominals = nominals,
              nTree = 100,
              casOut = dict(name = 'gbt_model', replace = True)
          )
```

A gradient boosting model with 100 trees was created on the training data:

Out[13]: **§ ModelInfo**

Gradient Boosting Tree for INSURANCE

	Descr	Value
0	Number of Trees	100.00
1	Distribution	2.00
2	Learning Rate	0.10
3	Subsampling Rate	0.50
4	Number of Selected Variables (M)	7.00
5	Number of Bins	50.00
6	Number of Variables	7.00
7	Max Number of Tree Nodes	31.00
8	Min Number of Tree Nodes	25.00
9	Max Number of Branches	2.00
10	Min Number of Branches	2.00
11	Max Number of Levels	5.00
12	Min Number of Levels	5.00
13	Max Number of Leaves	16.00
14	Min Number of Leaves	13.00
15	Maximum Size of Leaves	4585.00
16	Minimum Size of Leaves	5.00
17	Random Number Seed	0.00
18	Lasso (L1) penalty	0.00
19	Ridge (L2) penalty	1.00
20	Actual Number of Trees	100.00
21	Average number of Leaves	15.69

§ OutputCas Tables

	casLib	Name	Rows	Columns	casTable
0	CASUSER(student)	gbt_model	3038	54	CASTable('gbt_model', caslib='CASUSER(student)')

Calls the **gbtreeScore** action of the decisionTree action set to score the trained model on the validation data:

```
In [14]: conn.decisionTree.gbtreeScore(
             table = dict(name = indata, where = '_PartInd_ = 0'),
             model = "gbt_model",
             casout = dict(name="gbt_scored",replace=True),
             copyVars = target,
             encodename = True,
             assessonerow = True
         )
```

Out[14]: **§ EncodedName**

	LEVNAME	LEVINDEX	VARNAME
0	1	0	P_INS1
1	0	1	P_INS0

§ EncodedTargetName

	LEVNAME	LEVINDEX	VARNAME
0		0	I_INS

§ ErrorMetricInfo

	TreeID	Trees	NLeaves	MCR	LogLoss	ASE	RASE	MAXAE
0	0.0	1.0	16.0	0.346478	0.626526	0.217911	0.466809	0.673710
1	1.0	2.0	31.0	0.346478	0.612386	0.211418	0.459802	0.688979
2	2.0	3.0	47.0	0.346478	0.600191	0.205826	0.453681	0.704446
3	3.0	4.0	63.0	0.344756	0.588950	0.200720	0.448017	0.719473
4	4.0	5.0	79.0	0.317720	0.579801	0.196636	0.443436	0.736084
...	...	...	...	...	...	...	...	...
95	95.0	96.0	1505.0	0.257103	0.520122	0.173323	0.416320	0.959309
96	96.0	97.0	1521.0	0.258481	0.520224	0.173361	0.416366	0.961396
97	97.0	98.0	1537.0	0.258653	0.520356	0.173399	0.416412	0.960820
98	98.0	99.0	1553.0	0.259514	0.520597	0.173467	0.416493	0.962084
99	99.0	100.0	1569.0	0.260031	0.520686	0.173504	0.416538	0.961967

100 rows × 8 columns

§ OutputCasTables

	casLib	Name	Rows	Columns	casTable
0	CASUSER(student)	gbt_scored	5807	5	CASTable('gbt_scored', caslib='CASUSER(student)')

§ ScoreInfo

	Descr	Value
0	Number of Observations Read	5807
1	Number of Observations Used	5807
2	Misclassification Error (%)	26.003099707

elapsed 0.117s · user 0.62s · sys 0.00708s · mem 37.2MB

The validation misclassification is around 26%.

Loads the "percentile" action set that will be used to calculate quantiles and percentiles to be used in model assessment statistics and graphics:

```
In [15]: conn.loadActionSet('percentile')
         actions = conn.builtins.help(actionSet='percentile')

         NOTE: Added action set 'percentile'.
```

Assess and compare models:

```
In [16]: assess_input = 'P_' + target + '1'

         conn.percentile.assess(
             table = "gbt_scored",
             inputs = assess_input,
             casout = dict(name="gbt_assess",replace=True),
             response = target,
             event = "1"
         )
```

Out[16]: § OutputCasTables

	casLib	Name	Rows	Columns	casTable
0	CASUSER(student)	gbt_assess	20	21	CASTable('gbt_assess', caslib='CASUSER(student)')
1	CASUSER(student)	gbt_assess_ROC	100	21	CASTable('gbt_assess_ROC', caslib='CASUSER(stu...

elapsed 0.00795s · user 0.0258s · sys 0.0038s · mem 6.67MB

Two tables were created.
Prints a few rows from the **gbt_assess** table:

```
In [17]: conn.table.fetch(table='gbt_assess', to=5)
```

Out[17]: § Fetch

Selected Rows from Table GBT_ASSESS

	Column	_Event_	_Depth_	_Value_	_NObs_	_NEvents_	_NEventsBest_	_Resp_	_RespE
0	P_INS1	1	5.0	0.780727	291.0	224.0	291.0	11.133201	14.46
1	P_INS1	1	10.0	0.715290	291.0	216.0	291.0	10.735586	14.46
2	P_INS1	1	15.0	0.658898	291.0	196.0	291.0	9.741551	14.46
3	P_INS1	1	20.0	0.603189	291.0	186.0	291.0	9.244533	14.46
4	P_INS1	1	25.0	0.541653	291.0	160.0	291.0	7.952286	14.46

5 rows × 21 columns

elapsed 0.00144s · user 0.000237s · sys 0.00118s · mem 1.02MB

Prints a few rows from the **gbt_assess_ROC** table:

```
In [18]: conn.table.fetch(table='gbt_assess_ROC', to=5)
```

Out[18]: § Fetch

Selected Rows from Table GBT_ASSESS_ROC

	Column	_Event_	_Cutoff_	_TP_	_FP_	_FN_	_TN_	_Sensitivity_	_Specificity_	_KS_	..
0	P_INS1	1	0.00	2012.0	3795.0	0.0	0.0	1.000000	0.000000	0.0	..
1	P_INS1	1	0.01	2012.0	3795.0	0.0	0.0	1.000000	0.000000	0.0	..
2	P_INS1	1	0.02	2012.0	3795.0	0.0	0.0	1.000000	0.000000	0.0	..
3	P_INS1	1	0.03	2012.0	3795.0	0.0	0.0	1.000000	0.000000	0.0	..
4	P_INS1	1	0.04	2011.0	3787.0	1.0	8.0	0.999503	0.002108	0.0	..

5 rows × 21 columns

elapsed 0.0013s · user 0.00127s · sys 2e-06s · mem 1.02MB

To_frame converts the CASTable object to SASDataframe, which is like Pandas DataFrame:

```
In [19]: gbt_assess_ROC = conn.CASTable(name = "gbt_assess_ROC")
         gbt_assess_ROC = gbt_assess_ROC.to_frame()
         gbt_assess_ROC['Model'] = 'Gradient Boosting'
```

Computes confusion matrix:

```
In [20]: cutoff_index = gbt_assess_ROC['_Cutoff_']==0.5
         compare = gbt_assess_ROC[cutoff_index].reset_index(drop=True)
         compare[['Model','_TP_','_FP_','_FN_','_TN_']]
```

Out[20]:

Selected Rows from Table GBT_ASSESS_ROC

	Model	_TP_	_FP_	_FN_	_TN_
0	Gradient Boosting	1085.0	583.0	927.0	3212.0

Computes misclassification rate:

```
In [21]: compare['Misclassification'] = 1-compare['_ACC_']
         miss = compare[compare['_Cutoff_']==0.5] [['Model','Misclassification']]
         miss.sort_values('Misclassification')
```

Out[21]:

Selected Rows from Table
GBT_ASSESS_ROC

	Model	Misclassification
0	Gradient Boosting	0.260031

Creates ROC chart:

```
In [22]:  plt.figure(figsize=(8,8))
          plt.plot()
          models = list(gbt_assess_ROC.Model.unique())
          display(models)

          for X in models:
              tmp = gbt_assess_ROC[gbt_assess_ROC['Model']==X]
              plt.plot(tmp['_FPR_'],tmp['_Sensitivity_'],
                      label=X+' (C=%0.2f)'%tmp['_C_'].mean())

          plt.xlabel('False Positive Rate', fontsize=15)
          plt.ylabel('True Positive Rate', fontsize=15)
          plt.legend(loc='lower right', fontsize=15)
          plt.show()
```

['Gradient Boosting']

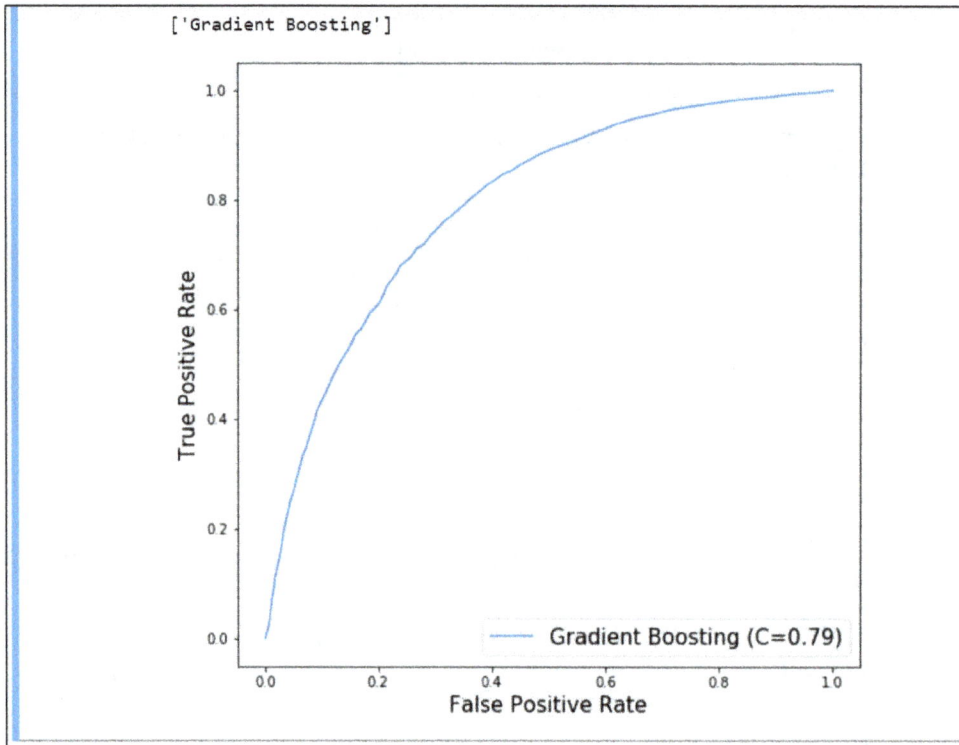

Returns table information for all the tables in this example:

```
In [23]: conn.table.tableInfo()['TableInfo'][['Name','Rows','Columns']]
Out[23]:
                     Name   Rows  Columns
         0       INSURANCE  19357       51
         1       GBT_MODEL   3038       54
         2      GBT_SCORED   5807        5
         3      GBT_ASSESS     20       21
         4  GBT_ASSESS_ROC    100       21
```

Ends the session:

```
In [24]: conn.session.endSession()
Out[24]:
         elapsed 0.000222s · user 0.00019s · sys 8e-06s · mem 0.227MB
```

6. Close the Jupyter Notebook.

End of Demonstration

Quiz

1. Transfer learning is beneficial for reducing labeling costs associated in retraining a model from scratch and to make classification rapidly adaptable in real time.
   a. True
   b. False

2. Using the correct target distribution is crucial in generalized linear models but not in a gradient boosting model.
   a. True
   b. False

Answers

1. True   2. False

Appendix A: Practice Case Study

Predicting Low Birth Weight

Context

Figure A.1: Data for Good—Analytics Helping Humanity

Low birth weight is a valuable public health indicator of maternal health, nutrition, health-care delivery, and poverty. Neonates with low birth weight have a more than 20 times greater risk of dying than neonates without low birth weight. In addition, low birth weight is associated with long-term neurologic disability, impaired language development, impaired academic achievement, and increased risk of chronic diseases including cardiovascular disease and diabetes.

Low birth weight is defined by the World Health Organization (WHO) as weight at birth less than 2500 g (5.5 lb.). Low birth weight continues to be a significant public health problem globally and is associated with a range of both short- and long-term consequences.

Data

These data focus on North Carolina births for 2000 and 2001. The original data sets include more than 120,000 births in each year and contain data on the ethnicity, age, education level, and marital status of the parents; prenatal medical care received; and information about the mother's reproductive history, including the number of previous pregnancies and live births (State Center for Health Statistics 2001, 2002).

The data set **births_2000.sas7bdat** represents an oversample (50% **LBWT=1**, 50% **LBWT=0**) of 17,097 records from the year 2000 to be used for training and validation. Plural births were filtered from the data. The percentage of low birth weight babies prior to oversampling is 7.2%. The data **births_2001.sas7bdat** represents an oversample (50% **LBWT=1**, 50% **LBWT=0**) of 16,687 records from 2001 to be used as a "future" test set. The percentage of low birth weight after removing plural births was also 7.2%.

LBWT	Low birth weight baby, defined as weight less than 2500 grams: **1**= Low birth weight; **0** = normal birth weight
FAGE	Age of father
MAGE	Age of mother
FEDUC	Education of father (number of years completed)
MEDUC	Education of mother (number of years completed)
TOTALP	Total pregnancies, including this one
PRENATAL	Months of pregnancy prenatal care began
BDEAD	Number of previous live births now dead
TERMS	Number of other terminations
LOUTCOME	Outcome of last delivery: **1** = Live birth, **2** = Fetal death, **9** = Not appl., unknown
YrsLastFetalDeath	Number of years since last fetal death
YrsLastLiveBirth	Number of years since last live birth
MARITAL	Marital status: **1** = Married, **2** = Not Married
CHILDREN	Number of previous children now living
ETHNICITYMOM	Ethnicity of mother (Values range from 0 to 9 and represent common groups in the U.S.)
ETHNICITYDAD	Ethnicity of father (Values range from 0 to 9 as above.)
CIGNUM	Average number of cigarettes daily
DRINKNUM	Average number of alcoholic drinks per week
DRINKER	Mother drinks alcohol: **1**-Yes, **0**-No
SMOKER	Mother smokes: **1**-Yes, **0**-No

Medical history for this pregnancy: **0** = No, **1** = Yes, **9** = Unknown, entire question blank

ANEMIA	Deficiency in the oxygen-carrying component of the blood
CARDIAC	Cardiac disease
ACLUNG	Lung disease
DIABETES	Diabetes
HERPES	Genital herpes
HYDRAM	Hydramnios or oligohydramnios (excessive amount or deficiency of amniotic fluid)
HEMOGLOB	Hemoglobinopathy
HYPERCH	Hypertension, chronic

HYPERPR	Hypertension for this pregnancy
ECLAMP	Eclampsia (seizures or convulsions in a pregnant woman that are not related to a pre-existing brain condition)
CERVIX	Incompetent cervix (abnormally weak cervix that can gradually widen during pregnancy)
PINFANT	Previous infant 4000+gm
PRETERM	Previous preterm or small
RENAL	Renal disease
RHSEN	Rh sensitization
UTERINE	Uterine bleeding

Obstetric procedures for this pregnancy: **0** = No, **1** = Yes, **9** = Unknown, entire question blank

AMNIO	Amniocentesis
ULTRA	Ultrasound

Data and documentation are reproduced with permission from North Carolina State Center for Health Statistics, Raleigh, NC, and the Howard W. Odum Institute for Research in Social Science at the University of North Carolina at Chapel Hill. These agencies are responsible for the original data only, and not for any content in the publication.

Analysis Questions

> Many questions in this case study do not have a definitive answer (for example, questions related to manual tuning or autotuning of the models are experimental in nature and there could be more than one solution for such questions).

1. **Building a Model Studio Project and Accessing the Data**
   a) Create a new Model Studio project and name it as NC_State. Choose the **births_2000** data as the analysis data. The data are available in the data folder.
   b) Configure the data such that you do not reject any inputs, even if they have a higher percentage of missing values. Ensure that any numeric variable with four or more levels remains defined with a level of *interval*.
   c) Split the data into two parts. Keep 70% for Training and the remaining for Validation.
   d) Assign the target variable. Do you have any variables that are rejected from the analysis?

2. **Assaying the Data**
   a) Have the data partitioned in your project? If not, execute splitting the data.
   b) Explore the analysis data and try to gain some knowledge about the variables by using both graphical and numerical methods. Select a subset of variables to provide a representative snapshot of the data.
   Hint: Add a Data Exploration node in the analysis.

    c)   Which one is the most important variable? Is the most important variable related to the health of the mother?
          Hint: Look at the Important Inputs chart.

    d)   What percentage of babies are low birth weight in this data?
          Hint: Look at the Class Variable Distributions chart.

    e)   What is the average age of mothers and fathers?
          Hint: Look at the Interval Variable Moments table.

    f)   Do any variables have missing values in these data? Which variable has the highest missing percentage? Although decision trees can handle missing values very well, do you think that the variable with the highest percentage of missing values should be rejected from the analysis? Why?
          Hint: Look at the Missing Values table in the Output window and then look at the Target by Input Crosstabulations chart.

    g)   Is there any association between parents' ethnicity and low birth weight? Should these variables be rejected from the analysis?
          Hint: Look at the Target by Input Crosstabulations chart.

3. **Creating a Decision Tree Model**
    a)   Create a decision tree model using all the default settings in Model Studio.
    b)   How many leaves does the model have? Up to how many leaves were originally grown on the tree?
          Hint: Look at the Model Information table in the Output window.
    c)   What are the rules defining segments with the highest and lowest probabilities of low birth weight?
          Hint: Look at the treemap chart and recognize the darkest and lightest Node IDs and then locate them in the Tree Diagram to see the rules.
    d)   Which are the most important inputs in predicting the low birth weight?
          Hint: Look at the Variable Importance table.

4. **Out-of-Time Testing of the Decision Tree Model**
    a)   You have used births data for the year 2000 to build the model and did the in-time validation of the model, no testing being done. Use the year 2001 data to do the out-of-time testing of the decision tree model.
          Hint: Use the Score Data node.
    b)   What are the ASE and MISC measures of the model's out-of-time performance? Are you getting comparable results for this out-of-time testing?
          Hint: Look at the Fit Statistics tables of the Score Data node and the Decision Tree node.
    c)   What is the advantage and disadvantage of this approach?
    d)   What if your out-of-time test performance is quite different from the in-time validation performance? What can you do?

5. **Improving the Decision Tree Model by Modifying Certain Tree Growth Options**
   a) Add one more decision tree in the pipeline and rename it as Tuned DT.
   b) Change certain tree growth options to improve the decision tree model. Do you think that increasing the tree depth, leaf size, and number of interval bins will benefit improving the decision tree model?
      Hint: Change maximum depth to 18, minimum leaf size to 70, and number of interval bins to 85. Explore other hyperparameter combinations. Be experimentative!
   c) How does this manually tuned decision tree perform in comparison with the default tree?

6. **Experimenting with the Pruning Strategy**
   a) How many leaves does the decision tree model have after pruning? Why have those many leaves been selected?
   b) Would it be acceptable to decrease the tree complexity by pruning the Tuned DT for reduced number of leaves, around 10 leaves fewer than what you have been getting? What are the two ways in which you can get this done in Model Studio? Apply any of these two methods and run the pipeline.
      Hint: You can explicitly mention the number of leaves you want in the decision tree or you can provide the cost-complexity alpha value.
   c) Compared to the previous tree, how does the resulting tree with reduced complexity (leaves) perform on accuracy? Is it worth reducing the tree complexity?
      Hint: Look at the Pruning Error Plot.
   d) How has the manually tuned tree performed as compared to the tree with default settings?
      Hint: Look at the results of the Model Comparison node.

7. **Creating a Default Forest Model**
   a) Create a forest model with all the default settings.
   b) How many trees are built to create the forest model? What is the out-of-bag sample proportion? What is the out-of-bag misclassification rate for this model?
      Hint: Look at the Model Information table in the Output window.
   c) Do you think that building a forest with one hundred trees is apt? Would increasing or reducing the number of trees be a good move?
      Hint: Look at the Pruning Error Plot.
   d) Which input is the most important predictor for low birth weight?
      Hint: Look at the Variable Importance table.
   e) Which inputs among the important ones are the most consistent across all the trees in the forest?
      Hint: Look at the Variable Importance table and use Sort and Add to Sort functionality in appropriate columns.
   f) How does the forest model compare to the best ASE and MISC values from previous decision tree models?
      Hint: Look at the results of the Model Comparison node.

g) Although the loss of predictive accuracy might occur, reduce the time for growing the forest model by changing some of the settings (for example, reduce the number of trees, and to compensate for fewer trees, increase the maximum depth a bit).

Hint: Change the number of trees to 25 and maximum depth to 25. Explore other hyperparameter combinations. Be experimentative!

h) Does this shallow forest have a comparable accuracy with the deeper one?

Hint: Look at the Model Information table in the Output window.

8. **Autotuning a Forest Model**

a) Create another forest model and use SAS Viya autotuning capability to autotune all the hyperparameters of this forest.

b) Has autotuning done some magic to improve the forest model? How do all the models in the pipeline perform?

Hint: Look at the results of the Model Comparison node.

c) In what ways is the autotuned forest different from the previous forest model?

Hint: Look at the Autotune Best Configuration table and compare it with the previous model.

d) Random forest models can be a little slower to execute than a single tree. However, they are generally particularly good for prediction. They can be difficult (or impossible) to interpret because they are ensembles of many different trees, based on different data and different variables. If model interpretation is inevitable and important, what can you do?

9. **Building a Default Gradient Boosting Model**

a) Create a decision trees gradient boosting machine with all the default settings.

b) Is the gradient boosting model deterministic or stochastic? How many trees does the gradient boosting model have? How many actual trees does the model have?

Hint: Look at the Model Information table in the Output window.

c) Should you further increase the number of trees?

Hint: Look at the Error Plot.

d) Is early stopping performed? What would be the validation misclassification rate of the next four trees beyond the actual number of trees in the gradient boosting model?

Hint: Look at the Fit Statistics table in the Output window.

e) Are there too many variables contributing to the model? Will decreasing the number of inputs to consider per split help in reducing the important variables?

Hint: Look at the Variable Importance table.

f) What is the learning rate used in the gradient boosting model? How does the learning rate play a role?

Hint: Look at the Model Information table in the Output window.

10. **Trying Alternative Settings in Gradient Boosting Model**
    a) Add another gradient boosting model in the pipeline and rename it GB Tuned.
    b) Change certain basic options and tree growth options to improve the gradient boosting model. Knowing that there could not be one single recipe to improve the default gradient boosting model, do you think that increasing the number of trees and sampling rate and modifying the learning rate a bit will benefit the gradient boosting model? How about decreasing the interval bins and tree depths?
       Hint: Change the number of trees to 125, learning rate 0.2, subsample rate to 0.85, maximum depth to 3, number of interval bins to 30, and number of inputs to consider per split to 25. Explore other hyperparameter combinations. Be experimentative!
    c) Although you have increased the number of trees, how many trees were there in the gradient boosting model? How does the average squared error behave with every tree added in the model?
       Hint: Look at the Error Plot.
    d) Based on average square error, which of the stochastic gradient boosting models appear to be better? How do they compare with the decision tree and forest nodes?
       Hint: Look at the results of the Model Comparison node.
    e) Which model do you choose for deployment? How do you select a model?

References

Allison, P.D. 2002. *Missing Data*. Thousand Oaks, CA: Sage Publications.

Bauer, E. and R. Kohavi. 1999. "An Empirical Comparison of Voting Classification Algorithms: Bagging, Boosting, and Variants." *Machine Learning* 36(1-2):105–139.

Blake, C., E. Keogh, and C. J. Merz. 1998. UCI Repository of machine learning databases. Irvine, CA: University of California, Center for Machine Learning and Intelligent Systems, Department of Computer Science. URL http://archive.ics.uci.edu/ml/

Breiman, L. 2001. "Random Forests." *Machine Learning* 45(1):5–32.

Breiman, L. 1996a. "Technical Note: Some Properties of Splitting Criteria." *Machine Learning* 24(1):41–47.

Breiman, L. 1996b. "Bagging Predictors." *Machine Learning* 24(2):123–140.

Breiman, L. 1998. "Arcing Classifiers (with discussion)." *Annals of Statistics* 26(3):801–849.

Breiman, L., J. H. Friedman, R. A. Olshen, and C. J. Stone. 1984. *Classification and Regression Trees*. Boca Raton, FL: Chapman & Hall/CRC.

Chen, T. and C. Guestrin. 2016. "XGBoost: A Scalable Tree Boosting System". Conference Paper, KDD '16: Proceedings of the 22nd ACM SIGKDD International Conference on Knowledge Discovery and Data Mining, San Francisco .

Dai, W., Q. Yang, G-R Xue,and Y. Yu. 2007."Boosting for Transfer Learning." ICML '07: Proceedings of the 24th international conference on Machine learning, Corvallis, OR.

Díaz-Uriarte, R. and S. A. de Andres. 2006. "Gene selection and classification of microarray data using random forest." *BMC Bioinformatics* 7(3):1.

Freund, Y. and R. E. Schapire. 1996. "A Decision-Theoretic Generalization of On-Line Learning and an Application to Boosting." *Journal of Computer and System Sciences* 55(1):119–139.

Friedman, J. H. 1999. "Greedy Function Approximation: A Gradient Boosting Machine." IMS 1999 Reitz Lecture.

Friedman, J. H. 2001. "Greedy function approximation: A gradient boosting machine." *The Annals of Statistics* 29(5):1189–1232.

Friedman, J. H. 2002. "Stochastic gradient boosting." *Computational Statistics & Data Analysis* 38(4):367–378.

Grzymala-Busse, J. W. and W. J. Grzymala-Busse. 2005. "Handling Missing Attribute Values." In: Maimon O., Rokach L. (eds) *Data Mining and Knowledge Discovery Handbook*. Springer, New York.

Hand, D. J. 1997. *Construction and Assessment of Classification Rules*. New York: Wiley.

Hastie, T., R. Tibshirani, and J. H. Friedman. 2001, 2009. *The Elements of Statistical Learning: Data Mining, Inference, and Prediction*. New York: Springer.

Kass, G. V. 1980. "An Exploratory Technique for Investigating Large Quantities of Categorical Data." *Journal of the Royal Statistical Society. Series C (Applied Statistics)* 29(2):119–127.

Kohavi, R., D. Sommerfield, and J. Dougherty. 1997. "Data Mining Using $\mathcal{MLC}++$ a Machine Learning Library in C++" *International Journal on Artificial Intelligence Tools* 6(4):537-566.

Liaw, A. and M. Wiener. 2002. "Classification and Regression by randomForest." *R News* 2(3):18–22.

Little, R. J. A. and D. B. Rubin. 2002. *Statistical Analysis with Missing Data, Second Edition*. Hoboken, NJ: Wiley.

Liu, F. T., K. M. Ting, and Z-H Zhou. 2008. "Isolation Forest." ICDM 2008. Proceedings of the Eighth IEEE International Conference on Data Mining.

Loh, W-Y. and Y-S Shih. 1997. "Split Selection Methods for Classification Trees." *Statistica Sinica* 7(4):815–840.

Loh, W-Y. and N. Vanichsetakul. 1988. "Tree-Structured Classification Via Generalized Discriminant Analysis (with discussion)." *Journal of the American Statistical Association* 83(403):715–728.

Loh, W-Y. 2002. "Regression Trees with Unbiased Variable Selection and Interaction Detection." *Statistica Sinica* 12(2):361–386.

Loh, W-Y. 2009. "Improving the Precision of Classification Trees." *The Annals of Applied Statistics* 3(4):1710–1737.

Lopez-Rojas, E. A., A. Elmir, and S. Axelsson. 2016. "PaySim: A financial mobile money simulator for fraud detection". In: The 28th European Modeling and Simulation Symposium-EMSS, Larnaca, Cyprus.

Maldonado, M., J. Dean, W. Czika, and S. Haller. 2014. "Leveraging Ensemble Models in SAS®Enterprise MinerTM." *Proceedings of the SAS Global Forum 2014 Conference*. Cary, NC: SAS Institute Inc. http://support.sas.com/resources/papers/proceedings14/SAS133-2014.pdf.

Martínez-Muñoz, G. and A. Suárez. 2010. "Out-of-bag estimation of the optimal sample size in bagging." *Pattern Recognition* 43 (1):143–152.

Morgan, J. N. and J. A. Sonquist. 1963. "Problems in the Analysis of Survey Data, and a Proposal." *Journal of the American Statistical Association* 58(302):415–434.

Murthy, S. K. and S. Salzberg. 1995. "Lookahead and Pathology in Decision Tree Induction." IJCAI-95: Proceedings of the Fourteenth International Joint Conference on Artificial Intelligence, Montreal. pp.1025–1031.

Murthy, S. K., S. Kasif, and S. Salzberg. 1994. "A System for Induction of Oblique Decision Trees." *Journal of Artificial Intelligence Research* 2:1–32.

Natekin, A. and A. Knoll. 2013. "Gradient boosting machines, a tutorial." *Frontiers in Neurorobotics* 7:21.

NC Department of Health and Human Services. North Carolina State Center for Health Statistics. "Statistics and Reports." Available http://www.schs.state.nc.us/data/archivedvitalstats.cfm#vol1.

Neville, P. G. 1999. "Decision Trees for Predictive Modeling." Cary, NC: SAS Institute Inc.

Neville, P. G. and P-Y Tan. 2014. "A Forest Measure of Variable Importance Resistant to Correlations." In Proceedings of the 2014 Joint Statistical Meetings. Alexandria, VA: American Statistical Association.

Oshiro, T. M., P. S. Perez, and J. A. Baranauskas. 2012. "How Many Trees in a Random Forest?" In Perner, P. (ed), Machine learning and data mining in pattern recognition, 8th international conference, MLDM 2012, Berlin, Germany, July 13–20, 2012. Proceedings.

Patil, G. P. and C. Taillie. 1982. "Diversity as a Concept and its Measurement (with discussion)." *Journal of the American Statistical Association* 77(379):548–567.

Probst, P. and A-L Boulesteix. 2017. "To tune or not to tune the number of trees in random forest?" https://arxiv.org/abs/1705.05654

Probst, P., B. Bischl, and A-L Boulesteix. 2018. "Tunability: Importance of hyperparameters of machine learning algorithms." https://arxiv.org/abs/1802.09596.

Probst, P., M. N. Wright, and A-L Boulesteix. 2019. "Hyperparameters and Tuning Strategies for Random Forest." WIREs Data Mining and Knowledge Discovery, 9(3):e1301.

Quinlan, J. R. 1986. "Induction of decision trees." *Machine Learning* 1(1): 81–106. doi:10.1007/BF00116251.

Quinlan, J. R. 1993. "C4.5: Programs for Machine Learning." San Mateo, CA: Morgan Kaufmann.

Quinlan, J. R. 1987. "Simplifying Decision Trees." *International Journal of Man-Machine Studies* 27:221-234.

Quinlan, J. R. 1993. *C4.5: Programs for Machine Learning*. San Mateo, California: Morgan Kaufmann.

Rao, J. S. and W. J. E. Potts. 1997. "Visualizing Bagged Decision Trees." KDD-97: *Proceedings of the Third International Conference on Knowledge Discovery and Data Mining* (Heckerman, D., Mannila, H., Pregibon, D., and Uthurusamy, R.,eds.). Menlo Park, CA: AAAI Press.

Ridgeway, G. 1999. "The State of Boosting." In Berk, K. and Pourahmadi, M. (eds.), *Computing Science and Statistics, Proceedings of the 31st Symposium on the Interface*. Fairfax Station, VA: Interface Foundation of North America, 31:171–181.

Ripley, B. D. 1996. *Pattern Recognition and Neural Networks*. New York: Cambridge University Press.

Schafer, J. L. and J. W. Graham. 2002. "Missing Data: Our View of the State of the Art." *Psychological methods* 7(2):147-177.

Scheffe, H. 1959. *The Analysis of Variance*. New York: Wiley.

SCHS: North Carolina State Center for Health Statistics. 2003. "Selected Vital Statistics for 2001 and 1997-2001." http://www.schs.state.nc.us/data/vital/volume1/2001/nc.html.

SCHS: North Carolina State Center for Health Statistics. 2012. "Selected Vital Statistics for 2000 and 1996-00." http://www.schs.state.nc.us/data/vital/volume1/2000/nc.html.

Segal, M. R. 2004. "Machine Learning Benchmarks and Random Forest Regression." Center for Bioinformatics and Molecular Biostatistics, University of California San Francisco, URL http://escholarship.org/uc/item/35x3v9t4

Shalev-Shwartz, S. and S. Ben-David. 2014. "Decision Trees." *Understanding Machine Learning: From Theory to Algorithms*. New York: Cambridge University Press.

Shannon, C. E. 1948. "A Mathematical Theory of Communication." *The Bell System Technical Journal* 27:379–423.

Smyth, G. K. 1996. "Regression Analysis of Quantity Data with Exact Zeroes." Proceedings of the Second Australia—Japan Workshop on Stochastic Models in Engineering, Technology and Management. Technology Management Centre, University of Queensland, 572–580.

Sutton C. D. 2005. "Classification and Regression Trees, Bagging, and Boosting." *Handbook of Statistics* 24:303–329.

Tweedie, M. C. K. 1984. "An Index Which Distinguishes between Some Important Exponential Families." In Ghosh, J.K., Roy, J (eds.). *Statistics: Applications and New Directions*. Proceedings of the Indian Statistical Institute Golden Jubilee International Conference on Statistics. Calcutta: Indian Statistical Institute. pp.579–604.

Van der Laan, M. J. 2006. "Statistical Inference for Variable Importance." *The International Journal of Biostatistics* 2(1):1–31. Article 2.

Jiang, W. 2002."On Weak Base Hypotheses and Their Implications for Boosting Regression and Classification." *The Annals of Statistics* 30(1):51–73. http://citeseerx.ist.psu.edu/viewdoc/summary?doi=10.1.1.134.9366

Wujek, B., P. Hall, and F. Güneş. 2016. "Best Practices for Machine Learning Applications." *Proceedings of the SAS Global Forum 2016 Conference*. Cary, NC: SAS Institute Inc.

Zhang, T. and B. Yu. 2005."Boosting with early stopping: convergence and consistency." *The Annals of Statistics* 33(4):1538–1579.

Zhang, H. and B. H. Singer. 2010. *Recursive Partitioning and Applications*. 2nd ed. New York: Springer.

Zhou, Z-H. 2012. *Ensemble Methods: Foundations and Algorithms*. Boca Raton, FL: Taylor & Francis.

Zhou, Z-H and J. Feng. 2017. "Deep Forest: Towards an Alternative to Deep Neural Networks." Proceedings of the Twenty-Sixth International Joint Conference on Artificial Intelligence (IJCAI-17), Melbourne, Australia

Zou H., and T. Hastie. 2005."Regularization and Variable Selection via the Elastic Net." *Journal of the Royal Statistical Society, Series B (Statistical Methodology)* 67(2):301–320.

www.ingramcontent.com/pod-product-compliance
Lightning Source LLC
Chambersburg PA
CBHW081046220326

41598CB00038B/7006